KEEPERS OF THE SPIRIT

NUMBER EIGHTY-NINE

Centennial Series of the Association of Former Students, Texas A&M University

KEEPERS OF

TEXAS A&M UNIVERSITY PRESS • COLLEGE STATION

THE SPIRIT

THE CORPS OF CADETS

AT TEXAS A&M UNIVERSITY,

1876–2001

John A. Adams, Jr.

Foreword by Ray M. Bowen

Library of Congress Cataloging-in-Publication Data

Adams, John A., 1951–
    Keepers of the spirit : the Corps of Cadets at Texas
A&M University, 1876–2001 / John A. Adams, Jr.
        p.   cm.—(Centennial series of the Association
of Former Students, Texas A&M University ; 89)
Includes bibliographical references and index.
ISBN 1-58544-126-0 (cloth)—ISBN 1-58544-127-9 (limited)
ISBN 978-1-60344-155-1 (paper)
    1. Texas A&M University. Corps of Cadets–History.
I. Title.   II. Series.
U429.T48A33   2001
355.2'232'071164242—dc21   2001000717

To the thousands of
Fightin' Texas Aggies
who as citizen-soldiers
made the supreme
sacrifice
in defense of
our great nation

I call, therefore, a complete and generous education, that which fits a man to perform justly, skillfully, and magnanimously all offices, both private and public, of peace and war.
John Milton, Of Education, *1644*

CONTENTS

Illustrations

TABLES

FOREWORD

Texas A&M University is a unique university within the United States. The people of Texas A&M University are passionate about what we do and how we do it. The loyalty of our former students, the dedication of our faculty and staff to our academic excellence, and our collective commitment to make the university better are the envy of many throughout public higher education. The source of this loyalty is an object of great speculation and discussion among individuals outside the Texas Aggie family. The history of the Corps of Cadets, as presented by John A. Adams, Jr., will help everyone understand this loyalty.

We are a product of a history that influences everything we do today. It is a history that has similarities with other public land-grant universities created during the latter half of the nineteenth century. The difference is that, while many others created during that period are not greatly affected by their history, we are dramatically affected. It is an exceptional graduate of Texas A&M University during the past forty-five years who does not identify with the small, all-male, military institution that existed during our first eighty years.

It is sometimes difficult to explain to a campus visitor why our military history is still such an important part of the university culture. In one dimension, it does not seem logical. We are a university of more than 43,000 students, one of the largest public universities in America. We are blessed with an exceptional faculty by any measure. Our Ph.D. graduates

are highly regarded and hold faculty positions in the best public and private universities in the country. Our faculty continues to receive extraordinary support, both public and private, for their scholarship. Our students are exposed to an education environment with great opportunity. They participate in internships throughout the world, from China to Washington, D.C. They study in University Centers in Mexico and Italy. More than 6,000 of our approximately 36,000 undergraduates participate as active members of faculty research programs. Within our large student body, the Corps of Cadets number a modest 2,000 members.

The answer as to why our military history is important today requires an understanding of the Corps of Cadets of Texas A&M University. This understanding needs to encompass the history of the Corps of Cadets as well as present-day knowledge about the Corps of Cadets. This book, *Keepers of the Spirit: The Corps of Cadets at Texas A&M University, 1876–2001* by John A. Adams, Jr., provides the necessary understanding.

The early history of the Corps of Cadets is actually a history of our university, the A&M College of Texas, in its early days. Every significant university event in our first eighty years is best understood in the context of the role of the Corps of Cadets. The altercation between President Thomas Gathright and the faculty over the selection of the 1879–80 cadet First Captain was so traumatic that all were forced to resign. The struggle after World War II to change the A&M College of Texas into a strong academic institution, while still maintaining an all-male military student body, created the energy to make A&M into the university it is today. This book, along with the earlier histories of Texas A&M University by George Sessions Perry and Henry C. Dethloff, provide great insight into the role of the Corps and, thus, our university, in its first eighty years.

When James Earl Rudder became president of Texas A&M in 1959, we were a troubled university. Our academic aspirations and our commitment to an all-male military university were in great conflict. The depth of this conflict is chronicled by John Adams in this work. He tells us about the efforts of Sen. Bill Moore, Aggie Class of 1940, to have women admitted to A&M in 1953, and how these efforts were thwarted by a fellow Aggie, Sen. Searcy Bracewell, Class of 1938. He shows us how the early efforts to admit women relate to efforts to make the Corps of Cadets optional during the middle to late 1950s, efforts that were finally resolved ten years later by President Rudder and the leadership of the Board of Regents. These troubling times produced the energy and determination that forever changed our university and its Corps of Cadets. Changes that resulted in a strong, comprehensive public university with an enrollment today of more than 19,000 women among its more than 43,000 students. It is a great university for which the Corps of Cadets remains one of the finest leadership laboratories in the country.

The author, a member of the Texas A&M University Class of 1973, is uniquely qualified to write about the Corps and the university during a period of profound change. These changes took place while our nation was experiencing the trauma of Vietnam and an alienation toward everything military. They were changes necessary to successfully bring women into the formerly all-male Corps of Cadets. They were changes necessary to maintain the relevance of the Corps in a post–Cold War United States.

The author mixes his special knowledge of "good bull" in the Corps with a sweeping view of people and events of the time. The author also helps us understand how the presence of the Corps of Cadets in this large civilian university has managed to implant into the rest of the university a culture of service and community leadership for its graduates.

The author allows the reader to understand the great aspects of the Corps and some of those aspects that trouble outsiders. The military environment of the Corps does not, in every detail, reflect contemporary military thinking about discipline and organization. Blended into the culture of the Corps is a strong element of a class or buddy system. The conflicts of a hierarchical military structure favored by the military establishment and a class or buddy system favored by the cadets have produced interesting challenges for the university over the years. A lesson one learns from reading this book is that these challenges have brought great benefits to the university, the state of Texas, and our nation. From the Spanish-American War to Desert Storm, officers commissioned from the Corps of Cadets of Texas A&M University have answered the call of our nation in times of crisis. They also have carried their talents into all other sectors of our nation. Through their corporate leadership, through their public service, and through their love of country, members of the Corps of Cadets make their university proud.

RAY M. BOWEN

PREFACE

I owe much to many.

Over the past two decades I have had the opportunity to delve into the past of one of the greatest organizations in the world—the Texas Aggie Corps of Cadets. In the process I have worked to capture the essence of the Corps. This story is one of which all Aggies have been a part.

I have been asked by many why I started this project. In many ways it began in the fall of 1969 when I entered my freshman or "fish" year in the Corps. The fish experience, for most, is the hallmark of being in the Corps of Cadets. The regimentation, styled within a framework that quickly transforms the new cadet from high school graduate to Texas Aggie, has been a constant denominator. Subjected to a seemingly endless indoctrination of campusology, mess hall etiquette, military formations, and a Spartan dorm life, the fish year experience forms the basis and the bond that set the Aggie Corps and its former students apart both on campus and after graduation. One uncommon thread has been the body of traditions and lore that has been passed down from class to class over more than a century. The Corps experience engenders pride, initiative, and steadfast loyalty to the institution, classmates, and one's "outfit" through a process of shared experiences, leadership opportunities, and access to a high-quality education.

We "fish" were expected to learn a litany of "famous facts" and campusology—referred to as "campo" by the cadets—and at every turn the lore of the Corps and Texas A&M was a significant part of our rite of passage to being in the Corps of Cadets. My outfit commander in Squadron One, Frank Montalbano III '70 from Beaumont, Texas, made sure the fish—without any doubt whatsoever—understood what it meant to be a Fightin' Texas Aggie. It was an exciting time, an experience that lasts forever. In late 1960s, the war in Vietnam captured national headlines; Gen. Earl Rudder was president of the university. Larger than life, Rudder affected an entire generation of students and alumni. He, more than any, was the gatekeeper as the A&M College entered the modern age as a leading national institution—Texas A&M University. His love for A&M and the Corps of Cadets was most evident.

Furthermore, through the years, I have marveled at how little history has been compiled on Texas A&M and the Corps of Cadets. The two-volume *Centennial History of Texas A&M 1876–1976* by Henry C. Dethloff published in 1976 was the first comprehensive look at the overall history of A&M and its agencies. While lesser works in 1895, 1935, and 1951 had preceded Dethloff's excellent work, none until Bill Leftwich's work in 1976, *The Corps at Aggieland,* had taken a direct look at the Corps of Cadets, and even that gave Corps history only in the form of anecdotes and an overview of a select number of key aspects of the Corps. Given the tremendous development of the research and extension programs at A&M I was amazed at the lack of institutional history or "memory" at Texas A&M. This again became apparent in the late 1990s during the Vision 2020 study on the future course of the institution. In the case of this work, I have included both detailed citations and an extensive bibliography. Thus, with the encouragement of Gen. Ormond R. Simpson '35, Richard "Buck" Weirus '42, Professors Dethloff and Allan C. Ashcraft '49, and many others, I endeavored to do what I think has never been done.

In the process I have interviewed hundreds of former cadets, students, and friends (and foes) of Texas A&M on all aspects of the Corps of Cadets. These interviews provided a tremendous foundation of insight and first-hand knowledge that ranged from 1904 to the present day. The hours of interviews with Joe Utay '08, A. J. "Niley" Smith '08, and J. V. "Pinky" Wilson '21 in the mid-1970s were priceless. The documentary history on the Corps is spread far and wide. One event that has hampered the chronicling of the data on the first three decades was the disastrous fire that gutted Old Main in May, 1912. Although official and unofficial records are available to fill in the early period, the records of the early college on, for example, enrollment statistics, Corps discipline records, and historical documents were forever lost. Thus, the numbers and data I have assembled are the best, given the information and resources available for this project. And to accomplish this, I have availed myself of as many sources as possible. I am indebted to the persistence and knowledge of both David L. Chapman '67 and Don Dyal at the Texas A&M University Archives, located in the Cushing Memorial Library, both leading experts on the documentary history of Texas A&M. Steve Smith and Angus Martin as well as the entire staff of the archives tolerated my ongoing search for as much information on the Corps and A&M as was possible to find. The new location of the archives in the Cushing Library is a testament to the truly critical importance of preserving and documenting the many facets and contributions of programs and individuals involved with the institution. Collecting, preserving, and enhancing the "institutional memory" of Texas A&M is paramount for research well into the future.

I am grateful for the auspicious assistance given by the staff of the Sam Houston Sanders Corps of Cadets Center. Joe and Pat Fenton '58, Lt. Col. Keith Stevens '71, and their staff provided tremendous help. The Sanders Corps Center is the showplace of the Corps history and should over time attract memorabilia and documents vital for a full appreciation of the lore and heritage of the Corps.

Special thanks are extended to Maj. Gen. Ted Hopgood '65, Commandant of the Corps of Cadets, as well as Gen. Don Johnson '55 for their encouragement and access to data on the Corps. Former A&M commandants Col. Tom Parsons '49, Col. Donald Burton '57, Maj. Gen. Tom Darling '54, Col. James Woodall '50, and Dr. J. Malon Southerland '65 provided timely observations and insight.

Many, many friends and colleagues have extended their suggestions, comments, and support: John H. Keck, David Perez, John A. Adams III, Calvin J. Green, Bob Boldt '66, Mike Casey '69, Carl Walker '73, Jim Kelly '52, Dick Hervey '42, Rick McPherson '62, Weldon Kruger '53, Walter Bradford '68, Joe West '54, T. L. and June Calvin, Buck Henderson '62, John Green '76, Louis Gohmert '75, Jerry Smith '70, and Mike Gentry '78. And a very special thanks is extended to John and Martha Adams for taking me to my first Texas Aggie Bonfire in November, 1952. To these and others who go unnamed, I am very grateful.

In my search for documents I received tremendous assistance from both Jerry Cooper '63, editor of the *Texas Aggie* magazine, and Bill Page '75, director of Government Documents at the Sterling C. Evans Library, Texas A&M. Archivist Linda Hanson Seelke at the Lyndon Baines Johnson Library in Austin, as well as the staffs at the Franklin D. Roosevelt Presidential Library in Hyde Park, New York; the Center for American History in Austin (formerly the Barker Center); and the National Archives in Washington, D.C., were also very helpful.

A special thanks is extended to those who took the time to review and comment on various sections of this work, and especially to Pam Johnson.

And as always Sherry, my wife, has endured my quest to complete this project. Without her encouragement and support this would not have been possible.

KEEPERS OF THE SPIRIT

INTRODUCTION

Let your watchword be DUTY, *and know no other talisman of success
than* LABOR. *Let* HONOR *be your guiding star in your dealings with your
superiors, your fellows, with all. Be true to a trust reposed as the needle
to the pole, stand by the right even to the sacrifice of life itself, and
learn that death is preferable to dishonor.*

Gov. Richard Coke, October 4, 1876

*In 1876 the College was inaugurated with six students . . . steadily
the number of students increased, until now the little squad of six has
grown into a splendid battalion.*

Texas A&M *Catalogue,* 1877–88

THE LEGACY and rich heritage of the Corps of Cadets at Texas A&M University have long been tied to the growth and development of both the nation and the state of Texas. One of the oldest surviving collegiate military-oriented programs in the nation, the Corps of Cadets has produced thousands of leaders for both the armed forces and private industry. The Corps, for decades the sole way of student life at the institution, was pivotal in the evolution of the spirit and traditions so integral to the development of Texas A&M and the Aggie esprit de corps. In war and peace, members of the A&M Corps have served with distinction. This presentation is a chronicle of the people and events that shaped the image and legacy of the Corps of Cadets at Aggieland.

The Aggie Corps of Cadets, a militarily oriented body of students established in 1876, traces its roots to a strong affinity with the broader concept and notion of the "citizen-soldier." Both the state and the nation have traditionally encouraged citizen input as well as the willingness to bear arms in time of need. However, it was not until the turbulence of the Civil War and the agony of the Reconstruction era that a mandated system of broad military training was formally legislated and implemented. To realize the citizen-soldier concept, and thus to have trained ample reserves in time of need, was a slow process that involved an indirect connection with those who envisioned a broad system that would expand the role, accessibility, and excellence of higher education. The catalyst for public land-grant institutions of higher education nationwide as well as the Corps of Cadets at Texas A&M was the Morrill Land Grant Act of July, 1862.[1]

The state of Texas at regular intervals since 1836 has initiated legislation that endorsed plans for the "establishment and endorsement" of a system of public education and higher learning in the state. "Such was the dream of the patriotic fathers of Texas," stressed Gov. Richard Coke at the inauguration of the college in October, 1876. Texas law set aside fifty leagues (221,420 acres) of "university" lands to support such future activities, and the state constitution endorsed the need to foster higher education. However, in the meantime, citizens of the state were more preoccupied with day-to-day life and in some cases Indian wars on the western frontier that dictated a reordering of priorities. The Civil War had done little to help that situation and further delayed the development of a state college in Texas.[2]

In other quarters, the concept of the land-grant college had been debated and fashioned as early as the 1850s. The driving force behind the land-grant concept was Justin Smith Morrill of Vermont. Morrill's first land-grant college act in legislative form was initially proposed in 1856 to establish a network of "national agricultural colleges" on the rigid model of the U.S. Military Academy at West Point. It failed to muster the needed votes. Ironically, pre-war attempts to pass the legislation failed because of Southerners' opposition, due in large part to their concern for states' rights. Furthermore, the concept of offering broad-based public higher education was foreign to the time-honored conventional wisdom of private universities. A universal, federally supported educational scheme in agriculture was further downplayed even though more than 90 percent of the nation's population lived and worked off the land. Morrill was persistent. Not until numerous modifications and the political shuffle in the U.S. House of Representatives during the Civil War did the land-grant act pass and become law with the signature of Abraham Lincoln on July 2, 1862.[3]

At the time of the passage of the Morrill Land Grant Act, this legislation was the first significant entree by the federal government into nationwide public higher education. Furthermore, the act included the Southern states that were in rebellion as well as future states to be carved out of the trans-Mississippi West. Thus, the agrarian states of the South, some of which did not wholeheartedly rush to implement the law, were in time to be a major postwar benefactor of this legislation.[4]

The land-grant act had two primary sections. First, it provided a system for the donation of federal public-domain lands to each state in the amount of 30,000 acres per representative and senator in Congress based on the census of 1860. Such lands and their disposition via sales were to form the funding and investment base to finance a college of higher learning for the benefit of teaching the "agricultural and mechanic arts." Texas, having no available federal public lands, received land scrip from the Department of Interior for 180,000 acres in Colorado. This land was sold in 1871 in quarter sections at $0.87 per acre and the proceeds invested in 7 percent gold (frontier) defense bonds. The dollars raised from these land sales formed the corpus that is the basis for the financing of public higher education in Texas to this day.[5]

Second, timing was critical to ensure receipt of the federal land grant. The Texas Legislature was compelled to act in order to acquire the land grant and funding from the federal government. On March 23, 1871, William A. Saylor, chair of the state Senate's Committee on Finance, introduced legislation to establish the "Agricultural and Mechanical College of Texas." Fully aware of the deadline, legislators passed the bill hastily. The governor signed it on April 17, 1871. Thus, the Texas Legislature, after numerous delays during the Reconstruction period, resolved to accept the mandate and obligation of the Morrill Act and provide support for a college by July 23, 1871. Under the terms of the Morrill Act (amended in July, 1865) there was a five-year deadline in which the state had to specify the establishment of a college to be opened by 1876. If the college was not opened, all benefits and grants under the Act of July 2, 1862, would expire.[6]

The final last-minute action by the Texas Legislature sealed the initial step to establish the A&M College. Section Four of the Morrill Act of 1862 embodied an equally important mandate. The Corps of Cadets can trace its roots to this section—the training of citizen-soldiers for national emergencies. Justin Morrill and Jonathan B. Turner stressed legislation that made "practical" higher education available to all "classes" of students, not just a select few. Public higher education was also the source of a broad-based training ground for a citizen-soldier reserve officer corps. The importance of such training was impressed upon President Abraham Lincoln during the dark hours of the Civil War. Lincoln insisted, with Morrill's assistance, that a reliable source be established in conjunction with the land-grant system to provide student instruction in "military tactics."[7] This section of the Morrill Act, which predates the formal Reserve Officers' Training Corps (ROTC), is often overlooked, yet the charge is unmistakable. To this end, nearly all of the designated land-grant colleges and universities across the nation have had or continue to have military training programs. Today, university-level training in military tactics is provided for by the ROTC program created in 1916–17 during the onset of American involvement in World War I.[8]

With the provisions of the Morrill Act accepted and acted on by the state of Texas, Republican governor Edmund J. Davis on April 24, 1871, named three special commissioners—Rep. Frederick E. Grothaus of DeWitt County, George B. Slaughter from Upshur County, and Sen. John G. Bell of Austin County—to begin the evaluation of possible sites for a campus.[9] Their instructions were to find a suitable location of not less than 1,280 acres in size, situated on a "well-drained position" yet near an ample source of usable water. Locations in Austin County (Bellville), Brazos County (Bryan), and Grimes County (Kellum Springs and Piedmont Springs) were considered. After looking at a number of locations, the commissioners narrowed the site selection to Brazos County. Bryan entrepreneur Harvey Mitchell and a local association of investors lured the site committee to a location south of the city that would meet their prerequisites. The local Brazos Valley economy was based primarily on the cotton and cattle industry and in the early 1870s had been given a significant boost with the completion of the Houston and Texas Central (H&TC) Railroad that connected the area with Houston as well as key markets to the north. Mitchell, a long-time resident and promoter in the Brazos Valley region, pledged 2,250 acres as a "donation" for the college if the site committee selected Bryan.[10] The committee accepted the offer and gave Mitchell forty-eight hours to provide title to the land. Mitchell moved rapidly, and on June 23, 1871, the deeds met the federal deadline for filing a donation of lands with the Brazos County clerk for 2,416 acres in the name of the Agricultural and Mechanical College of Texas. Today the campus exceeds 5,214 acres and is one of the largest such operations in the world.[11]

In the 1870s, Brazos County was "considered one of the poorest counties" and among the most remote locations in the state.[12] The campus, isolated except for the north-south rail connection via the Houston & Texas Central, was accessible primarily by horseback or stagecoach. The bridge spanning the Brazos River west of Bryan to Austin and San Antonio was in questionable condition. Travel westward was safest by making rail connections either in Hearne or Hempstead. And

for many years cadets, faculty, and visitors, as well as detractors of the A&M College, cited as its worst attribute the remote and isolated location. The Austin *Tri-Weekly Statesman* in late May, 1872, flatly concluded, "The present location of the College is a sure guarantee that no good will ever be derived from it." In time the very location would shape and enhance the status of the fledgling college.[13]

With the decision to locate the campus six miles from Bryan made, the legislature appropriated funding in early 1872 to begin construction. After some delay the "main" building was begun in August, 1873, and was completed and inspected by Governor Coke in October, 1874. Old Main, as it became known, was a grand hall on barren prairie dotted with post oaks and mesquite. In late 1874, a second structure, Steward's Hall (later named Gathright Hall), was funded with an initial $25,000 grant to serve as a multipurpose facility to include apartments for the faculty and president, student dormitory rooms, kitchen facilities, and a mess hall. Construction of faculty residences, other support buildings, and a crude road system continued throughout 1875 and into early 1876. The Texas Legislature had appropriated, by late 1876, a total of $187,000 in capital improvements for the college. At this juncture, the governor appointed a new five-man board of directors who met in Bryan to begin a search for a president, faculty, and staff.[14]

The year 1876 was to prove a banner year. The opening of the college mirrored historic events taking place not only in Texas, but across the nation. The year marked the centennial of the country, the patenting of the telephone by Alexander Graham Bell, and nationwide mania over a game known as baseball. Texas concluded passage of the draft of

its new state constitution. The state was grop-
ing to put its economy back on track. The
promise of accessible higher education greatly
appealed to those who intended to chart the
future of the state. None was more dedicated
to this purpose than Gov. Richard Coke. He
had seen the worst of times and fully believed
that the region was entering a new era.[15]

The original college board of directors
was actively chaired by Governor Coke and
met twice in Bryan during June, 1875. After
considerable discussion, the board extended
a unanimous offer to the former president of
the Confederacy, Jefferson Davis, to be the new
college's first president. Confederate sentiment
and tradition ran strong in Texas. The wave of
postwar militarism and nostalgia that swept
the South was an enduring legacy of the bygone days of the Lost Cause—a nostalgia that
perceived links with the Old South as well as
the notion that soldierly virtues were the hallmark of moral values and the martial spirit of
an honorable man and worthy citizen-soldier.
The relationship between the Lost Cause and
higher education in the South is unmistakable.
According to one historian, "Militarism now
found expression exclusively in more abstract,
mythical terms—honor, patriotism, duty,
respect for law, sacrifice, and even piety. The
southern military tradition's claim that military
service and education were intimately associated with these virtues was one element of the
tradition that emerged intact from the ravages
of the war and in turn had a profound impact
on higher education in the postwar South."[16]

Davis, a West Point graduate and wounded
veteran of the Mexican War of 1846–48, was
one of the few Confederates imprisoned for
rebellious actions during the Civil War. He
had in fact strongly objected to the federally sponsored Morrill Land Grant Act on

the grounds of states' rights. Nonetheless,
he greatly appreciated the courtesy of being
singled out by the Texas A&M Board. Davis
declined the presidential appointment, instead
highly recommending Mississippi educator
and school administrator Thomas Sanford
Gathright for the new position of president.[17]

During the remainder of 1875 and into 1876,
a three-member subcommittee of the board
assembled a list of projected candidates for
president and faculty to initially fill six staff
positions. At the July, 1876, board meeting in
Austin, the board organized the college into
seven departments. Courses to be offered were
also recommended and patterned along the
early-nineteenth-century model of classical
studies and belles-lettres. Such curriculum
stressed "literature that is an end in itself
and not practical, or purely informative."
The courses were clearly not oriented toward
instruction in the practical aspects of "agriculture and mechanical arts (engineering)." The
college was thus initially organized around
the model of other such institutions of higher
learning of the day. Only time, experience, and
a better focus on the A&M College mandate
and mission would change this initial organizational scheme.[18]

Though Gathright at first declined, he
later accepted the offer to move to Texas at an
annual salary of $3,000.[19] To complete the
faculty William A. Banks of Virginia was
selected as professor of mathematics; C. P. B.
Martin, D.D. (doctor of divinity), was named
professor of chemistry, natural science, and
practical agriculture; John T. Hand was appointed professor of ancient languages; and
Alexander Hogg assumed the position of professor of pure mathematics. The board selected
Virginian Robert Page Waller Morris as professor of applied mathematics, mechanics, and

military tactics. Morris, due in large part to his "militarist" background (he was an 1872 graduate of the Virginia Military Institute), at the age of twenty-three became the first Commandant of Cadets.[20]

The new faculty and board on July 23, 1876, agreed to print five thousand promotional pamphlets outlining the general goals, courses, expenses, review of the instruction in "military tactics," and regulations of the new college. Because the population of the state in 1876 was more than one million, all expected a large response on opening day. The *1876–77 Annual Catalogue* contained a brief overview of the "practical instruction in the school of the soldier" and military duties that the college would expect of the cadets: "The drills are short, and the military duty involves no hardship. The military drill is a health-giving exercise, and its good effects in the development of the physique and improvement of carriage of the cadet is manifest."[21] President Gathright and the board selected the first Monday in October as the opening date.

In a sequence of errors and misinformation, thousands of new Aggie Corps freshmen have for decades learned and recounted a slightly inaccurate story: that the opening of the school was delayed for lack of students. The error was apparently introduced in the first list of campusology students were required to memorize; it appeared in numerous publications: the cadet handbook, the YMCA *Students' Handbook, The Cadence,* then *The Standard,* and, subsequently, in the widely distributed (yet unofficial) maroon-covered three-by-four-inch pocket pamphlet *Aggie Facts and Figures* compiled by College Station entrepreneur and bookstore owner J. E. "Old Army" Loupot, Class of '32.[22]

One of the first "famous facts" or "campus-

ology" addressed the opening and inauguration of the college in 1876: "A&M was opened on September 17, 1876, with only six students appearing. Therefore, the opening date was postponed until October 4, 1876, at which time 40 students enrolled."

This legendary inaccuracy in the above dates and sequence of events can be traced to the confusion and memoirs of one of the so-called "first six" students, Charles Rogan '79. It is quite possible that Rogan is not the only one responsible for the legend of the "first six." The records of the first days of the college are sketchy at best and based largely upon the recollections of students and local townspeople in nearby Bryan. Most of the official college records were lost in a tragic fire that destroyed Old Main in May, 1912. Rogan, during the college's semi-centennial preparations, recalled, "The six who reported for matriculation on September 17, the day the College was to have opened, included, besides myself, Archie McIver of Caldwell, Burleson County; H. E. Vernor, and a boy by the name of Smith, a son of the old Col. D. Porte Smith, the first college physician, both of whom resided at Bryan. Ed Shands of Austin was another, and a son belonging to Professor J. T. Hand."[23] Others who are said to be among the "first six" include Aubrey L. Banks, son of Professor William Banks, and John Crisp from Columbus, who by their own account, claimed to be the first and second students to register for classes at Texas A&M.[24] Reflecting on the opening day of the college, Alexander Hogg, math professor, in 1890 gave the following statement to the state's leading newspaper, the *Galveston Daily News:* "Six was an important number in the affairs of the school; for there were six directors, six professors, [six buildings], and *only* six students for a period of six weeks."[25]

Charles Rogan, in a letter dated February, 1928, went into great detail about the opening, stating that the "college was advertised to open on or about September 17th, 1876." However, when, according to Rogan, the school did not open, he waited around until October 2 to register. No such record of this advertised September date exists. Although the date is in question, the number of the original students to enroll is substantiated as six in the annual report of the A&M Board of Directors to the Texas Legislature in 1879.[26]

Gathright and his staff reported for work to prepare the new college on September 1, 1876. Students began to matriculate on Monday, October 2, and by Wednesday, between six and twenty students had arrived to be present for the formal inauguration presided over by Gov. Richard Coke. The *Galveston Daily News* reported that "three or four hundred citizens were in attendance along with a handful of students."[27] Having pushed the final funding through the state legislature and chairing the new college's board of directors, the governor was both enthusiastic and eloquent in his inaugural remarks. The vision of a truly broad-based and available "Texan college" for public higher education had become reality. Coke highlighted the accomplishment in detail, stressing that "Texas is preparing to embrace and be worthy of the great destiny which the big years of the future have in store for her."[28]

After reviewing the goals and mission of the Agriculture and Mechanical College of Texas to prepare "young men of Texas for the high duty of the American citizen," the governor concluded his address with an optimistic perspective on the future of the new institution and its alumni: "In time, these halls will become classic, and the strong men of Texas, the men who will control the destinies of the State and direct her government, and her material development; men from the farm, the shop, the counter, the bench, the senate and the forum, who have been prepared for life's great struggle here, will, after we have been gathered to our fathers, meet in these halls and with grateful hearts amid the scenes and struggles of their youth, in poetry and song, and in silvery eloquence, chant the praises of their Alma Mater."[29]

The narrative that follows is a chronicle of the events and personalities that shaped the history and legacy of the first 125 years of the Corps of Cadets at Texas A&M. The story of the Corps is written by thousands of cadets from all walks of life. The training, education, and traditions passed from generation to generation have shaped the image and destiny of the institution and its former students. In peace and war, the citizen-soldiers who came to Texas' first venture into public higher education have left their indelible mark as citizen-soldiers on their alma mater, the state of Texas, and the nation.

1

THE HOWLING OF WOLVES

The military system of school government . . . tends to develop in the student a high sense of personal honor and moral responsibility, and to give him those habits of regularity, promptness, self-reliance, and respect for proper authority, which go far to make the good citizen and the successful man of business. It thus becomes a potent factor in the formation of true character.

John G. James, A&M President, 1879–80

Many parents imagined that the military discipline at A. and M. would prove a one-all for their wayward sons, and many of the latter came with their minds set on getting just as much fun as possible out of their banishment.

William A. Trenckmann, Class of 1879

THE ORGANIZATION of the Corps of Cadets began shortly after the festive inauguration of the college on October 4, 1876. The mandate of the Morrill Land Grant Act of 1862 to include "military tactics" proved both attractive and challenging to the early faculty as they worked to maintain order on the new campus. In the late 1870s, memories of the "Lost Cause" and the military heritage of the South were keenly engraved in those entrusted to set the mold for the Agricultural and Mechanical College of Texas. The population of the state exceeded one million in 1876. The "Corps," as it was called from the first days of classes, would be the pivotal campus organization. Remote in location and eager for continuity and purpose, the new college slowly adopted the veneer of military bearing and discipline. It mattered little that neither Congress nor the U.S. Army in the years between 1862 and 1916 made any serious attempt to define guidelines and details of what the original 1862 legislation meant. From its earliest days, the Corps of Cadets at Texas A&M, though organized along military lines, was a quasi-military organization interwoven with the larger mandate of the parent institution to train and educate young Texans. Thus, during its formative years, the Corps proved a tremendous catalyst in providing a common denominator to a diverse student body.[1]

Robert P. W. Morris, appointed primarily as a math professor, was designated by the A&M Board of Directors and President Gathright to serve as the first Commandant of Cadets and the interim instructor of military tactics. This additional duty was viewed as a temporary arrangement until a regular officer of the U.S. Army could be detailed to the campus. As a recent graduate of Virginia Military Institute (VMI) and the youngest faculty member, with the most "recent" military experience, Morris, often addressed by the honorary title of "major," was given the task of drafting a set of regulations to govern the Corps and organize the new cadets along military lines. He pursued his assignment as Commandant with a great deal of energy and foresight. Given the remote conditions of the campus and lack of any precedent, he worked to instill a sense of purpose and esprit de corps at the fledgling college. Gathright and the faculty of six agreed that the military discipline system would provide more stability. Though well liked by the students, Morris—the youngest Commandant in the institution's history at age twenty-two—faced a tremendous challenge.[2]

The campus in 1876 was situated on a bare prairie and consisted of two buildings—Old Main and Steward's Hall or Gathright Hall—along with a number of scattered faculty cottages and outhouses. The new college did not offer adequate amenities. R. W. Guyler '79, who came to campus his fish year by ox-drawn wagon, recalled, "There was no such thing as running water, no electricity, no streets, and no sanitary toilets." Kerosene oil furnished light for studying. Students were required to pump and carry water from nearby shallow wells and were issued axes to cut and carry their own firewood. Cadets were left to their own imagination for extracurricular activities. Thus the military regime was deemed useful by the faculty to keep the cadets out of mischief and to maintain discipline. Cadet barracks were located on the upper floors of both buildings. Meals were served under the direction of the college steward and business manager, Gen. Hamilton Prioleau Bee, on the first floor of Gathright Hall. The fifty-four-year-old former Confederate general was the embodiment of the "Lost Cause." A veteran of the Mexican War and former speaker of the house in the Texas Legislature, at the close of the Civil War in 1865 he went to Mexico before returning to San Antonio in the mid-1870s. Credited as an excellent administrator, he proved a positive influence during the formative days of the college.[3]

During the early days rabbits, deer, wild or stray cattle, and mustangs roamed freely across the campus. One faculty member recounted that it was not uncommon to see a pack of wolves roaming the campus. One new cadet who had arrived late to enroll was attacked by wolves during the day "in full sight of the main building!" The environment of the new college was raw and rugged, and the

The treeless, windswept campus was a spartan location. On the left is Gathright Hall and to the far right is the majestic Old Main. This picture was taken in the late 1870s from the present location of the Fish Pond. Courtesy Texas A&M University Archives, Cushing Library

students were equally tough and rowdy. A former state senator reflecting on the conditions at A&M wrote that he "had rather give his boy a pony, six-shooter, bottle of whiskey and deck of cards and start him out to get his education than send him to A&MC." Equipped in that manner in 1876 the senator's son or anyone else's would have made friends quickly. Campus life was both simple and challenging in 1876.[4]

The rustic conditions and view of early campus life are reflected in a memoir drafted by Mrs. William A. Banks, wife of the first English professor:

COLLEGE IN 1876

In the early days of the A. & M. College our diversions were few, an occasional walk in the late afternoon, a favored one being "down the Line" which was only a long rope stretched from the main College building to a stake post located where the last & fifth brick residence now stands. Only the main building & the Mess Hall which is now called the Gathright Hall was finished & occupied by Profs & Cadets. The College opened the first day of its work with six students, six Prof. & six members of the board, though by Feb 1st 1877 we had 300—as they were coming in every day. This line remained intact from 1st of Sept 1876 till March 1877 as only a small quantity of material for building was brought to the grounds at intervals so it was the 1st of June before all the residences were finished. Our beloved Commandant [Morris] being yet a single man, he did not require a residency during all this time.

The Pres. Mr. Gathright was so anxious

to see the good work go on, that he would often suggest a "walk down the line" to see if a few more bricks had arrived. It was no uncommon sight at that time to see a pack of wolves leap out in front of us at the sound of foot steps, from among the tall rank weeds that encompassed the campus grounds, when taking these walks, so it was not considered safe for ladies or children to go alone.

I was informed by Pres. Mr. Gathright he would like to have a lady guest at the College to spend time with us and as his family had not come, he having yet no house, he would be pleased if I would entertain her in my quarters on the second floor of the now Gathright Hall. Hospitality had always reigned in my nature, so I readily a[s]sented to his wishes—this guest was Miss Mary Hunt of Ala. now Mrs. Afflick of Brenham, Tex. There was only one little child on the campus for a long time, this little one being then Katie Lee Banks & was called by the Profs. the "child of the College". There was too a stout boy of under 14 yrs, Duncan Martin who had come to enter College but was too young. This boy was lonely for need of playmates and wandered one sunny afternoon far out from the building, chasing butterflies—the little lonely girl sat at my window watching him with strained eyes fearing he would be completely lost among the tall strange weed—finally a cry of distress from the boy reached the ears of the little girl & she ran to inform the father of his boys dangers for she saw the wolves jumping fiercely on him and biting him again & again. The father immediately summoned the aid of the Com. and Corps of Cadets to the boy's rescue & the wolves leaped rapidly away out of sight at the coming of the body of men. We were next anxious as to the seriousness of the lad's injuries & had him brought into my quarters where his wounds were properly dressed, after which he was safely ensconced to his Father's keeping and all was soon quiet again.

Our guest had now arrived & we had puzzled our brain how we should amuse her during her visit. Soon after her arrival supper in the Mess Hall was announced. The introduction of Profs. to our guest & many other little pleasantries passed the time at the Faculty table. After the Cadets had left the Mess Hall some of the Profs, our guest & I lingered over our cups of tea, engaged in delightful conversation when suddenly there came from the last cadet who had gone out a terrible cry of "oh I'm killed, I'm killed". Immediately the Com and Cadets were again called, the cadet having turned back again & had fallen on the cement floor of the Mess Hall porch. These wolves had returned & had crept in stealthily behind him & had bitten him severely on his body—he was carried to comfortable quarters where he rec prompt attention. The wolves then made their next appearance with their faces pressed closely against the Mess Hall windows all along on one side sniffing & snorting as if they meant to come through—just at our table & you may imagine our fright as the horrible mouths of these animals looking in on us as if they would devour us in an instant. It is needless to say that the ladies were panic stricken & did not hesitate to depart taking our flight above stairs. Our little girl was inconsolable sobbing & crying saying "she could not get up the steps".

The Com. in the meanwhile gave orders for the Cadets to, form in line that a thorough search would be made for these marauders—In an instant at the sound of footsteps they had scampered away and never came so near again, though for three years about dark they would assemble in the small skirt of the woods near the last brick residency & render that weird howl till the wee hours of the night. After reaching our apartments & composing ourselves we decided we had been more than amused for one evening. Our guest is a well known poetess & this grand & courteous old man Gen Bee who was with us then requested Miss Hunt to write this evenings entertainment for him which she did in fine style & the next evening was spent in listening to her account of it with much merriment & laughter.[5]

Newly arriving students had only to be fourteen years of age, knowledgeable in simple arithmetic, fractions, English grammar, and geography. To be admitted they were expected to read in the "fifth reader," be able to write legibly and—if necessary—present testimonials from home. Pre-1880 admissions to the college were based on a system similar to West Point and the Naval Academy, whereby the student body would be filled by political appointment. The plan called for three students from each state senatorial district, one from each congressional district, and two appointed by the U.S. senators from the state at large. This system was to ensure both a broad representation from across the state as well as a quick means to fill the ranks. The selection process lasted for only a short while and was abandoned when it could not be properly implemented. President Gathright interviewed each student prior to enrollment and class assignment. Each was given a room and a copy of "the bible"—a twenty-eight page set of the *Rules and Regulations*. These rules were also printed in each annual catalog

and provided the first degree of strict regimen and discipline to which most new students had ever been exposed. The catalog further detailed the academic structure of the college and fees. The new college offered four areas of study: agriculture, mechanics and engineering, languages and literature, and military tactics. The cost for a first-term cadet was $128.50, which included a matriculation fee of $20.00, board ($12.00 per month), two sets of uniforms ($40.00 and a laundry fee of $6.00), fuel and light charge of $3.50, and a catch-all $5.00 charge for "incidentals." The second or spring term cost only $68.50 total. The total annual fee of $197.00 was payable in four installments: $64.00 on the first day of classes and three equal payments of $44.33⅓ on the first day of December, February, and April. Books and supplies were estimated to range from $10.00 to $15.00, yet such expenses could be "worked out" if a student was short of cash. In the early years, rooms were free of rent and included a bed, table, chair, basin, and water bucket.[6]

According to the 1876 *Catalogue of the State Agricultural and Mechanical College of Texas*, each cadet was to be issued two uniforms; one set was a frock of cadet gray with one row of college buttons, gray pants, and a forage cap trimmed in black. Major Morris fashioned the cadet uniform after that of VMI. A second set of fatigues was provided for everyday wear. Uniforms were required to be worn at all times during the term. Each student was required to have seven shirts, two pairs of shoes, towels, sheets, four pairs of drawers, four undershirts, and seven collars. The college *Catalogue* stressed that the dress code would be adhered to with no deviation—no fancy cravats or hats. Mustaches, whiskers, and long hair were prohibited.[7]

William A. Banks, A.M.
Prof. of Modern Languages
and English Literature

John T. Hand, A.M.
Prof. of Ancient Languages
and Literature

C. P. B. Martin, D.D.
Prof. of Chemistry, Natural
Sciences and Practical Agriculture

R. P. W. Morris
Prof. of Applied Mathematics
Mechanics and Military Tactics

Alexander Hogg, A.M.
Prof. of Pure Mathematics

Thomas S. Gathright, A.M.
President, and Prof. of Mental and Moral
Philosophy and Belles Letters

The first faculty of six was headed by Thomas S. Gathright of Mississippi. Professor R. P. W. Morris taught math and served as the first Commandant of Cadets. Courtesy Texas A&M University Archives, Cushing Library

As the Corps began to settle into daily routine the Commandant and faculty took great efforts to set up a discipline policy primarily based along military lines. Though not the regular Army, the appearance and demands of militarily oriented procedures clashed with those students not willing to adjust to the new environment. Neither cadets nor faculty knew fully how to proceed. The balancing act was how to maintain proper military standards while at the same time realizing that these "uniformed cadets" were nothing more than inquisitive, restless teenagers. Thus, a fine balance continues throughout the Corps of Cadets history to the present day.[8]

For most cadets, this was their first ad-

TABLE I-I

Corps of Cadets Enrollment, 1876–91

Year	Peak Enrollment	Graduates
1876–77	106	—
1877–78	261	—
1878–79	248	—
1879–80	144	2
1880–81	127	1
1881–82	258	12
1882–83	225	8
1883–84	108	14
1884–85	142	11
1885–86	170	11
1886–87	176	10
1887–88	214	17
1888–89	207	19
1889–90	279	14
1890–91	316	16

Source: *Annual Catalogue* of Texas A&M, 1876–1891

venture away from home. The mere trip and exposure to the new college, though rather bleak, was in those days the major event in a new student's life. Very quickly, over the first decade, changing conditions on campus, faculty turnover, and discord put the very existence of the college in question. However, those early cadets, although at times rowdy, gained a fierce love and devotion to their alma mater—which is seldom generated in such a short time period. The fierce loyalty of the cadets set a pattern of obligation, with many returning often to campus in times of unrest and controversy as well as to meet in reunion.[9]

Discipline was a prime concern of the new administration. The college *Rules and Regulations,* issued in late 1876, were rather detailed both on the administration of the college as well as conduct of the cadets. The twenty-eight-page document was subdivided into eight articles or sections. Article VIII, entitled "Discipline," comprised more than two-thirds of the *Rules.* The early premise, not lost on future military organizations, was to set rules and "delinquencies" or demerits and use such a framework to create uniformity and discipline. Scarcely was a topic or offense not outlined. Demerits were set as to the "degree of criminality" of the offense, with the most severe being classed as "offenses of the first class" or ten demerits, down to "seventh class offenses," which resulted in one demerit. The demerit files were maintained by name in order of the total demerits received during a term, with those having the least number being placed first. This card box system quickly set the cadets apart. Fifty demerits resulted in a reduction of rank, and 150 in the first term or 250 for the two terms combined would result in the cadet's being declared deficient in conduct and "forthwith sent to his home by the President." Any cadet who received fewer than ten demerits was to be recognized publicly on commencement day as "distinguished in conduct." This honor eluded most cadets.[10]

The detailed rules were primarily for those cadets who did not live up to the admonishment of the *1877 Annual Catalogue:* "Discipline has been administered with rigorous impartiality, tempered with a wise discretion, and the elevated tone and high moral sense pervading the Corps vindicate the wisdom of the administration. It would be hard to find in any institution, a collection of young gentlemen of better feeling, and generally, of more refined and considerate conduct."[11] A sample of the 157 rules under Article VIII is testimony to the extent of the 1876–77 *Rules.* The detail and content is based on Morris's experience as a cadet at VMI and Gathright's concern about—and some say fear and distrust of—the "military features" of the college.[12]

ARTICLE VIII

DISCIPLINE

I. The students of the College shall constitute a military corps, and be subject to military discipline, under the immediate command of the Commandant of Cadets. . . .

IV. No student shall drink, bring, or cause to be brought within the student's limits, or have in his room, or otherwise in his possession, wine, porter, or any spirituous or intoxicating liquors, brandied fruits, or viands, upon pain of being dismissed. . . .

VIII. No student shall play at cards, or any game of chance, or bring, or cause to be brought upon the premises of the College, or have in his room, or otherwise in his possession the cards or other materials, used in these games, on pain of being dismissed, or otherwise less severely punished.

IX. No student shall cook or prepare food, or have cooked provisions in his room, or given an entertainment there or elsewhere, without permission.

X. No student shall be allowed to keep a waiter, horse, or dog. . . .

XV. Any student who shall insult a sentinel by word or gesture, shall be dismissed, or be otherwise less severely punished. . . .

XVII. No student shall visit during the hours of study, or between Taps and Reveille, or be absent from his room at those times for any purpose without permission from the proper authority. . . .

XXIX. Any student who shall wantonly abuse the person of any other student, by playing unjustifiable tricks upon him, shall be dismissed, or otherwise less severely punished

according to the nature of the offense. . . .

XLI. In all details of military duty, the rules and regulations for the government of the Army of the United States are to be observed so far as they are applicable to, and do not conflict with, the regulations adopted for the government of the College. . . .

XLIII. The students shall be organized into one or more companies, according to number, and the officers shall be appointed by the President upon the recommendation of the Commandant. The selection shall be made from those students who have been most active and soldier-like in the performance of their duties, and exemplary in their general deportment.

In general, the officers will be taken from the first class, and the non-commissioned officers from the second and third classes.

XLIV. From the first of March to the first of December, there shall be an infantry drill every day (while the college is in session), when the weather is favorable (Saturdays and Sundays excepted), at such regular hours of the day as the Commandant of Cadets may appoint, and for the residue of the year at his discretion.

XLV. There will be a dress parade at retreat, the weather permitting, according to the form prescribed in the General Regulations for the United States Army.

XLVI. There shall be an inspection of the battalion under arms, every Sunday morning, when the weather permits, according to the form prescribed in the General Regulations of the United States Army. . . .

LV. No student shall lend or exchange his arms, or accoutrements, or use those of another student. . . .

LIX. Every student, on rising in the morning, shall make up his bed, neatly, hang up his extra clothing, and arrange all his effects in the prescribed order. . . .

LXII. No student shall throw water from any room in buildings into the hall, or into the galleries, or spit on the floor, or sit in windows. . . .

LXVI. No student shall play on any musical instrument on Sunday, nor during study hours on any other day. . . .

LXXIV. No student shall throw missiles or stones of any description in the vicinity of the buildings. . . .

SUPERINTENDENTS FOR DIVISIONS

LXXXII. Each division of quarters shall be placed under the superintendence of the Cadet Officers of the corps, who shall be charged with immediate care of its police, and the preservation of general good order therein. . . .

XC. The Commandant of Cadets shall make a minute inspection of the rooms, furniture, &c., of the cadets as often as he shall deem necessary for the preservation of the buildings and furniture, and for the maintenance of discipline. He will also make occasional inspections for quarters after taps. . . .

OFFICER OF THE DAY

XCIV. The Officer of the Day shall be detailed from the roster of cadets acting as officers. . . .

CIII. He shall be held responsible for all public property deposited in the guard room. . . .

SENTINELS

CVI. There being perhaps no better test of discipline than the manner in which the duties of sentinels are performed, cadets should understand the honor and responsibility of a sentinel on post. . . .

CXV. Any student who shall neglect his guard duty by deserting or sleeping on his post, or in any manner impair the security of the public property, by neglect of the duties developing upon the guard, by law, shall be dismissed, or otherwise less severely punished. . . .

MESS HALL

CXXVIII. The senior Cadet Captain is superintendent of the mess hall and the other cadet officers in the order of rank are assistant superintendents. It shall be their duty aided by the non-commissioned officers to preserve order therein and to enforce the mess regulations, and to march the corps to and from the hall in a military and soldier like manner. . . .

CXXXII. Every cadet shall march to and from meals, except the officers of the day and the senior corporal of the guard not immediately on duty. The latter shall precede the corps to the mess hall, shall report any cadet who may enter the hall before the corps, and shall remain until the relief arrive. . . .

EXCUSES

CXLVI. Any cadet officer having fifty demerits for one quarter, will forth with be reduced to the ranks. . . .

SOCIETIES, ETC.

CLXI. No society shall be organized among the cadets without a special license from the President; nor shall any assembly of cadets be held for this or any other purpose without his express permission. . . .

REGULATIONS

CLVII. A copy of these Regulations will be deposited in each of the student's rooms for the safe keeping of which the orderly will be held responsible; and it is made the duty of the Commandant of Cadets to see that this regulation is complied with.

There is evidence that President Gathright did not fully agree with the provisions for military training and regimentation mandated by the Morrill Act and supported by the A&M Board of Directors. Gathright had not attended a military school and had avoided serving in the Confederate army during the Civil War. His background in classical education (in what he deemed a more relaxed and refined atmosphere) called for an adjustment to the regimen of rules and military procedures set forth by Morris with his overall approval. At times the military discipline and procedure irritated Gathright, and it was to become the source of future friction at the college. In one incident a cadet sentry at his post on guard outside the president's office halted Gathright and "threw [aimed] his gun on him" and abruptly demanded to know who he was and what his business was. P. L. Downs recalled, "The boy was a green country lad but had his instructions to stand guard and let no one pass that did not give the password or have permission from the Commandant. This action greatly irritated Gathright and he never got over it."[13]

The first student body was organized along military lines into a battalion and divided into two companies of cadets—A and B. Cadet officers were selected for "proficiency in drill and deportment." Each cadet company was "officered" by one captain, one first lieutenant, one second lieutenant, and a number of sergeants and corporals. The appointments to these positions were reviewed and conferred by the faculty upon the recommendations of the Commandant of Cadets. Such positions were honorary and effective for each collegiate year unless otherwise rescinded or forfeited for misconduct. The Commander of Company A served also as "First Captain" or Cadet Corps Commander. Alva P. Smyth of Mexia was selected as the 1876–77 Commander of

Company A and First Captain. Company B was led by Charles Rogan of Giddings.[14]

The adjustment to campus life by both cadets and faculty proved slow. By December, 1876, nearly fifty cadets were enrolled. The spartan first semester was celebrated with a Christmas of "feasting and merry making." Those students who were unable to go home were treated to a surprise dinner hosted by the Commandant and the ladies of the campus. An unknown cadet provided the following account to the *Galveston Daily News:*

Quickly but steadily we marched along to the resounding notes of fife and drum. As we entered the mess hall door we were dumbstricken. We could hardly believe our eyes; for there we stood before tables that fairly groaned under the many good things heaped upon them. But we were not long in coming to ourselves again. It is waste of time to state what happened there; suffice it to say that in less than an hour we, instead of the tables, were groaning.

May we never forget this our first Christmas day spent in the halls of the A. and M. College of Texas.[15]

By June, 1877, the number had doubled to 107. The first year ended with a cadet parade and brief ceremony, at which Major Morris was praised for his training of the Corps. The cadets adjourned until October 1, 1877. Early students P. L. Downs, W. A. Trenckmann, and G. W. Hardy fondly recalled in Ousley's *History* that Morris was a "spirited Virginia Gentleman of rugged virtues." Though Gathright was never completely comfortable with the military emphasis at the college, Morris was a devoted militarist in the sense that he believed in the wisdom and the efficiency of the military discipline. As the first Commandant of Cadets, he deserves much credit for the roots of the esprit de corps that would become a hallmark of the Cadet Corps.[16]

During the summer recess Gathright and the staff worked to continue to upgrade the campus. There was a concern that the small campus could not withstand a large increase in cadets. It was estimated that 160 students could be comfortably housed. With the assistance of Governor Coke, who resigned as governor in December, 1876, to become a U.S. senator, a regular U.S. Army officer was requested and assigned to the A&M campus in January, 1878. With the new Commandant came 200 rifles and accoutrements, 1,000 rounds of ammunition, and two field pieces with caissons. In his annual report to the governor, Gathright indirectly expressed reluctance on the degree of regimentation at the college: "Whilst the military feature is required, and is beneficial in discipline, etc. the popular complaint against this department—'too much military'—can be met and appeased without detriment to the College."[17]

The popular Major Morris returned to full-time instruction in the math department when Capt. George T. Olmsted, Jr. (USA),

TABLE 1-2
Corps Organization, 1876–77

Major R. P. W. Morris, Commandant
Cadet 1st Lieutenant C. A. Burchard, Adjutant

| Alva P. Smyth | Charles Rogan |
| Captain, Company A | Captain, Company B |

Corps Strength:
December, 1876: 48   June, 1877: 107

became the first regular Army Commandant of the Corps and the only employee of the college not on the state payroll. A debate ensued among the faculty on the academic status of the newly assigned officer, yet it was soon resolved. The federal statutes that allowed an officer to be detailed to the college stated that the officer "should be a recognized member of the faculty, with equal vote, and not simply a prefect of discipline."[18] Yet the Army was indifferent to establishing formal guidelines and a curriculum of instruction in "military tactics." Thus, the expertise and enthusiasm of the pre–World War I Commandants assigned to A&M was critical to the level of cadet training and discipline. Cadets were equally concerned with the assignment of a federal officer but for different reasons. One cadet noted upon first sight of the new Commandant, "He is about seven feet tall and he looks like he could take any boy here back of the neck, and hold him out and slap his jaws!"[19]

Major Morris departed A&M in 1880 to become a lawyer, relocated to Duluth, Minnesota, and, after serving in a number of local elected offices, was elected to the U.S. Congress and served from 1896 to 1903. He concluded his career as the U.S. district judge for the District of Minnesota and died shortly after delivering a stirring campus commencement address to the Class of '24, attended by the largest gathering of the "Grand Old Gentlemen" from the 1876–79 cadets in reunion.[20]

Captain Olmsted proved highly capable and made a marked impression on the cadets. His primary responsibilities included the daily drill and instruction in the manual of arms as well as overseeing the discipline program. However, for the most part, only basic military tactics were taught on an informal schedule—unlike the more vigorous programs at VMI and West Point. Cadet training and discipline were instilled by room and uniform inspection as well as formations in front of Old Main for roll call in the mornings and the evenings. Cadets were expected to be in uniform at all times, stand guard duty on a regular basis, and not leave the campus without proper permission. Furthermore, few activities or amusements were available to the cadets. In the late 1870s cadets attended the State Fair in Houston; however, Bryan—only six miles from the campus—was usually off limits.[21]

The worst fears of Gathright and his staff occurred in the fall of 1877 as the student enrollment began at 200 in October and exceeded 250 by Christmas. A third cadet unit, Company C, was added, and sophomore cadet Charles Rogan was selected First Captain of the Corps for both 1877–78 and 1878–79. Housing conditions became so acute that students slept in the hallway of Old Main, the lobby of the president's office, and in the mess hall. The seriousness of the overcrowding was vividly captured by cadet William Rufus Nash of Columbus, Texas, in an early October, 1877, letter to his mother: "You must not be surprised to see me at home any time. I am getting tired of this sort of business. They have got Waddy and myself in a room with nineteen boys, and some of them are the worst sort of ones. The President says he will move us as soon as he possibly can."[22] One bright spot was the addition of Bernard Sbisa in January, 1878, as the chief steward and dining hall manager in charge of all meals. The overcrowding attracted public attention from not only concerned parents but also the statewide press who hastily concluded that the administration did not

One of the earliest known pictures of an A&M cadet. Robert Fulton Jones entered the college in September, 1877. Courtesy Sam Houston Sanders Corps Center

know how to manage a college properly. Texas A&M's prosperity and growth were almost its undoing.[23]

Persistent housing problems, a shortage of funding, and perception in the press that A&M was not fulfilling its mission plagued the college until the early 1880s. The first college newspaper, *The Texas Collegian*, attempted to depict a rather normal college environment. To the consternation of adults, however, idle students passed the time in odd or rowdy ways; an item appearing in the newspaper in 1879 noted that "two cadets attended the hanging at Bryan on the 2nd inst. at the request of their parents."[24] Esprit de corps was vital to the cadets, even if generated by illicit actions such as a mass student exodus to the Brazos River for a swim. Outings to the river, some seven miles west of the campus, were forbidden yet proved one of the few cadet adventures that could interrupt the monotony of campus life. However, the concern over the direction of the college was sparked in mid-1879 by the visit of the state legislative review and investigative committee, a lack of planning for housing, and an unforeseen disturbance among the Corps and faculty that erupted due to "grave dissensions and bitter animosities."[25]

In an effort to ease the housing problems, Gathright appealed directly to the legislature for new funds to build additional barracks. Once approved it would be eighteen months before the needed space was completed. In the meantime, an altercation developed between the faculty and President Gathright over the selection of cadet John C. Crisp '80 as the 1879–80 First Captain and Commander of Company A. This bizarre incident, embroiling students, faculty, the alumni, and the governor of Texas, was to rattle Texas A&M to its young foundations. Cadet Crisp had worked his way up through the ranks of the Corps from corporal, to first sergeant (1877–78), to captain of Company B (1878–79) and was in line for promotion to the top cadet position as First Captain. He was an officer of the prestigious student literary organization known as the Austin Society and had no demerits, a good academic record, and the backing of his peers in the Cadet Corps.[26] However, in June at the annual staff meeting for the selection of cadet officers for the ensuing school year, the faculty (though Crisp was recommended by both the Commandant and the president) voted to reject cadet Crisp's promotion. In a short time it was revealed that Crisp was an unwitting casualty of an internal faculty dispute between President Gathright and Prof. Alexander Hogg. Though a victim of petty internal politics and bickering, Crisp did not easily abandon his appointment as First Captain.[27]

In the summer of 1879 Crisp conducted an intensive letter-writing campaign to inform his classmates and to assemble testimonial letters. This campaign to line up support proved very successful with both the cadets and the former students or ex-cadets. Captain Olmsted, while on vacation with his family in New Braunfels, wrote Prof. Louis L. McInnis in late September prior to the opening of the fall semester of his concern over the divisive nature of the events at the college and concluded that beyond a doubt there is "an air of distrust . . . when will it be over?" Reflecting the concern of the rapidly deteriorating situation, Gathright, sensing the worst, advised a few friends that "there is evidently a conspiracy against me" and the "very future of the College" is jeopardized.[28]

Upon the cadets' return to campus, the Corps of Cadets on November 16, 1879, petitioned Gov. Oran Roberts to step in personally to settle the festering dispute and approve Crisp to his rightful position as First Captain. In a detailed appeal to the governor, Crisp was adamant in his own defense.[29] Roberts responded by calling a hearing of all involved before the A&M Board in Bryan in late November. All activity on the campus and in nearby Bryan came to a halt for three days as the board interviewed and questioned Gathright (ill at the time with pneumonia), each faculty member, numerous students, and cadet Crisp. The proceedings received detailed statewide attention and coverage in the press.[30]

Governor Roberts, a former Texas Supreme Court chief justice and Confederate infantry officer during the war, wasted little time in sorting through the issues. Responding to Roberts's comments on the college in the *Galveston Daily News,* Rufus C. Burleson on November 20 wrote a special delivery letter to the governor in Bryan observing that "more importance to brass buttons and a little idle show and a captain of cadets than the grand aims of an A.& M. College" has allowed "petty wrangling and unprofessional bickering" to undermine the general "design" of the college. Such advice helped seal the governor's decision. With swift action, some say pre-planned, the governor on November 22 called for the resignation of the president and the entire faculty. Only Captain Olmsted and subsequently Professor McInnis were spared the axe; the remainder departed the campus by December 1, 1879. John Crisp also permanently departed the college, never to complete his studies. In the decade that followed, Crisp became a real estate and livestock broker in Uvalde and, in 1886, editor and manager of the Uvalde Publishing Company. He regularly corresponded with Professor McInnis, whom he held in high regard, and requested advice on the necessary arrangements for his brothers to attend A&M in the late 1880s. Though embittered, he resigned himself to his misfortune in a letter to McInnis in early August, 1887: "Glad to see old A&M C. progressing, I shall never lose interest in her. Though I have probably been counted as an enemy to the College by some, I have never allowed to pass when I could put in a word for her. In one respect . . . the misfortune was perhaps no greater for the school and for me."[31]

Unrest at Texas A&M in a presumed "military" setting concerned all but the closest observers of events and tradition in the South. Military training, loyalty, and discipline in Southern colleges was viewed as an excellent means to foster positive character traits in young men—the framework of a worthy citizen-soldier. So too was there a sense of fairness and individualism. Military historian Rod Andrew, Jr., in *Long Gray Lines: The Southern Military School Tradition, 1839–1915,* captured the essence of the Crisp affair as well as periodic student protests or "strikes" that would surface at Texas A&M and other military schools across the South:

> Militarism by any definition, insists on obedience and respect for lawful authority. But the American military tradition—especially the southern version of that tradition—included a heritage of individualism, personal autonomy, and rebellion *against* authority. The mythical Revolutionary, as well as Confederate, soldier was a hero who had taken up arms not to enforce obedience but to assert personal autonomy and reject illegitimate claims of established authority. The American heritage of rebellion against tyranny, often expressed in military terms, created serious problems for southern colleges that relied on military discipline and that thrived on the military tradition. These schools were prone to frequent disturbances. Whenever the cadets began to feel that the discipline was becoming too irksome or arbitrary, there was the danger of rebellion by an entire class or student body.[32]

Thus, the Crisp affair was a major turning point for the Corps of Cadets and the college—the faculty had been eliminated, the president discharged, and the cadet system of promotion and ceremony held hostage by those who could not otherwise settle their differences. However, those who hailed the college's demise underestimated the "little school" on the Brazos. The Corps stood firm to its military regimen and due to the statewide publicity actually grew in number. Captain Olmsted was able to expand the training program and enhance discipline with the employment of a VMI graduate, Capt. J. W. Clark. In an effort to support the college, the governor made a commitment to assist the institution financially if the academic program was revamped to teach "practical" courses. Just prior to adjourning for Christmas break, college officials welcomed reports of additional funding from the legislature. Hopes were high for a calmer 1880 spring semester.[33]

The campus unrest was indirectly responsible for a larger involvement by the former students, or "ex-cadets" as they were called, of the college. Loyalty to the Corps and college has always run deep among the alumni and was no exception in the earliest years. In the fall of 1879, a group of former cadets met in Houston at the Bachelor's Hall and informally established the Association of Ex-Cadets. The former students held meetings throughout the balance of the year and called for an on-campus gathering in June, 1880. This group became the forerunner of the Association of Former Students chartered in 1926. The ex-cadets expressed grave concern over the faculty dispute; they were pleased when the governor moved rapidly to fill the void left by the departed Gathright staff. The early leadership and officers of the association were among the best cadets from the first four years of the college. William M. Sleeper of Waco was selected as president, and W. A. Trenckmann, P. L. Downs, and E. E. Fitzhugh were elected vice presidents. A driving force behind the organization was Edward B. Cushing '80, who served as chief secretary. Beginning in 1880, a day was set aside for alumni activities at each graduation in June. And all looked forward to cooperating with the Corps, the Board of Directors, and the new college administration.[34]

The board moved swiftly to restaff the college administration and faculty. A Virginia Military Institute graduate (1866), a cadet veteran of the Battle of New Market, and former president of Texas Military Institute at Bastrop, Texas (1867–70), John Garland James was hastily selected by the board in November, 1879, as the new college president.[35] In an open letter "To the people of Texas" the board confidently reported, "A new president and professors of ripe experience, vigorous manhood and of acknowledged ability and reputation in the state were chosen and in charge. One hundred and thirty students are in attendance and we hope and believe the number will be speedily increased."[36] To put an end to the unrest in the Corps and extend a gesture of goodwill, all the cadets were marched into a meeting with President James to shake hands and "make the new beginning."[37] No stranger to conflict, the former Confederate colonel in the Army of Northern Virginia acted quickly to assemble a new faculty that would stress the practical training indicative of an agricultural and mechanical college. Governor Roberts's stress on the importance of teaching the basics of agriculture was also a major concern of the state Grange, a powerful farm movement that wielded tremendous influence in the state legislature and in the popular press. The governor's instructions to James were based on the idea he often espoused that "civilization begins and ends with the plow." Thus he advised the college to teach the cadets to farm. Roberts was a "strong believer in industrial education that would train the hand as carefully as the head." The strict interpretation of the "letter and spirit" of the Morrill Land Grant Act was a radical swing toward a "vocational" orientation for all instruction at the college. This back-to-the-basics emphasis was equally as unbalanced as Gathright's narrow focus on the classics. James's struggle to do the right thing was painfully taxing yet key to the college's finding a leveling stage for a practical course of instruction. To restaff the faculty, James encouraged Henry H. Dinwiddie, a fellow Virginian, close friend, and Civil War veteran to become professor of chemistry. James R. Cole, a native of North Carolina, was selected as professor of English. Charles P. Estell became professor of ancient languages, and Captain Olmsted remained as Commandant. In a surprising move, the board approved L. L. McInnis for a temporary position in the math department. In June, 1880, he was retained on a permanent basis. Dinwiddie, Cole, and especially McInnis would figure prominently in events of the college over the next two decades.[38]

James was pleased to report to the board and the statewide press that the problems of '79 did not result in undue damage to enrollment of cadets in the Corps of Cadets. With the departure of Crisp, William H. Brown of Navasota was selected First Captain. The Corps was to prove resilient to controversy and change during the early years. The distraction of the Crisp affair was but one incident to the cadets compared with the ongoing problems of poor water, bad winter weather, overcrowding that never seemed to diminish even with the addition of new barracks, and the constant threat of sickness and influenza epidemics. Though there were many hardships, one barometer of cadet life has for years been the letters home to family and friends about the daily routine. None is more poetic and descriptive than the pen of fifteen-year-old cadet Robert W. Guyler in May, 1880. His letter was written in response to a stranger's letter, a possible family acquaintance from Galveston:

I received your letter today. I was very much surprised for I saw on the envelope that it was mailed in Galveston. I was more surprised when I saw that it was written by a Mr. Poindexter whom I have never seen or heard of before but it was none the less welcome as it has been a good while since I received one from home. I am well and getting along fine in my studies which are Latin, English Composition, & Grammar, 1st Academic Algebra, and 2nd Academic Arithmetic. We have six study hours, first hour comes at nine o'clock right after prayers, then all the boys who have a class that hour have to assemble in their different section and are marched into their respective section rooms by a boy who is appointed to this office by the Prof. who heads his section. At every hour the drum beats and the Prof. have to disband their sections and new ones take their place. There is Church every Sunday in the Chapel. Dinner is at one o'clock. School turns out at four. 4:15 We have drill when the weather is fit, right after drill comes dress Parade. I am a marcher, then comes Supper. 6:40 is study call then you can not visit one another's room under no pretense whatever & cannot absent yourself 10 minutes from your room on pain of being reported. 9:30 is Tattoo when you can go to bed, it has just

The absence of paved roads turned the streets into deep mud pits, passable by only oxen and mule-drawn wagons. The wet conditions in the early years contributed to the rash of illnesses among the cadets and faculty on the fledgling campus. Courtesy Texas A&M University Archives, Cushing Library

beat not five minutes ago. 10 is Taps after which no one can leave his quarters or burn his light without permission. Reveille comes very early, I do not know how soon but it is very early for I hate to get up. Today was my birthday. I am fifteen years old. The first letter I ever received from a preacher was on my fifteenth birthday. There was a regular fist & skull fight here today between two of the boys. Some of the big boys parted them, one got a bloody nose & the other got a black eye. The weather is beautiful, had a heavy shower but of short duration Wednesday, the ground being dry soaked it up as fast as it fell, nearly all of the boys were in swimming in the tank tonight. There is 65 boys in the school & forty were in the tank so if the Pres. tried to do anything with us he would have to punish the whole school & then we

would kick and he could not do a thing, as it is nearly Taps I will close for tonight and finish tomorrow.

May 22nd / 80. We had target shooting yesterday evening only twenty five of the best shot. I was not in it. A good many of the boys are out hunting dewberries as there is a good many round the College. The College has a library from which any of the boys can draw books provided he return them in two weeks. The College also takes all the best papers that are published from all parts of the World, printed of course in our language. Several of the cadets are sick in the hospital. One of them was so sick he had to go home. We have a few fleas. We do not want any more. Examination comes on the 23rd day of June & expect a nice time not so much in anticipation of my examination as I do of going home. Why did you ask me did I ever hunt a water duck, did you ever see a land duck if so what species? I have hunted everything but never killed anything larger than a crow in the game line, you must excuse this bad place in the paper I did not know it was on the paper until I was through . . . write often give my love to all.[39]

One vestige of the old "classical" system was the creation of a number of student organizations that stressed the arts and literary pursuits. It is interesting that within this rustic frontier atmosphere, cadets organized the Stephen F. Austin Literary Society and the Calliopean Society, named for the goddess Calliope, the muse who presided over heroic poetry and eloquent literature. These organizations were both exclusive and secretive. At times they gathered jointly to debate key issues of the day. Their existence was approved by the president in hopes of keeping the cadets occupied. The elitist attitudes of the two societies resulted in a subtle response by the other cadets. The first of a long list of unauthorized underground clubs and groups began to organize among those cadets not included in the approved societies. The "Stay Out Late" or SOL Club was one of the earliest of such groups. The high jinks continued throughout the spring and up until June.

The first titled degrees of civil engineer were conferred to Cadet First Captain William Harrison Brown and cadet Louis John Kopke of San Felipe, Texas. According to the 1880–81 college *Catalogue* the Corps at Final Review June, 1880, totaled 144 cadets.[40]

The early 1880s continued to be a difficult time for the college and the Corps. President James made every effort to ensure the cadets were provided for and well housed. The

The son of Gen. Sam Houston, Temple Lea Houston, is one of the most notable "frogs" to enter the Corps at midterm. Very popular with his fellow cadets, he was active in a number of student organizations. Courtesy Texas A&M University Archives, Cushing Library

agriculture program began to experience a modest growth. Additional funding, though slow in coming, was approved for the college. Cadet enrollment dropped from between the 125–50 level to 90 in 1881, and the cadet battalion was reduced from four cadet units (Companies A, B, C, and D) back to two—A and B. Concerned with the conditions at the college, a special Texas legislative committee visited the campus and noted in a March, 1881, report, "It should be remembered, that its [the college's] retrograde tendency is due, in great measure, to the serious misunderstanding with the faculty" and not the cadets. The committee further dispelled charges that the military training under the Morrill Act was given "too much prominence," responding resoundingly that "the military feature is of no disadvantage." The Corps and college were on the right track.[41]

To economize, James recommended that the number of uniforms be reduced and that the "dress hat" and parade uniform be eliminated. For agriculture work and classes, civilian clothing was permitted for the first time. Captain Olmsted optimistically reported to the board that the Corps had begun to have a better grasp of military protocol and that the "military system is used here simply as a means of discipline and wholesome restraint." Cadets were drilled in the procedures found in *Upton's U.S. Infantry Tactics* and regularly held target practice and classes in signaling. Six-pounder howitzers with caissons and limbers, provided by the U.S. government, were used to train an artillery battery. The Commandant recommended that a "military department" be formed to enhance cadet training and to give permanence to the detachment. Olmsted cautioned that without a "permanent" program the emphasis could be reduced and "gradually cut down until there shall be nothing left to it." In mid-1881 Captain Olmsted was reassigned to Fort Leavenworth, Kansas, and replaced by an interim Commandant. Olmsted met with tragedy in August, 1884, when he was ordered to be tried by general court-martial for embezzlement. He was found guilty and, though he appealed directly to President Grover Cleveland, was given his dismissal from the Army.[42]

Olmsted's permanent replacement, Lt. Charles Judson Crane, arrived by way of stagecoach in College Station in the dead of a cold wet Texas winter in January, 1882. Given the six-month absence of a full-time Army officer, Crane found discipline low and the Corps in need of immediate leadership. However, it was not the status of discipline or lack of excellence in military drill that plagued the campus. Instead, the living conditions on campus began to undermine attempts to increase the size of the cadet corps. In his annual report to the president, Lieutenant Crane urged that the conditions be improved to ease the overcrowding. The Commandant reported, "There are on hand 130 single beds, 26 double beds, 174 mattresses, 114 chairs, 15 mirrors, 66 tables, 70 pillows, 25 buckets, 62 wash stands, 37 wardrobes and about 15 lamps." These supplies proved inadequate as enrollment in the spring of 1882 exceeded 250 cadets. President James expressed grave concern the "we [A&M] will be full to overflowing."[43]

Living and health conditions continued to be a problem. From November, 1882, through early 1883, a dozen cadets and a faculty member died of either influenza, measles, dysentery, and/or related disorders. Rumors, despite the efforts of the college administration to the contrary, began to surface statewide that the deaths were due to the poor location of the college. The *Galveston Daily News,* the *Austin Statesman,* and the *Houston Post* fueled the debate at every opportunity. Such innuendo was timely given the efforts of the newly opened University of Texas in Austin to expand its operations and secure a larger portion of the higher education funding from the state legislature. Located only blocks from the state capitol, the new university would continually use its proximity to the capitol to garner increasingly more state funding. At

A&M conditions did not improve, as the Corps dropped to 108 cadets. In March, 1883, President James resigned and his friend Col. James R. Cole was selected as acting president.[44]

Notwithstanding the harsh conditions on campus, Lieutenant Crane noted that "the cadets were eager to learn." The faculty allowed the new Commandant ample time and leeway. Crane reported, "I began at the beginning, with instruction in the School of Soldier without arms, and I carried it up to include all battalion exercises, also ceremonies." Instruction included the basics of artillery and infantry drills, a limited amount of target practice, and the basics of sentinel (guard) duties, outposts, and picket service. Guard mounting was strictly conducted to help ensure discipline and as a means to alert dorm occupants in case of fire. Once the basics were in place Crane substituted Saturday morning drill with instruction in the use of an old three-inch muzzle-loading field piece. The young cadets were amazed at the use of artillery, firing at ranges up to 300 yards. The roar of the cannon fire broke the normal Saturday morning campus silence.[45]

Crane modified the cadet rule book with regulations that mirrored those of the U.S. Military Academy. Hoping to develop a sense of professionalism, the more promising upperclassmen were provided instruction in advanced military tactics and fencing. Knowing that the public visitors and parents enjoyed the pomp and ceremony, the Corps, during commencement weekend in June, performed all sorts of drills, target practice exhibitions, and witnessed the awarding of a prize to the best drilled cadet and the best shot. Crane's duty at A&M ended in November, 1883, with reassignment to Fort Sill, Oklahoma. The Corps presented him with a farewell present, and as he traveled through Texas to his new post, A&M cadets met him in both Austin and San Antonio to provide transportation and a tour. A career Army officer, Crane would return to Texas A&M in September, 1917.[46]

James's resignation was primarily the result of frustration. He had grown weary of the constant struggle to improve the campus with limited resources and the impending political struggle to place the campus in a positive light in both Austin and in the press statewide. A shortage of funding prevented a systematic approach to improvements, and conditions on campus bordered on the primitive or at best Spartan. Even though the president was able to secure a single telephone connection to Bryan, the campus throughout the decade lacked running water for bathing and had no electricity or central heating. However, according to Henry C. Dethloff in *A Centennial History of Texas A&M University,* the Corps persisted in spite of James's departure and the seemingly negative views of the cadet lifestyle. The military regimen was unmistakably popular with the students, agreeable to the faculty, and quite acceptable to most political and public figures, who saw in the Corps a close association with southern military tradition and the bygone glories of the Confederacy and the "Lost Cause."[47]

After departing College Station, James assumed the presidency of the City National Bank in Wichita Falls in 1884 and remained a staunch supporter of the college. He corresponded on a regular basis with state officials on the status of the college as well as with faculty and friends at the campus. In many respects he was better able to assist A&M and encourage its development from his vantage point away from the college. Securing a handle on the growing pains of the college would continue to be a problem as A&M was now pitted against the university scheduled to open in 1883 in Austin for both funding and public approval.[48]

The balance of the decade of the 1880s proved to be equally unsettling for the college and the Corps. The prime distraction, over which the cadets had no control, was the week-by-week fight over the continuing status of the college in the eyes of the legislature and the public. The statewide press continued to have a field day pitting A&M against the new university in Austin. Though it was suggested that A&M should be closed and converted into a "lunatic asylum," there were also those who questioned the wisdom of the legislature and Governor Ireland to create hastily a new campus in Austin and a medical branch in Galveston.[49] The net result for A&M was that sparse state funding would now be further subdivided. The Corps during the 1883–85 period numbered more than 200; its organizational structure was once again divided into four cadet units. The official report to the governor and the state legislature concluded, "The cadets in a great measure govern themselves."[50] Notwithstanding, the rustic campus was prone to numerous events of excitement that drew statewide attention, as illustrated in the *Galveston Daily News* in 1884:

About a week ago the students rooming in the mess hall building were thrown into a state of consternation at the sound of a terrible explosion in the third story of the building. Two students who live in Bryan were permitted to return home from college of Friday night and remain with their parents until Monday morning. So last Saturday a week ago, while in Bryan, a fellow student knowing they were in town, went into their room to be to himself, where he could study in quiet and repose. It was cold. He determined to make a fire. The student put paper in the stove, laid in kindling, struck the match, and before he could say Jack Robinson he was thrown violently against the door, some eight feet from the stove. The stove was carried to the ceiling and the window sash lay in shattered pieces on the ground below, a distance of sixty feet. The student was somewhat mutilated. His hair was singed off close and he was terribly burned about the face and eyes. The cause of the explosion was a pound of gunpowder which the young Bryan men had placed in the stove to prevent a[n] accident of any kind that might happen while absent from their room.[51]

Newly appointed Commandant Lt. John S. Mallory worked to maintain proper military discipline, always mindful that the cadets were first and foremost students. Mallory recommended that adequate bathing facilities be provided to improve on the once a week chance to bathe and urged construction of plank walks to connect buildings in order to prevent the cadets from getting their feet muddy and wet. The Commandant theorized that wet feet caused the rash of cadet sickness on the campus during winter. Furthermore, he believed that cadets should be relieved of cutting their own firewood, that cadets should *not* be allowed to receive "sweetmeats and food of any kind" from home—due to the fact that the mess hall "fare is abundant," and that the uniform issue should be improved to combat the bad weather. However, such timely improvements were slow in coming. Faculty turnover, poor funding from the legislature, and a wavering Board of Directors did not diminish the confusion.[52]

Though Cole was confirmed as president in June, 1883, he remained in the job only a few months. Due to a surprise move by the board, the position of president was abolished in favor of creating a newly elected position known as "chairman of the faculty." Cole was not consulted on this change. Under the new scheme the chairman would have reduced responsibilities due to the creation of the new administrative staff position of "business manager." Thus, in reality, the power of the chief executive was diminished. In late July, 1883, the faculty elected Hardaway Hunt Dinwiddie, a veteran of the Battle of New Market, as the first chairman of the faculty. Dinwiddie was valedictorian of the VMI class of 1867 and adjutant of the cadet regiment. His presence at Texas A&M brought to four—along with Morris, Cole, and James—the number of VMI former cadets to influence the organization and development of the A&M Corps of Cadets during its first decade. Cole, who in later years felt he was eased out of the presidency to make way for this more progressive chairman–business manager scheme, departed quietly to become superintendent of schools in Dallas.[53]

Dinwiddie had been a close observer of both the campus and state political intrigue since he first came to the college in 1879. Thus he quickly took the initiative to publicize the college and the Corps positively to the legislature and the state Grange. He took care to cultivate the support of Grange leadership, particularly Archibald Johnson Rose, the grand master and an agrarian reform crusader. Rose was originally from North Carolina and

had served as a major in the Confederate army. After the war he settled in Salado, Texas, and was a founding member of the Grange of Patrons of Husbandry in 1873. In 1887, Gov. Lawrence Sullivan Ross appointed Rose to the A&M Board of Directors. Though a strong agriculture advocate, Rose also strongly approved of the Corps of Cadets. To generate support and create goodwill for the college, Rose was invited to review the Corps and deliver a commencement address in the spring of 1885. Efforts to make Rose a friend of the college would pay multiple dividends for Texas A&M throughout the balance of the 1880s and well into the 1890s. Rose's initial skepticism about Texas A&M turned, and he became an ardent advocate and supporter of the college, so much so that Gov. John Ireland appointed him to the board of the University of Texas in order to neutralize his influence on behalf of A&M. On the question of merger with the University of Texas, Dinwiddie publicly stated support for the new institution but opposed any notion of unity of the two campuses, citing that they had fundamentally different missions. Dinwiddie emphasized the unique role of the land-grant college by hiring a new professor of agriculture, George W. Curtis, and a professor of engineering, Roger H. Whitlock.[54]

Although the college catalogs during the late 1880s generally de-emphasized the role of the Corps and military training, the organization remained popular with the cadets. The lifestyle continued to be spartan, and there were few extracurricular activities. The number of cadet units varied from two to four as Dinwiddie and the faculty worked to maintain a cadet enrollment between 150 and 200. Students came and went almost at random. One cadet recalled that in 1885, "the railroad

The state's agrarian roots and the Southern military tradition were important twin factors in the early years of the college. This drawing is from the cover of a 1885 campus invitation. Courtesy Texas A&M University Archives, Cushing Library

had no depot, trains did not stop unless someone was going to get off or on, when someone was going to get off there would be two toots and the boys would yell 'fresh fish.'" Dinwiddie kept careful track of the departures. Other than "home sickness," the prime reason for dropping out of the college was "because they [cadets] could not get money."[55] It had long been hoped that the state legislature would provide a tuition supplement or stipend that would allow more students to attend the college. Such measures never materialized. The class of 1885 graduated ten cadets. Thereafter, the "course of study" at

the college was changed from three- to four-year degree programs. Lieutenant Mallory, during his tour at A&M from 1884 to 1886, worked to streamline the Corps's procedures and foster cadet leadership. Mallory was in large measure successful because of the able cadet leadership of First Captains Robert M. Rutherford '85 of Seagoville and Frederick E. Giesecke '86 from New Braunfels.[56]

In October, 1886, Lt. Guy Carleton replaced Lieutenant Mallory as Commandant. Capitalizing on the successes of his predecessor, Carleton urged the cadets to avoid spontaneous, nonproductive activities in favor of the rigors of the military. Yet, as one cadet reported in a March, 1887, letter home, "When the Commandant was absent the [cadet] Capt's would have sham battles, you see we march at intervals of five yards, not in step and can carry our piece any [way] we want and have lots of fun." While noble in their intent, the cadets, ranging in age from fourteen to eighteen, spent much more time off the drill field engaged in horseplay and prohibited activities, such as secret trips to far-off Bryan, clandestine beer busts in the nearby wooded areas, or an occasional free-for-all brawl that usually resulted in numerous demerits and restriction to campus. Practical jokes and games of dominoes helped waste idle time. However, pause was taken on April 21 to "fire a salute for the battle of San Jacinto."[57]

Out of the organized chaos began to emerge a number of organizations and traditions that continue to the present day. In late 1887 with the encouragement and approval of Lieutenant Carleton, who some accused of wanting to turn "our school into a second West Point," an honor guard and drill team was formed of upperclassmen. The daily routine was well in place by the mid-1880s:

6:00	reveille sounds followed by roll call
6:30	room inspection, swept and in order
7:00	breakfast
8:00	guard mounting and lan guage recitations
9:00	recitations in all departments and lab work
1:00	dinner
2:00	shop, field and lab work
5:00	drill, three times per week
6:00	retreat
7:00	supper
8:00	study call
9:45	tattoo—prepare for bed
10:00	taps—"when the lights must go out"[58]

This organization was first named the Scott Guards and later the Scott Volunteers in honor of the college business manager, Thomas M. Scott. During the June, 1888, commencement exercises, the Scott Volunteers made their first public appearance to honor the visit of the governor—Lawrence Sullivan Ross. The drill presentation in white duck uniforms and pith helmets with spikes on the crown impressed and pleased all. Little did the governor know that this organization would in time permanently carry his name. The *Austin Daily States-man* echoed its approval by claiming that the events at the college were the "most interesting exercises in the whole history of the institution." With the graduation of sixteen cadets (seven in agriculture and nine in "mechanical arts"), the newspaper concluded, "the prospects of the College are unclouded and it has entered upon a career of undeniable usefulness to the state." These were welcome words after more than a decade of trial and turmoil.[59]

Due to the untimely death of Dinwiddie in December, 1887, L. L. McInnis was selected as chairman of the faculty in January, 1888. As a survivor of the unrest of 1879, he more than anyone knew the risks and rewards of the top position at the college. However, his knowledge of the board and his many friendships placed him in a keen position to advance the college. Having taken part in the trial and error of the establishment of an "agricultural and mechanical" program, he was able to use such first-hand knowledge to implement a broad-based and successful academic program. He secured the approval of the Board of Directors to create a more streamlined group of academic departments and thus added numerous new courses. Furthermore, by 1888 the college had completed the transition to a four-year degree-granting program.[60]

During the last years of the 1880s the Cadet Corps averaged 210 cadets per school term, organized into a battalion of four companies. Discipline remained stable with few flagrant violations. Lieutenant Carleton actively ensured proper deportment both on and off the drill field. He conducted frequent room inspections and prepared lectures in practical military tactics for each class. Yet, like his predecessors Mallory and Crane, he expressed concern that only limited time was allotted to teach "rudimentary" drills and tactics. Too much time was being taken in the "setting up" of drills and not their actual execution. By 1887, the War Department started a regular inspection of college military programs, but it refused to recognize the military graduates of A&M with any standing except "the empty honor of enrollment in the Army Register." Thus, in spite of four years of military instruction, graduates had no direct opportunity for a military career either in the national or state armed forces.[61]

The senior cadet officers for the Class of 1887. Variations in the cadet gray uniforms were not uncommon due to the variety of tailor shops that provided the uniforms for the Corps. Sitting left to right: Cadet First Captain John B. Hereford; Deputy J. H. Freeman; Adjutant H. C. Hare; Cadet Captains T. B. West and G. H. Rogers. Standing left to right: Cadet Lieutenants F. L. Fordtran, J. H. McNair, E. Gruene, L. E. Allen, and E. R. Knolle. Courtesy Texas A&M University Archives, Cushing Library

This sense of frustration was compounded by the overcrowding, the constant cycle of sickness on the campus, and the meager funding to support the college programs. Such seemingly simple items as a new flag for the battalion, tactical manuals, and proper uniforms suited for both the winter and summer months were not easily obtained. Though the Commandant's salary was paid by the federal government, he had little or no discretionary funds. Cadet housing was expanded in 1887 with the construction of Pfeuffer Hall and Austin Hall, each housing seventy-five students. Plans were well under way for a forty-one-room, three-story dorm, later to be named Ross Hall.[62]

Commandants during this period found themselves in a unique position to influence and mold the Cadet Corps, yet with few resources: no staff, training materials, or supplies. Each Commandant, regardless of his parent regular Army unit (Crane: Second Cavalry; Mallory: Second Infantry; and Carleton: Twenty-fourth Infantry) was dispatched on temporary duty status out of Fort Leavenworth, Kansas. This central point gave each an opportunity to compare notes on not only the Texas A&M program but also other land-grant college detachments throughout the South. In addition to a lack of resources, there was a void of direction and purpose for "military tactics" at the colleges under the mandate of the Morrill Act. The conflict of "discretion" constantly put military training at odds with the academic pursuits of the various land-grant colleges.[63]

Lieutenant Carleton, in his final report to the chairman of the faculty at Texas A&M in 1889, captured the tenor of the dilemma: "It is unfortunate that the wording of the said act simply requires instruction in military tactics, leaving the amount of time and study to be devoted to it entirely at the discretion of the college authorities. The result of this indefinite requirement is evasion of the spirit of the law in many of the colleges benefitted by it, a deplorable lack of uniformity in the amount and kind of instruction afforded the students of the different colleges, and, generally a defeat of the purpose of the law in question."[64]

Notwithstanding, Carleton hastened to add that he could assure the faculty and board that A&M "provides more liberally for the military instruction of its students than any other institution in the South . . . yet, even here the time devoted to this subject is inadequate to attain the results which I believe were contemplated by the law." Reflecting the observation and concern of previous Commandants, Lieutenant Carleton outlined the measures (at the time bold recommendations) needed to enhance not only the Corps but the overall standing of the college. First, a fully staffed Department of Military Science and Tactics should be established and properly funded—unlike the current arrangement in which meager funds are provided for only "necessary repairs" of equipment. Next, the Commandant should have more authority

for the maintenance of discipline and enforcement of the rules and regulations. And finally, because the time allowed for military training is inadequate, a full five hours per week should be allocated for "the important work of the military department."[65]

Possibly sensing a change in the mood on the campus and across the South toward military training, Carleton concluded that his observations were "with full acknowledgement [sic] of the existing feelings among certain classes of citizens that the military feature of the Agricultural and Mechanical College should not be made too prominent." His perceptions and concerns about the mood of the press, the farmers movement, and the state legislature were well founded. Nevertheless, it must be remembered that his focus was on the enhancement of the Corps of Cadets. To this end, the Commandant also recommended that a height requirement for cadets' admission into Texas A&M be set at a minimum of five feet.[66]

Carleton's concern in no way diluted the overall positive association of Texas A&M with the military tradition of the citizen-soldier, both in the South and the nation. Those who helped to mold and shape Texas A&M during the formative years were staunch devotees of the conservative southern military traditions personified by the struggle for Texas independence and by the memory of the Great War between the States, referred to by many at the time as "the late unpleasantness." Regularly during the early years, April 21—San Jacinto Day—festivities were observed with much fanfare. Faculty and students had regular contact with the Confederate Veterans Association and its leaders whose attachment to the days of the Lost Cause were personified by an ongoing sense of obligation to its memory. In honor of the passing of Confederate President Jefferson Davis, a solemn day of mourning, to include a memorial service and a cannon salute every hour from 6 A.M. to midnight, was observed on December 11, 1889. The Corps did not strive to copy overtly the old Confederate militarist of bygone days nor to perpetuate its myth but instead reflected on its legacy as a sense of obligation and high duty.[67]

Lieutenant Carleton was replaced by Lt. William F. Scott of the First Cavalry in September, 1889, but he remained on campus a year, until August, 1890. Lieutenant Scott, a career soldier, was both surprised and perplexed by the conditions he found at the college. The campus was, for the first time, lit by electric lights via a line from Bryan. Each room had "immense" light with a sixteen-candle power lamp.[68] The new "illumination" did not ease the drudgery of campus life, although it did allow for a measure of free time. Scott reduced drill to only three times a week from the daily routine of his predecessors. Resigned to his remote duty, the Commandant did continue the tradition of unannounced inspections. One cadet recalled that "one night Lieut. Scott was inspecting—he found some of the boys making candy—he sat down and waited until the candy was done, he helped himself to the biggest share of it and [was] out without saying a word."[69]

Cadet mischief in the late 1800s equaled any that had existed at the college. Though burdened with drill, classes, and detail work in laboratory and the agricultural shops, the cadets always found time for fun. The evening "water fight" so often practiced in the 1960s and early 1970s in the cadet quadrangles dates from the 1890s. "We had long roil last night and the boys ran out with buckets of water to put out (the fire I mean), all the fire there was was some papers in front of the main building some of the boys threw the water on each other and some of them a bucket full on Lt. Scott and then we all run back and got in bed."[70]

Lieutenant Scott never completely gained organizational control of the Corps of Cadets and campus life. In like fashion, the faculty and administration also remained unsettled. The office of the chairman of the faculty was abolished in June, 1890. In August, 1890, shortly after the faculty unrest, Scott was abruptly notified that he was "posted" or reassigned back to the regular Army. In the interim, Prof. Frederick E. Giesecke '86 was selected as the acting Commandant. At the age of twenty-four, he was the first former cadet of the Texas A&M Corps to hold the position of Commandant of Cadets.

A. J. Rose, chairman of the A&M Board, wrote Lieutenant Scott while he was away on leave at Leavenworth, Kansas, that his services would no longer be needed, saying, "the interests of the College demand a change."[71] Concerned that his military record at the War Department would be impacted, Scott insisted on a formal hearing to clarify the grounds of his hasty dismissal prior to the end of his tour of duty. No such meeting was held. However, Rose sent Lieutenant Scott a laudatory letter. Faculty unrest and rumors of hazing caused Scott to demand redress from the "shabby and unwarranted treatment" by "the damned scoundrels" on the board.[72]

A stalwart alumnus and professor of drawing, Giesecke was Cadet Corps Commander of the Class of 1886. Although Commandant for fewer than sixty days, he was to have a lasting impact on the college in a number of roles over the next five decades. His brief tenure as Commandant allowed him to offer first-hand observations on the status of the cadets and their needs. In early October, 1890, 2nd Lt. Benjamin C. Morse, Twenty-third Infantry, arrived as Commandant and professor of military science and tactics. Morse, an energetic West Pointer, reviewed both Giesecke's and the faculty's recommendations. He had his hands full.[73]

The first decade and a half (1876–90) drew to a close somewhat fittingly in 1890 with a massive influenza outbreak in January and February. The dismissal without notice or reason of the chief college administrator—L. L. McInnis, chairman of the faculty—was followed by the appointment of English professor William L. Bringhurst as acting president on July 1, 1890. College Board Chairman Archibald Rose lamented that "things are [still] not in the best shape at the College." This reference is primarily directed at the lack of administrative leadership, inadequate political contacts in Austin, and the constant shortfall of funding needed to put the college on a sound footing. By 1890, the Corps of Cadets was the predominant organization on campus. The military regimen was ingrained in the institution. With the assistance of the Commandant, the faculty and staff viewed the regimentation as a positive means of instilling discipline, organizing a diverse number of students, and fulfilling federal and state requirements based on the mandate to provide military training and leadership in partnership with higher education.

Thus, what was needed was a man of vision, with proven leadership ability, political savvy, and a high stature among the people of Texas to revive and enhance the standing of the college. Few could have filled such a rigorous profile better than Gov. Lawrence Sullivan Ross, who was offered the presidency of the A&M College of Texas on July 1, 1890.

2

THE ROSS YEARS

The young men of the State can acquire at this institution a knowledge that will prepare them to achieve the highest and best results in any station through the reliable factors, education, industry, and a proper moral instruction, by the application of plain moral precepts to every act of life. In addition to this, the military feature of the College is of transcendent importance, though probably not fully appreciated.

Lawrence Sullivan Ross, August 8, 1890

Through the sheer force of his personality, Ross saved A&M.

San Antonio Express, November 5, 1950

FEW MEN HAVE LEFT their mark on the state of Texas and Texas A&M as did Lawrence Sullivan Ross. A tough-minded self-made man, Ross was a true product of the American frontier. He was a man of action—a Texas Ranger and Indian fighter during the late 1850s. He rose from the rank of private to brigadier general in the Confederate army (1861–65) and was twice elected governor of Texas (1887–91). A lifelong champion of public higher education, he was selected president of the A&M College of Texas in mid-1890, thus forgoing further political office at either the state or national level. Sul Ross proved pivotal to the survival, growth, prestige, and future of the fledgling A&M College and the Corps of Cadets.

Ross's very nature seemed to attract attention, as his life closely paralleled the growth and development of Texas. His character was strongly enhanced and influenced by his environment and the training given by his father, Shapley Prince Ross. That Lil' Sul—or Sully, as he was fondly called—was a fighter and leader was no accident. In 1963, grandson Lawrence Sullivan Ross Clarke '21 related the following episode to R. Henderson Shuffler. During Governor Ross's inaugural ball in 1887 at the Menger Hotel in downtown San Antonio, a disturbance erupted at the main entrance as the honored guests were being

Lawrence Sullivan Ross as a general in the Confederate army, circa early 1865. The former Texas Ranger and former governor of Texas became the sixth president of Texas A&M in 1891. His popularity with the cadets and political savvy placed the college on a solid footing. Courtesy Texas A&M University Archives, Cushing Library

received. Upon arriving at the front door, the elder Mr. Ross, a nimble seventy-six-year-old former Indian fighter and pioneer in the truest meaning of the word, was asked to present his letter of invitation.

"Don't have one!" snorted the old man.

"If you don't have an invitation, you can't come in here," he was informed.

"The Hell you say!" the elder Ross replied. "If I don't get in there, I'll call my boy Sul out, and we'll take this place apart!"[1]

He got in.

Sul Ross, more than any other leader of his day, bridged post–Civil War Texas with the goals and aspirations expressed by Gov. Richard Coke to settle, develop, and educate a new generation of young Texans. His resilience and initiative, won through conflict and action, placed him at a unique juncture of the state's history.

Ross was raised eighty miles north of College Station on the Little River in Milam County and later moved to nearby Waco. Shapley Ross had moved his family from the Iowa Territory to Texas in 1839–40 in response to promotional information on the new country called the Republic of Texas. While growing up on the still untamed frontier, Ross was frequently encouraged by his mother to obtain a formal education. He first attended Baylor at Independence, Texas, south of Bryan on the Brazos River in 1856, and chose to transfer to Florence Wesleyan University in Florence, Alabama, in 1857. During his summer vacation, he was put in command of a troop of friendly Indians and scouts from the Brazos Indian Reservation. This group joined the Second U.S. Cavalry in the late summer of 1858 under the command of Bvt. Maj. Earl Van Dorn in a campaign against the Comanches along the Canadian River and in the

Wichita Mountains. The experience gained during the Indian Wars would prove invaluable, but it almost cost Ross his life. During the final charge against the Comanches he was shot through the shoulder with an arrow and within minutes of that wound he was shot point-blank by a charging Indian using a .58-caliber Springfield carbine. Quick action by a friendly Indian scout, Caddo John, saved Ross's life. The bullet had entered the chest and exited the back between the shoulder blades, miraculously not damaging any vital organs. Caddo John probed and cleaned the wounds, applied some homemade medicine, and moved Ross to the rear to be treated by an Army surgeon. Though also wounded, yet not as seriously, Van Dorn dictated a letter of commendation detailing the bravery and leadership of Captain Ross. He strongly recommended to the secretary of war that Ross be given a direct commission as an officer in the U.S. Army.[2]

Ross never accepted the regular U.S. Army commission. After a period of recovery from his wounds in Waco, he returned to graduate from Florence Wesleyan in 1859 and then returned to Texas to be appointed an officer in the Texas Rangers in 1860. Ross married Elizabeth Dorothy Tinsley and settled in the Waco area. With the outbreak of the Civil War and the secession of Texas from the Union on March 2, 1861, he left the service of the Rangers and enlisted as a private in a Confederate company organized and commanded by his brother, Capt. Peter F. Ross. Within months he was promoted to major and then colonel, ultimately commanding the Sixth Texas Regiment. For his conspicuous bravery and leadership at Corinth, Mississippi, he was promoted to the rank of brigadier general in 1863 at the age of twenty-three. During the

course of the war Ross fought in 135 engagements in ten states, captured thirty stands of Union colors, and had five horses shot from under him.[3]

Judith Ann Benner, in her detailed biography of Sul Ross, concluded, "Ross can be considered as an illustrious example of the American *citizen-soldier* who serves his country when needed and then returns to civilian life once the crisis is past."[4] Upon returning to Texas, Ross applied for and received presidential pardon from the federal government and quickly stepped forward to help lead efforts toward rebuilding the state. During the years after the war, in addition to managing a small plantation on the Brazos River, he was a sheriff in McLennan County, elected a senator in the Texas Legislature, and served as a member of the Texas Constitutional Convention in 1875. After a number of attempts to encourage him to run for statewide office, he agreed to head the 1886 Democratic ticket for governor. Following the departure of two-term governor John Ireland (1882–86) the field for the Democratic nomination opened with five candidates. Ross's stature and statewide appeal proved formidable against the opponents: A. M. Cochran, the Republican candidate, and Prohibitionist E. L. Dohoney. Touted in the media as the "little cavalryman" and a "war horse," his first gubernatorial victory in late 1886 was marked by the largest majority of popular vote to that time. Ross captured 73 percent of the popular vote in 1886 and easily won reelection for a second term in 1888.[5]

Ross's appointment as president of Texas A&M had roots long before his 1891 inauguration. During the Texas Constitutional Convention, while a member of the legislature and as governor, he lobbied for strong, practical—

or what he called "popular"— educational programs for the citizens of the state. During his administration as governor, not only were funds increased to A&M but also to the other state schools. He championed the establishment of the medical school in Galveston, the first of its kind in the state. As early as 1887, Ross had met with the A&M Board of Directors concerning expenditure of revenues in the university fund. Competition for state funding between the two institutions was fierce. A vocal faction in Austin noted that the growth of the University of Texas could best be accomplished by limiting support for Texas A&M. Ross's understanding of the fund, as well as his involvement with the directors of the University of Texas and the directors of Texas A&M, in time proved to be invaluable to the continued existence and growth of the A&M College.[6]

In 1888, Ross worked carefully with the chairman of the faculty, Louis L. McInnis, in order to gain legislative approval of a plan calling for funding to begin a comprehensive building program. The plan provided the first sizable funds to be given the college since its opening in 1876–77. The appropriation of $41,500 included (1) $20,000 for a dormitory and assembly hall; (2) $2,000 for furniture; (3) $9,000 for repairs; (4) $2,000 to improve the water supply; (5) $2,500 for fencing; (6) $1,000 for barns; (7) $4,500 for lab and shop equipment; and (8) $500 for work stock and farm implements. As a token of the college's thanks, the governor was invited to take part in the June, 1888, commencement exercises for sixteen students: seven in agriculture and nine in civil engineering. During his visit it is quite possible that, in addition to a warm welcome and tour of the campus and surrounding community, informal discussions were held to probe his possible interest in becoming president of the college. No governor since Richard Coke had shown as much concern in the A&M College as Ross. Having known Coke as a boy growing up in Waco, Ross had looked up to the transplanted Virginian as both a role model and mentor. Henderson Shuffler, in a unique monograph, *Son, Remember . . . ,* written for the seventy-fifth anniversary of the college in 1951, observed: "Sul Ross looked the part of the scion Virginia gentlefolk, which in reality Richard Coke was. Coke's big frame fit the picture one would envision of the son of hardy Shapley Ross."[7]

In 1888 and 1889, even with the improved funding and relative stability of the school, there were those in Austin who would rather see the college closed. The friction, of course, continued to revolve around the university fund and the question as to whether or not Texas A&M was "really needed." Ross again was helpful in prevailing over those in Austin who had desired to discontinue what opponents termed "the cancer on the Brazos." During the summer of 1890 he was offered the presidency of A&M at the salary of $3,500 per year and formally accepted on August 8, to be effective February 1, 1891.[8]

In his letter of acceptance to the members of the A&M College Board of Directors, Ross acknowledged the "high honor" of being selected president, concluding with a reflection on the "development of self-reliant manhood" and the Corps of Cadets: "The military feature of the College is of transcendent importance, though probably not fully appreciated."[9]

Upon arriving at College Station, Ross found no "cancer," but a college in need of his reputation, leadership, and administrative ability. Noting that "everything looked cheerless" in February, 1891, he wrote to Maj. H. M. Holmes, his former gubernatorial aide and close friend: "I have this morning assumed charge at the A&M C. So far as I can tell, my presence is very acceptable to all parties and a short talk made to the young men this morning seemed to take very well. . . . The school is full of sick, 15 cases of measles and a 'la grippe' prevails to an alarming extent everywhere in this part of the state. I find the work in the details somewhat annoying, but not difficult, and I will soon have things running smoothly here. I think I shall like it."[10]

Quite possibly his acceptance of the presidency of A&M—thereby rejecting encouragement to stand for the U.S. Senate—saved the college. The Cadet Corps of 200 and the small faculty and staff had weathered challenges of establishing the new institution for much of the period since the opening of the college. The conditions that greeted Ross confirmed the seriousness of the situation. There was no running water, and what was available in the wells (cisterns) was contaminated. There was a shortage of housing (including a suitable residence for the Ross family), a disgruntled faculty, an overindulgent Board of Directors, and an unruly student body. Ross lamented, "I am settling down to a steady gate . . . the change was radical, but not greater than I anticipated."[11]

Notwithstanding the dismal conditions, Ross was immensely popular with the cadets as well as their parents. A true hero and near legend throughout Texas, he was the embodiment of the highest virtues of the honorable men of the Lost Cause. It was said that parents sent their sons not to "the A&M College," but to "Sul Ross," a fact that Ross proudly acknowledged: "Their sons come on my account."[12] The college and its new

president were at a defining moment. Few were more suited for the task of setting the college on a solid footing, yet problems had to be overcome rapidly. At the heart of his plans for the college was the enhancement of student life and the creation of an atmosphere that fostered esprit de corps among not only the cadets, but also the faculty, former students, friends of the college, and the board.

Lt. Benjamin C. Morse (USA) arrived on campus on September 27, 1890, to assume the duties of the Commandant following the interim stint by Frederick E. Giesecke. Newly issued orders from the War Department expanded the duties and authority of the Army officer at each land-grant location. All training was to be carried out in a formal fashion. The designated "professor of military science and tactics" or PMS&T and Commandant received new guidance not included in the original Morrill Act. In recognition of this expanded role, the board approved a yearly stipend of $700 "for his services taking charge of College discipline."[13] Encouraged at the prospect of General Ross becoming the college president, the young West Point graduate began one of the first reorganizations of the Corps. The initial structure of four companies (A–D) would remain intact for organizational purposes. In order to create more leadership positions he divided each company into two platoons, thus creating eight units. Additionally, Lieutenant Morse established what would be the origins of the Cadet staff or Corps staff by selecting three senior cadets designated "Cadet First Lieutenants," one each to serve as adjutant, quartermaster, and private secretary to join with the First Captain (still the Commander of Company A) as well as three juniors to fill cadet noncommissioned staff slots. This plan was agreed upon by the president, who appointed cadet officers upon the recommendation of the Commandant. During the fall of 1890 and early 1891, in addition to adding to the time dedicated for drill, the first roots of the now famous Aggie Band were sown with the organization of a "drum corps" consisting of four drums and four fifes. This small unit was placed under the leadership of a paid drum major, R. R. Fisk. In September, 1893, Joseph Holick was hired as the Corps bugler and director of a nine-piece cadet "string band."[14]

Ross, in the meantime, concentrated his efforts on the enhancement of the facilities and entrusted cadet discipline to the Commandant. By the fall of 1891, as a result of a special legislative appropriation, construction started on a new dorm—eventually to be named Ross Hall—a major addition to the mechanical engineering shops, and the president's residence. The new construction was helpful in spite of the fact that overcrowding persisted throughout the 1890s. Existing dorms, Austin (1888) and Pfeuffer (1887), had an intended rooming capacity of about 150. Rooms designed to hold only 2 cadets routinely housed 3 to 4 cadets per semester. Enrollment in 1890–91 was 279 cadets. Ross was concerned about the negative publicity in state newspapers regarding both the overcrowding and the poor health conditions. Improvements in campus life would prove significant to a more positive image of the college in the media as well as overcome meddling from the board and government officials in Austin. Ross's previous political and legislative experience in Austin helped fend off critics of the college. He noted that the "devilment" caused by "petty annoyances with the boys [cadets]" and outside observers could best be overcome with improved conditions.[15]

Valuable allies, both then and now, were the former students of the college. During the 1880s, a fledgling group, known first as the Association of Ex-Cadets and then the Alumni Association of the A&M College of Texas (the term "former student" would not come into wide use until the 1920s), increasingly demonstrated more interest in events at the college. Among the early active leaders were F. E. Giesecke '86 and Walter Wipprecht '84, both prominent ex-cadets, who helped organize on-campus events, such as the annual commencement dinner, of interest to the former cadets. Giesecke used the arrival of Ross to the campus as a rallying event encouraging all to join and be a part of the new spirit and purpose at the college. Correspondingly, Ross knew the potential of such an organizational effort due to his lifelong active association with the Confederate Veterans Association.[16]

Concerned with discipline and the reduction of freshman hazing, Lieutenant Morse also helped revise a number of Corps rules and regulations. In addition to inspection on Sunday morning, the Commandant required, for example, a written explanation for violations of the rules and an adherence to policy. Educating the new cadets on military procedure and protocol was critical to both discipline and morale, since "a large number of the delinquencies committed is through ignorance rather than intent." Such items included enforcement of the rule that all new cadets would be "at least 5 feet in height" and that cadets would wear the proper uniform at all times and bathe at least once a week. The standard uniform was the VMI style: cadet gray jacket and trousers. However, due to shortages in material and tailors, cadets at first wore a varied mix of styles, sizes, and colors. Uniforms were available from tailor shops in Bryan and often were not obtainable for new freshmen until November or December of the first semester. With the addition of an electrical line from Bryan in the late 1890s, it was hoped an on-campus tailor shop would ease the uniform problems.[17]

The improved morale fostered by Lieutenant Morse proved timely. By April, 1892, with the easing of overcrowding due to the addition of the "electrified" Ross Hall (not yet named), the sick list of cadets also began to diminish. By graduation of 1892, 326 cadets helped host a grand banquet for the returning A&M alumni and ex-cadets. Improvements to Corps housing and other campus building projects continued throughout the summer of 1892. As conditions improved, so too did favorable media coverage of the college. The growth of the college, along with discussion about the course curriculum comprised of "liberal and practical education," began to raise questions as to the purpose of the college. Debate over the viability of public higher education continued to be a statewide issue.[18]

Like anyone long active in state politics, Ross (and therefore the college) were not without critics. The 1893–94 governor's race between the incumbent, Gov. William C. Hogg, and challenger Charles A. Culberson proved a volatile arena in which to debate the merits of the A&M College. Hogg was a vocal supporter of higher education in Texas. While not against Texas A&M, he primarily endorsed the younger program at the University of Texas in Austin. To reduce funding for the A&M program would allow more funding for the university. Furthermore, there were special interests in Austin

By the early 1890s the cadet uniform became more decorative with the addition of the long coat, sash, and three rows of buttons. Courtesy Texas A&M University Archives, Cushing Library

who wanted not only more resources for the University of Texas but also funding for the expansion of the medical school in Galveston. Thus, an attack on Ross was an attack on A&M. Ross was criticized by outsiders for developing an "imported Yankee Republican faculty" and charged by the *Texas Farmer* magazine with turning the college into a school of "elaborate military peacockery."[19] His prestige did not prevent him from being needled at every turn by the legislature, Hogg, and the anti-A&M newspapers statewide. Ross often expressed his concern over the public perception of the college and its students: "The College has poor reputation for health in general esteem and this will likely prove detrimental for a time to its prosperity."[20]

Ross privately expressed his concern that a Hogg victory could damage the college. If reelected Hogg would pack the A&M Board of Directors with anti-A&M appointments, which was a cause for concern: "If Hogg gets in he will probably decapitate me or make my position so unpleasant that I have to retire . . . the College has been greatly improved in my respect, and no one will be able to rob me of that credit any way."[21] The cadets were generally unimpressed by the political entanglements and duly lampooned Governor Hogg by electing him an honorary member of the Fat Men's Club. Much to the delight of Ross and the alumni of the college, Culberson was elected the seventh governor of the state.[22]

Charges of "military peacockery" and unrest on the campus proved more political than factual. Ross and the Commandant had quickly gained the respect of the cadets, and in turn the Corps responded to both the leadership and the training. At least yearly, a representative of the U.S. Army inspector general's office conducted a formal inspection and evaluation of each university offering military training. The official report of Maj. R. D. Vroom, referring to his inspection of the college in May, 1893, offered a tremendous glimpse of the campus and Corps in the mid-1890s:

The President of the College is ex-Governor L. S. Ross. The government is vested in a Board of Directors, consisting of five members, appointed by the Governor of the state. The members of the Board are appointed from different sections of the state and hold office for six years or during good behavior and until their successors are qualified. The College is non-sectarian. The number of College buildings is twenty-four, including Professors' residences and principal farm buildings. The requirements of law are met, the institution being prepared to teach annually 225 students. The military department was established October 4th, 1876, when the College was formally opened, and the detail has been continued since. The military course is popular with the students, satisfactory to the Faculty and receives proper support from the College authorities. In determining class standing, or relative standing on graduation, the military course is given the same weight as other departments. The Professor of Military Science and Tactics is First Lieutenant, Benjamin C. Morse, 18th Infantry, who has been on duty at the College since September 27th, 1890. The military professor resides at the College and is provided with quarters on the same terms as other professors.

He is a member of the Faculty, with all the rights, privileges and authority of other heads of departments. In addition to his other duties, he performs those of Commandant of Cadets. The military organization is a battalion of four companies. The battalion staff consists of one First Lieutenant and Adjutant, one First Lieutenant and Quartermaster, one First Lieutenant and Private Secretary, one Sergeant Major and one Quartermaster Sergeant. Each company has one captain, one first and one second lieutenant, one first sergeant, four sergeants and five corporals. There is no separate artillery company, but a select company for special drill, known as the "Ross Volunteers." There is a drum corps service which is voluntary. A band is not maintained. Officers and non-commissioned officers are appointed by the President, upon the recommendation of the Commandant of Cadets. Commissioned officers are appointed from the first or senior class; sergeants from the second, and corporals from the third class. The uniform consists of grey blouse, gray [*sic*] trousers with black stripe and grey forage cap. Rank is designated by shoulder straps with chevrons, as in the United States Army. The national color is carried by the battalion. The number of students in attendance at date of inspection was 215, of whom 213 were in the military department. All of the military students in the military department live at the College and all but three are over fifteen years of age. The discipline of the students is very good and is maintained by the military department throughout the College. Breaches of discipline are punished, by confinement to rooms and guard room; extra hours of duty; demerits, suspension and dismissal. The aptitude of the students for military instruction is excellent. The following have shown special aptitude for military service, Cadet Captain B. C. Parsons and W. H. Mitchell, and Cadet Lieutenant and Adjutant J. W. Hawkins, all of whom are desirous of entering the Army as second lieutenants by civilian appointments. Their general standing in studies is good and they are apparently physically sound.

The College campus furnishes an excellent drill ground. There are no facilities for indoor drill. The ordinance and ordinance [*sic*] stores on hand, all of which are the property of the United States, consists of 230 Springfield cadet rifles, caliber 45, and 230 sets of accoutrements, and are properly cared for under the direction of the military professor. The rifles are in excellent condition, but the accoutrements are old and worn and should be replaced. None of the stores are unfit for use. The full allowance of ammunition has been drawn from the United States. The Commandant's office is in the main College building, and was found to be in excellent condition. The quarterly report has been regularly rendered to the Adjutant General of the Army and copies of all reports and correspondence are retained for transfer to the officer's successor.

A morning report book is kept for each company and a consolidated morning report book for the battalion. A guard report book, order book, and delinquency book are also kept. I received a view of the battalion of cadets, which was commanded by First Lieutenant B. C. Morse, 18th Infantry. The ceremony was well rendered and the battalion presented a fine appearance. The review was followed by an inspection of the battalion. The uniforms were neat and well fitting and the arms in excellent condition. The accoutrements were clean, but the leather is old and not susceptible of polish. The number of cadets present at the inspection was 178; absent, 37. The battalion was subsequently drilled by the Commandant. The drill embraced most of the movements in the school of the battalion and was highly credible. The battalion drill was followed by company drills, each company being commanded by its cadet captain, and the exercises concluded with a very handsome drill by the select company. The excellent condition of the battalion reflects great credit upon the very capable instructor, Lieutenant Morse. I visited and inspected the dormitories, Mess Hall, kitchen, etc., all of them were found to be in good order.[23]

By 1893–94, the campus was beginning to look and operate like a college. Striving to develop an identity for the campus, inquiries into the college were informed that this was "College Station," not Bryan. New additions to the campus further enhanced living conditions. An electrical lighting power plant replaced the single line from Bryan. An ice house, a "steam" laundry, a 60,000-gallon above-ground water tank, and natatorium (indoor pool) were completed. In the first four years of the Ross administration, Lieutenant Morse had worked well with the president and the cadets to restore a sense of esprit de corps. Increasingly cadets took pride in the uniform. Interested in a more colorful and fancier look, senior cadets petitioned the Board of Directors to allow them to wear "regulation white pantaloons and helmets as a dress uniform." The governing board denied the request on the grounds that such a combination would be "injudicious."[24] The Cadet Corps of 313 graduated 34 seniors in the Class of '94, among them the president's son, Frank R. Ross, cadet lieutenant and quartermaster. In his final written comments, prior to reassignment to an active Army unit, Lieutenant Morse commented on the course of progress at A&M in the annual catalog to prospective students: "The military system is a means of enforcing discipline and securing regularity in the performance of academic duties, and tends to inculcate in the students that habit of truthfulness and manliness of character which characterize young men as gentlemen."[25] And thus quite possibly his comments were part of the origin of the epitaph left to Ross— Soldier, Statesman, and Knightly Gentleman.

It is little known or seldom stated, but Ross both before and during his presidency at A&M was courted heavily by state politicians to run for a third term as governor, possibly to run for the U.S. Senate, or to accept an appointment to the then powerful Texas Railroad Commission. A popular man in his mid-fifties, he was still quite capable of holding further public office. The lure of public elected office was great, but after reviewing his personal letters it is not hard to ascertain that he felt he could do the most good as president of the A&M College of Texas.[26]

Not only did Ross improve solid funding for the campus, he was president during many significant events affecting the college, its students, and former students. Among the many events and activities that took place during his administration were the fielding of the first intercollegiate football team, which came to the campus in the early 1890s; the redesign and acceptance of the Aggie Ring in 1894; the organization of the band and orchestra; the forming of a cadet Glee Club—later to be known as the Singing Cadets; the reorganization of the Ross Volunteers; and the participation of the cadets in numerous Corps trips to Houston, San Antonio, and the San Jacinto battlefield on April 21 (in time to be known and observed at A&M as Aggie Muster). Enrollment doubled between 1891 and 1898; the *Catalogue* and curriculum were upgraded and revised; and the first student annual, the *Olio* (1895), and the campus newspaper, the *Battalion* (1893), were published. The Association of Ex-Cadets formed its first clubs statewide, established an alumni placement bureau to assist new graduates with jobs, and created the class agent concept. Each graduating class elected a representative—or class agent—to assist the association with reporting of the activities and whereabouts of the class. Each class since the 1890s has continued the tradition of the class agent. This is of course not to imply that Ross was single-handedly responsible for each of the above, but he did inject an atmosphere of stability and leadership that made these events possible. He encouraged the student body, as well as the faculty and former students, to organize and retain a common identity with their alma mater. His sense of esprit de corps proved to be that extra ingredient needed during the 1890s to develop A&M into a school of solid reputation.[27]

For the first time in the college's history the struggling institution had a seasoned chief executive who championed both the enhancement of higher education and the need for the regimented training offered by the Corps of Cadets. The Cadet Corps underwent major changes and growth during the 1890s. Ross's emphasis on education—primarily an expansion of the faculty and curriculum—and cadet discipline was augmented by increased funding from the Texas Legislature and a better acceptance of the college statewide. Ross personally interviewed each new potential student to ensure there was an understanding of the mission of the college. In presentations to the former students during his tenure, he challenged A&M College graduates to excel in their "chosen profession" to the betterment of their alma mater. Thus, the esprit de corps of campus days was transformed into an active legend by vocal ex-cadets or former students, often demonstrated by rousing after-dinner toasts and spirited "wildcats" or "yells" (the cadets never cheered!)—which predated yells used at athletic events—created for special events and commencement:

Rip-rah-ree! A. and M. C.
Hoopla-hoopla-zip-boom-zip
boom-rah-rah-ree! [28]

The Aggie class ring is one of the enduring symbols of the institution. First designed in 1894, the ring has had only slight modification over the past century. This 1898 ring cost $15.00. Courtesy Texas A&M University Archives, Cushing Library

CHANGING ROLE OF THE COMMANDANT

During the 1890s the Corps of Cadets began to operate along a more rigid military regimen. While President Ross was preoccupied with funding from the Texas Legislature to enhance campus facilities, the Office of the Commandant was given greater latitude in the day-to-day training and discipline of the student cadets. In an agreement with the U.S. Army, regular Army officers were first assigned to the college as Commandant and PMS&T in 1878. Of the first ten Commandants of the Corps, eight were graduates of West Point. However, none were more effective than those assigned to A&M during the 1890s. Lts. Benjamin C. Morse (August, 1890–September, 1894), George T. Bartlett (October, 1894–April, 1898), and Harry B. Martin (1898–99) proved instrumental in redefining a more ordered approach to military training. However, the process was not an easy one due to periodic faculty unrest.

Lieutenant Morse, for example, had been suspicious of the strange circumstances surrounding the departure of Lt. William Scott in August, 1890. Scott was unwittingly a victim of an internal faculty dispute over the administration of the college, which resulted in the termination of L. L. McInnis. Knowing the young Army lieutenant was expendable, the board laid the blame on Scott for a portion of their petty bickering.

The peacetime Army in the late 1800s was small, only about 25,000, and thus the active officer corps were well known to each other. This was the case in 1890 when newly appointed Commandant Morse and former Commandant Scott exchanged letters and views on the course of action for the A&M Cadet Corps. These young Army lieutenants and captains were careful of their active duty actions and were duly concerned that campus politics and experiences while on detached duty in College Station would not harm their careers. By early fall of 1890 at issue was how active and vocal the Commandant would be in college activities and discipline beyond his assigned duties to teach military tactics. Scott strongly recommended to Morse that he petition the board and president to provide an additional stipend (above his Army salary) if the Commandant was to take on added duties not within his role as professor of military science and tactics.[29]

Any such controversy regarding the Commandant or his mistrust was dashed with the arrival of Governor Ross as president. Ross was quick to recognize the very problems of disorganization for which he was sent to College Station to rectify. Lieutenants Morse and Bartlett were to work closely with Ross to edify the Corps. What Morse would start,

Lieutenant Bartlett would build on and enhance beginning in the fall of 1894. The new college president and Commandant began to make a number of changes. In addition to an interview with the president, each new prospective cadet was subjected to a brief entrance exam. In an era before standardized testing, a sample of the admission questions were published each year in the college catalog. Questions covered arithmetic, English, and history topics. With overcrowding a constant problem, the admissions testing helped in identifying those most suitable to be considered for college. The exam of the 1890s would challenge some of today's entering freshmen:

1. Name the two Austins who were founders of Texas.
2. What hard fighters fell at the Alamo?
3. Who commanded the *Alabama?*
4. In what battle did Stonewall Jackson fall?
5. Spell correctly: eez, seez, pleez, neez, neese, poleese, acheev, beleef, looz, brooz.
6. Name the rivers of Texas.
7. Write a half-page account of your trip to this place.
8. If by selling land at $30 per acre I lose 25 percent, at what price must I sell it in order to gain 40 percent?
9. What is meant by a centimeter?
10. Write a brief composition on the "Resources of Texas."[30]

The fall of 1894 was to be an exciting period for the college and the Corps. The student body grew to its largest size ever, 372 cadets. More new students applied than could be accepted, and housing continued to be a serious problem. Notwithstanding the overcrowding, a popular campus sport caught the imagination of both the cadets and the alumni. On October 20, Texas A&M and the University of Texas played the state's first intercollegiate football game before a crowd of nearly 800 in Austin's Hyde Park. The final score was 38-0 in favor of the university's "Varsity." Failing to set up a rematch in College Station, the cadets wrapped up their first gridiron season with a 14-6 win over Galveston Ball High School on Thanksgiving. The team during the first couple of seasons never numbered more than fifteen players—average weight 152 pounds. The new Commandant hoped to capitalize on the football enthusiasm.[31]

Lieutenant Bartlett, an 1881 graduate of West Point, was thirty-eight years old when he arrived in College Station. As the oldest Commandant to date, he was also the most experienced in the demands and leadership of a military-style educational environment. In addition to lessons learned during his student days at West Point, he had served as Commandant and professor of military science at the Pennsylvania Military College (1885—88) at Chester and as superintendent of the Military Institute (1888–90) at Hamilton, Virginia. With Ross firmly in control of the college and the political lobby in Austin by 1893–94, the arrival of Lieutenant Bartlett on campus proved fortuitous.[32]

Discipline improved and the Corps continued to grow. More importantly Bartlett had a firmer grip on what was needed to train future potential officers for the Army. He increased drills and was the catalyst for the formal organization of a ceremonial, eighteen-member cadet band. A professor of music, George W. Gross, was hired in 1895 to develop and lead the cadet band and orchestra, considered by early 1897 the "finest amateur organization [band] in Texas."[33] New distinctive uniforms consisting of cadet gray jackets, white trousers, a three-inch white belt, and a white pith helmet were adopted. Gross added a novel feature for the 1897 commencement, long before the band became a marching unit, by developing a musical arrangement and drill for the Ross Volunteer Company. Finally, band leader Gross composed the first song about Texas A&M, titled "A. & M. Cadets," hailed as a "piece of music destined to become the popular two-step of the season."[34]

Lieutenant Bartlett developed a tremendous rapport with the cadets, in an era when teenagers did not associate with authority figures. Having been raised on the rough Kansas frontier Bartlett had acquired tremendous insight into the motives and backgrounds of the college's young cadets. Cadets ranged in age from fourteen to eighteen years and had received only the most basic high school education prior to arrival at A&M. His insistence on excellence, military bearing during training, and discipline were pivotal for those in the college during the late 1890s. Bartlett's maturity and command of the Cadet Corps proved timely. The *Galveston Daily News,* often critical of the A&M College and its "military features," noted in October, 1897, "one of the remarkable features of the institution is the discipline enforced. It is marvelous."[35]

It was Bartlett who arranged and set protocol for General Ross's funeral. In time of need, the campus and community invariably called on the military contingent in the Office of the Commandant to handle such affairs.

The 1894–95 football team coached by F. D. Perkins posted a record of 1-1. Their first game was with the University of Texas and the second, an A&M victory, was over Galveston Ball High School, 14-6. Courtesy Texas A&M University Archives, Cushing Library

This know-how and attention to duty and protocol would repeat itself in the years to come with the deaths of President L. L. Foster, Gen. James Earl Rudder, and Dr. Eli Whiteley.[36]

All was not pomp and ceremony or drill and polish in the 1890s. On February 13, 1895, a heavy snowstorm resulted in an all-out Corps snowball fight among the cadet units, resulting in sixty-one cadets being placed on the sick and injured report. In May of the same year, the Corps made its first official Corps trip to Houston as the honor guard for the grand reunion of the Confederate Veterans Association. Ross, a lifelong leader and advocate in the Confed-

erate Veterans, was eager to put the A&M cadets, totaling 337, on public display. The Corps annually in April visited Houston and the San Jacinto battleground until the turn of the century. On campus, Mrs. Ross was active in assisting the cadets to organize a Mother's Day observation, the precursor to the weekend-long "Parents' Day" festivities in the 1970s.[37]

Far and away the high point of the first two decades of the A&M College was the enthusiasm and feeling of accomplishment obtained by the institution and the Corps by the spring semester of 1897. The class of '97, twenty-six members strong, was showcased at the four days of commencement exercises in early June. The *Battalion* proclaimed that the college and the Corps had "never shown off better."[38] The festive events included chapel services, inspections, a parade, competitive drill, orchestra music furnished by the cadet band, a baseball game against the "boys from Bryan," three evening banquets, the annual meeting of the Alumni Association, and a special drill performance by the Ross Volunteers. All came to the college to "have a good time."[39]

Legislative visitors from Austin, alumni officers, A&M board members, and special guests were surprised at the progress of the college. They discovered that, contrary to representations made in various statewide newspapers and the *Farm and Ranch* magazine, the college presented a positive impression. The level of alumni organization and emphasis led by E. B. Cushing, Judge Charles Rogan, Frank Reichardt, and Salis Hare, Jr., was striking. Addressing the annual alumni banquet in the Old Mess Hall, Maj. W. R. Cavitt of Bryan, a member of the college board for more than a decade, lamented the limited capacity of the college to house cadets, claiming, "we want 1000 young men . . . and we can get them. We must ask the legislature for assistance, and if they won't help us let us elect legislatures who will!" The Corps of Cadets had demonstrated excellence in the face of much hardship, confirmed by the 1896–97 U.S. Army inspection report that cited the college as "the finest military institution in the south."[40]

Total Corps strength in 1897 numbered 222 cadets, as compared to 338 cadets at the Military Academy at West Point and 166 at VMI in Lexington. Enrollment in the Corps of Cadets and four years of military training prior to 1897 afforded no direct opportunity

for a commission in the regular Army, due in large part to the fact that West Point graduates met the needs of the small standing army. Notwithstanding the fact that there was no formal program to provide commissions for the armed services, Lt. George Bartlett strongly recommended six graduating cadets in the Class of '97—Charles C. Todd, Albert J. Kyle, Ben F. Bryan, Henry M. Eldridge, George M. Shires, and Hermon H. Ueckert—to both the adjutant general, the State of Texas, and the adjutant general of the U.S. Army in Washington as "having shown aptitude for military service."[41] However, the only one selected was Lieutenant Todd; he became the first regular Army officer commissioned directly from Texas A&M. During the balance of the decade, four cadet units—A, B, C, and D—made up the Corps organizational structure.[42]

However, it was a senior cadet who would provide the most stirring address during the '97 commencement. Charles Todd, First Captain of the Corps and Commander of Company A, presented a valedictory address on the "duties of cadetship" and the importance of a solid education—proclaiming the class of '97 had indeed obtained "the strong armour of education . . . in order to withstand the onslaughts" of the world. Yet it was his praise for Governor Ross that so eloquently captured the genius and catalyst behind the growth and success of the college. Ross had persevered with the college when others denigrated or abandoned the institution: "Honored president, it is you we have gone to in our darkest hours of trouble; when we might have given up in discouragement had it not been for your encouraging advice. We have not forgotten that only a few months ago you were offered, and refused a position probably more pleasant. It is not the nature of a Texas boy to forget an action of this kind, and though you may never have a monument of marble and precious stone, your image is so engraved upon the hearts of the students of this college as to live forever."[43]

On January 3, 1898, after two days of deer hunting in cold and wet conditions in the Navasota River bottom, General Ross contracted a severe case of influenza and died suddenly of congestive heart failure. The loss marked the end of an era for the young college that had come of age. The old warrior, politician, and educator left his indelible mark on the A&M cadets during the 1890s.

THE SPLENDID LITTLE WAR

As the century drew to a close Aggies would be called on for the first time to serve the nation in time of conflict. Shortly after a terrific explosion and the sinking of the battleship USS *Maine* in the Havana harbor on the night of February 15, 1898, the United States declared war on Spain and its possessions in Cuba and the Pacific. Although the Spanish-American War, involving Cuba and the Philippines, was short-lived, it emphasized the reasons that the Morrill Act was drafted and passed in 1862—to provide citizen-soldiers and officers in time of national need. The U.S. Army was initially unprepared. The call for volunteers, even in this period of limited conflict, dramatically highlighted the need to have a solid base of trained citizen-soldier reserves in time of need to fill the ranks of the armed services, which grew from a small regular force of 25,000 in April to more than 225,000 volunteers by November, 1898. Texas

Edwin J. Kyle '99 holds the claim of being the only cadet to serve as Commandant of Cadets, serving from March to May, 1899, due to the absence of an active duty officer. He relinquished his duties upon the arrival of Capt. Frank P. Avery. After the turn of the century Kyle became a distinguished professor of agriculture. Kyle Field bears his name. Courtesy Texas A&M University Archives, Cushing Library

A&M was no exception as the campus was caught in the grip of war fever in the spring of 1898. After nearly two decades, the war proved the first genuine opportunity for A&M alumni to serve on active duty. On campus, cadets petitioned the president to organize a volunteer regiment. Although no Aggie regiment was organized, "the good effects of military instruction" resulted in many of the older senior cadets resigning from the college to join more than

100 former students of the college who volunteered for service in what Teddy Roosevelt called a "splendid little war." A majority of the Texas A&M volunteers were given direct officer commissions in the First, Second, Third, or Fourth Texas U.S. Volunteer Infantry (USV), mobilized by Gov. Charles A. Culberson in late 1898. At war's end all but a few were discharged without further military obligation.[44]

The small core of regular Army officers on campuses nationwide were ordered to active duty units. Officers serving as college military instructors were key to the initial mobilization. Lieutenant Bartlett was recalled and promoted to major upon his departure from A&M. All regular Army personnel were ordered to active Army units for the duration of the war. On active duty, Major Bartlett was one who championed the young officers from colleges other than the service academies. Though a staunch West Pointer, Bartlett was quick to recommend many A&M cadets he had known as Commandant to regular Army positions.[45]

The U.S. Army determined that due to the "continued exigencies of the military" it was inadvisable and impossible "to detail an officer on the active list" to Texas A&M.[46] In April, 1898, to address the vacancy, Lt. Charles C. Todd '97, former Cadet Colonel of the Corps and a newly commissioned Army officer, was named to assume the duties as the acting Commandant in the absence of an active duty Army officer. At twenty-four years of age, Todd became one of the youngest Commandants in the history of Texas A&M and the second alumnus to oversee the Corps following F. E. Giesecke. For his two months of service as Commandant of Cadets, the A&M Board authorized Lieutenant Todd

to be given full pay and benefits totaling $28.80.[47]

As a result of the good showing by Texas A&M alumni and students and recommendations such as Bartlett's, the inspector general of the U.S. Army noted that "so complete was the service rendered by the [A&M] graduates of the College during the war," he recommended that some of the "best of these students" be given appointments in the regular Army. Though critically wounded in the Philippines in 1899, Lieutenant Todd was one of the few former A&M cadets to remain on active duty. In addition to assisting many A&M graduates, Bartlett departed A&M in 1898 to direct the logistical support for the invasion of Cuba and retired a major general in November, 1918, after extensive duty in England, France, and Greece during World War I. In 1948, the Texas A&M Class of 1898, on the occasion of their fiftieth reunion, drafted a special resolution of thanks to General Bartlett, expressing their gratitude for his service to the Corps of Cadets and Texas A&M.[48]

THE ROSS LEGACY

The Ross years at A&M, seven in all, would prove a watershed for the college and the Corps of Cadets. The legacy of the Ross mark on Texas A&M and its traditions is unmistakable. From the rough-and-tumble days of the 1890s, the Ross presidency solidified the position of Texas A&M as an up-and-coming institution of higher learning. In the case of the Cadet Corps, Ross, with the help of a more active, professional approach from the Commandants, instilled a sense of purpose, pride, and esprit de corps that comes out of

TABLE 2-1

Official Annual Fall Corps Enrollment,
1891–1919

Year	Enrollment	Year	Enrollment
1891–92	331	1905–1906	369
1892–93	293	1906–1907	488
1893–94	313	1907–1908	579
1894–95	349	1908–1909	577
1895–96	320	1909–10	711
1896–97	282	1910–11	833
1897–98	325	1911–12	895
1898–99	344	1912–13	868
1899–1900	382	1913–14	772
1900–1901	345	1914–15	775
1901–1902	432	1915–16	732
1902–1903	354	1916–17	847
1903–1904	349	1917–18	874
1904–1905	393	1918–19	1,284

Source: Office of the Registrar, Texas A&M University

both adversity and promise of future greatness. The training of the Corps at Texas A&M during the 1890s was ably represented in the war record of those who served in 1898–99, as well as in the growing clout of the former students working in business, engineering, and agriculture. Furthermore, the foundations of Texas A&M's rich traditions and lore became steadfast.

The Aggie cadets who attended the college and graduated during Ross's tenure as president considered themselves special. They banded together in what came to be known as the Sul Ross Group. Later, under the auspices of the Association of Former Students at the yearly alumni reunion programs, the group expanded to include all those A&M graduates who had been out of the university for fifty years.

THE ROSS VOLUNTEERS

In 1905, the Ross Volunteers were given their permanent name in honor of General Ross. One of the oldest student organizations in the state of Texas, the original company was formed in 1887 and called the Scott Volunteers in honor of Col. T. M. Scott, who at that time was the business manager of the college. The purpose of the organization was to recognize the "most military men" in the college and organize them into a crack ceremonial unit. When General Ross arrived as president in 1891, the name was changed to the Ross Volunteers. Upon his death, however, the name was changed to reflect each new administration that followed: the Foster Guards (1898–1901) and the Houston Rifles (1902–1905).[49]

When he assumed the presidency in September, 1905, Dr. H. H. Harrington, son-in-law of Governor Ross, requested that the name be permanently changed to the "Ross Volunteers." Members of the company, often called the "Rosses" in the 1890s, were designated by white duck uniforms with gold and white cords. A special tin helmet was designed for the RVs but was discarded for the lighter weight white military cap. For a brief time in 1907–1908 the RVs adopted a uniform reminiscent of Teddy Roosevelt's Rough Riders— gray trousers, blue shirts, boots, and a large Stetson hat. In 1909, Volunteers returned to the traditional white uniform. Membership for the most part was limited to the junior and senior classes, although during World War I nonmilitary students and a few sophomores were admitted. Most notably, J. V. "Pinky" Wilson, author of the "Aggie War Hymn," was admitted as a sophomore in late 1916. In 1942, the company was disbanded for the

TABLE 2-2

The Original Sul Ross Classes, 1891–1901

YEAR	GRAD & SPEC	SR	JR	SOPH	FISH	TOTAL
1891–92	5	38	49	145	94	
1892–93	4	17	69	100	103	
1893–94	3	34	56	100	100	
1894–95	4	33	66	136	133	
1895–96	3	30	64	130	126	
1896–97	7 (4)	30	43	85	185	297
1897–98	3 (13)	24	36	113	148	337
1898–99	4 (12)	24	28	134	134	356
1899–1900	6 (14)	25	72	143	136	396
1900–1901	2 (41)	21	61	122	135	382
1901–1902	3 (44)	33	57	112	218	467

Source: The numbers are compiled from the annual catalogs of the College and cross-referenced with other available records and reports. While there is steady growth in the total size of the Corps, due mainly to larger fish classes each year, there is also a very dramatic attrition rate over the four-year academic period.

duration of World War II; it was reactivated by the Commandant, Col. Guy S. Meloy, Jr., in 1948. Since the reactivation, membership has been limited to juniors and seniors, reaching a maximum strength of company at 144 in 1972–73 and subsequently numbering 90 members in the 1980s and 1990s.[50]

Senior members of the company constitute a special 21-member firing squad chosen by their peers for Silver Taps and the annual April 21 Aggie Muster ceremonies and serving at the pleasure of the Commandant for special functions. Since their earliest days, the RVs have been the honor guard for the governor of Texas and the official escort for the inauguration of the governor every four years. While various appearances have been made nationwide, the most long running has been the official escort each year of His Majesty King Rex at the Mardi Gras parade in New Orleans.[51]

The initial tribute to Ross was a memorial plaque placed in the college chapel. During the two decades following Ross's death, cadets and former students collected funds, along with a special appropriation from the Texas Legislature, to erect a fitting memorial to the general. Cadet Todd's 1897 vision became reality in 1919 when a life-size bronze statue, the only one on the main campus until the modern era, was dedicated in front of the Academic Building to the renowned Texas soldier, leader, and educator.[52]

In later years, Ross's grandson, L. S. Ross Clarke '21, expressed his displeasure with the veneration of his famous grandfather. He recalled that, when he was a cadet, upperclassmen required him to keep an umbrella in his room. Every time it rained (which can be often in the Brazos bottom) Clarke was required to grab his umbrella and rush out to hold it over the statue of Ross in front of the Academic Building. "Before it was over that year,

I got pretty damned tired of grandpa," he said.[53]

Looking west, the Ross statue remains at the center of campus activity.

And learning the inscription on "Sully" has been required campusology material for new Corps fish to the present day.

> *Lawrence Sullivan Ross*
> *1838–1898*
> *Soldier, Statesman, and Knightly*
> *Gentleman*
> *Brigadier General, C.S.A.*
> *Governor of Texas*
> *President of the A&M College of Texas*

3

RATTLED
TO A FRAZZLE

*The advantages of military discipline as a means of governing a
student body cannot be questioned . . . ambition and high aim in life
require a strict military discipline to keep down rowdyism.*

Capt. Andrew Moses, Commandant, 1908

*There are more students in tents than were enrolled in the College in
1906. The student body is the largest under military discipline in the
world.*

Col. Robert T. Milner, President, 1910

THE OUTPOURING of grief state-wide for Governor Ross was tremendous. Cadet Corps Commander Charles C. Todd '97 gave a moving tribute to the thousands that attended services in Waco, Ross's boyhood home. The funeral bier, surrounded by four former Texas governors—Charles A. Culberson, Oran M. Roberts, John Ireland, and Joseph D. Sayers—was covered with flowers and the red, white, and blue flags of the Confederate stars and bars. In death, as in life, his legacy and contribution to the fledgling institution continued to focus attention on the A&M College and the Corps of Cadets. In the interim Roger H. Whitlock, professor of mechanical engineering, was named president pro tempore. By early spring, 1898, a full-scale search was conducted to identify a successor, concluding in June with the selection of Lafayette L. Foster, one-time Speaker of the Texas House of Representatives and a former member of the Texas Railroad Commission. Politically savvy, he had been a member of the A&M Board of Directors in 1887–91 due to the practice of having the state's commissioner of agriculture serve as an ex-officio member.[1]

Foster, a lay Baptist minister, faced the same major challenge—too many interested students and not enough space. Although there was not enough dorm space on campus to meet the demand and interest in the college, in the fall of 1898 more than 400 students were confirmed for entrance. Meanwhile, despite the close of the Spanish-American War after calling all its officers into active units, the U.S. Army had not yet begun to reassign officers to college campuses. Faculty member Capt. Harry B. Martin '95 followed Lt. Charles Todd's brief stint as interim Commandant.[2] Martin faced identical problems encountered by his regular Army predecessors. The Corps of Cadets grew to 391 by early 1899, even after running newspaper stories statewide that the college was full and unable to accommodate any additional cadets. In spite of these stories, hundreds appeared to attempt enrollment. For five years, the college had turned away potential students, leading the board to beseech the legislature strongly to provide urgent funding to enlarge the facilities, that "those turned away are as much entitled to share in the benefits it confers as those who are here is an incontrovertible fact . . . they pay taxes . . . and it is their right in common with others to enjoy its educational advantage."[3] The "right" to attend the college was vastly overshadowed by the poor state of on-campus facilities: questionable drinking water, no sew-age system, and constant threat of fire due to wood-burning stoves.[4]

Captain Martin in early spring, 1898, returned to his regular teaching duties. In the interim, prior to Final Review, Edwin Jackson Kyle '99, cadet First Captain, functioned as the Commandant. A well-known campus leader during in his cadet days, Kyle would become a stalwart in developing Texas A&M after the turn of the century. Col. John C. Edmonds, the Commandant designate by the A&M Board of Directors, arrived on campus in late fall of 1898. Colonel Edmonds was born in 1847 in Alexander, Virginia, and enlisted in Confederate service at the age of fifteen and fought in the Confederate army under Mosby's command. After the war he graduated from VMI in 1872 and moved to Texas in 1878 to teach. He served as president of Austin College and nearly a decade as mayor of Sherman, Texas. In 1898, he was appointed a "colonel" in the Fourth Texas Regiment, with hopes of duty in Cuba. His appointment as Commandant of Cadets at A&M was questioned by some who believed that "a citizen as commandant" with no federal military experience could not adequately deal with regular Army officers. The concern primarily involved the War Department and its review of Corps training procedures. However, Edmonds's interim role as Commandant apparently did no harm to the Corps or its ability to be prepared for the annual state and federal inspection.[5]

Notwithstanding a colder than usual 1898–99 winter, the Corps received high marks and praise from a visitation committee sent by the Texas Legislature to survey the status of the college. By demonstrating proficiency in the military features of the Corps, the cadets were on their best behavior.

Cadets were authorized to wear the Army-style campaign hat for the first time. Shortly thereafter, Gov. Joseph D. Sayers invited the entire Cadet Corps to attend the governor's spring reception on May 26 in Austin. A special chartered train, the normal method of moving large groups before the age of the automobile, decked out with A&M flags and banners allowed both cadets and faculty a break from campus duties. In June, the governor returned to A&M to confer degrees to twenty-three cadets in the Class of '99.[6]

Except for the first gridiron game with the varsity of the University of Texas on Thanksgiving in Austin, the 1900–1901 college session was without special note. With two minutes remaining in the game and Texas leading 11-0, the game was halted due to bad weather and darkness. Back on campus, the two college cannons that were "called away" in April, 1898, for the war in Cuba were returned to the Corps, much to the delight of the cadets who looked forward to staging "a right decent sham battle" for commencement.[7] Fall, 1901, enrollment was 458, 68 more than the record enrollment of 1899–1900. However, tragedy jolted all those associated with the college with the announcement of President Foster's sudden death—most likely from complications of pneumonia—while visiting in Dallas on December 2, 1901. Foster was buried on the A&M campus, in a newly designated cemetery he had dedicated only a year earlier for use by the college. Located between what is today Duncan Dining Hall and Dorm 9 in the Corps dorm quadrangle, the cemetery was moved a mile southwest of the campus in 1938 to make room for the new cadet dorm area.[8]

Professor Whitlock again stepped in as acting president until summer, 1902, when Dr. David Franklin Houston, selected from a field of nearly twenty candidates "after hours of balloting," was named president. Houston, the former dean of the faculty at the University of Texas, while more an academician than a politician, proved better able to take up where Ross left off. One observer noted, "His predecessors, Governor Ross and Colonel Foster, have indeed done well by the college; however, Professor Houston is the first president we have had who was specially trained for the job."[9] With the Corps now numbering more than 400, immediate measures were needed to address the overcrowding and needs of the college. Inheriting a college in transition, Houston had a two-pronged approach to the demands on the college to expand housing and needed improvements: first, obtain the needed appropriations for facilities and new faculty with "proven scholarship and training"; second, manage the enrollment demands by increasing the minimum age for admission from fourteen to sixteen years of age and raising the scholastic requirements for admission.[10]

The enrollment management plans at first appeared to work. In the fall of 1902, the Corps of Cadets had a slight decrease in enrollment (the first time in seven years); however, late-arriving cadets in October and November swelled the Corps total back above 400.[11] Cadets were given an off-campus diversion to San Antonio in one of the first official football weekend Corps trips and parade in support of the Aggie squad, which played Texas at the old International Fair Grounds to a 0-0 tie on October 2. The cadets paraded up East Commerce, by the Alamo, and back on Houston Street. By the turn of the century, football became a major part of cadet life. President Houston encouraged the cadets to "raise [football] to the highest standards." Each of the four cadet units fielded a Corps intersquad team, providing a good recruiting source for the varsity team.[12]

Back on campus, it was reported that the new age limit policy resulted "a marked improvement in the quality and attitude of the student body."[13] However, the *Longhorn* lamented that the shift in admissions policy "has resulted in an almost complete change in the spirit and general attitude of the corps . . . the addition of one year [two years] to the age for entrance has robbed the corps of much of its youthful appearance and character."[14] Capt. Frank P. Avery was assigned as the new Commandant in the fall of 1902, coinciding with the selection of President Houston. Though enrollment fluctuated, and efforts were made to limit new students, Houston—a progressive educator and endless visionary—envisioned growth, more faculty, and a vastly expanded campus able to house at least 2,000 cadets. However, in 1903–1904, room was available for only 408 students. A few temporary tents were needed for the overflow. The fiscal year 1903–1905 legislative appropriation for the college was $266,070, and although it was the highest sum to date, it was still more than $100,000 short of what was needed.

Dr. Houston was keen to survey and cultivate the backing of the former cadets, knowing that their accomplishments in industry and business were important for the image of the college. The alumni could also be instrumental in efforts to urge the legislature to increase funding. To recognize their accomplishments, the annual college catalog began to carry a special section on the Alumni Association, listing by name, academic course

of study, occupation, and residence those who had been conferred a Texas A&M degree. Organized in 1889, the association would play a critical role in the survival of the Corps and the college. In 1902, after a quarter of a century of existence, of the estimated 3,000 cadets who had matriculated, the college boasted 378 "graduates."[15]

Houston noted the accomplishments of the alumni as proof and clear justification of the need to expand the scope of the college. An idealist for his time, Houston was hastily lured away by the University of Texas in 1905 to be its president. After a short period in Austin, he became a member of President Woodrow Wilson's cabinet (1913–20), serving in two different posts, first as secretary of agriculture and then as secretary of the treasury.[16] Such administrative changes, except for the newly assigned Commandant, Capt. Frank P. Avery, were scarcely noted by the average A&M cadet. The lone cadet battalion, comprised of four cadet companies, added an additional element of "artillery" as well as an expanded cadet band in the fall of 1902 under the direction of George S. Tyrell. The Commandant, with the approval of the faculty, encouraged the code of conduct be enhanced with a viable honor system. However, mischievous and rowdy cadets stretched the intent and spirit of the code to its limits.[17]

Always starved for amusement and extracurricular activities, cadets were allowed an occasional visit to a carnival or circus in Bryan or Hearne to break the monotony. However, an altercation between the A&M cadets and the Indians in the traveling Buffalo Bill's Wild West Show and Circus headed by Col. William F. "Buffalo Bill" Cody, resulted in the cadets being temporarily confined to campus and banned from off-campus visits to amusement shows. One account called the altercation at the Bryan show the "great massacre of '02 and the last unheralded" frontier battle:

Hard pressed for cash, adventurous students had an overwhelming desire to see the show gratis. Well, without much bickering the students marched in on the Indians in the middle of their performance just about the time the braves were demonstrating a stagecoach hold-up in a grandstand battle. Hearing all the ruckus at the gates and having some inkling about the big rush the students had threatened if they were not invited in to see the show gratis, the Indians abandoned their scene and rushed to attack the invaders. To say the least, as told by both sides, the scrap was terrific. The show, of course, was abandoned but the last unheralded Indian frontier battle went on far into the night. It took place mostly around the show grounds, but extended into the Bryan freight yards and thence all the way to College. Due to superior Indian armament consisting mostly of tommy hawks, ponies, lariats, bows and arrows with blunted points, quirts etc as against rocks, sticks and fists the students were gradually forced into full retreat. By the wee hours they were back on the campus trying to recoup but their reorganization plans fell thru when the rest of the students preferred to sleep on rather than fight the Indians.[18]

Yet by morning, the cadets were planning a counteroffensive. To avoid any further altercations with the spirited students, Colonel Cody re-routed the baggage train traveling from Bryan to Brenham in order to thwart a cadet ambush of his circus train. To calm the standoff, Colonel Cody met with President Houston and the ranking cadet officer, L. W. Wallace '03, and the truce was honored.[19]

On campus, sports began to fill the gap for much-needed activities to the relief of the president and Commandant, who hoped that the cadets would reserve their energy for the gridiron. Intramural sports were gradually introduced, but it was the popularity of intercollegiate football that soon captured the interest of the students. The cadets and alumni were elated with the tremendous success of the 1902 football team that not only beat the varsity of the "state university" in Austin (long before they were known as the Longhorns), but also went on to be crowned the Southern Champions. Organized football on campus was first played on the drill field and in the spring of 1905 then on an improved portion of an old vegetable patch managed by the professor of horticulture, E. J. Kyle '99.[20] The first awarding of the "T" for Texas A&M to student athletes by the Athletic Association began in 1905–1906, discontinuing the practice of recognizing students with a "C" for College. To support the growing college athletic program the cadets began to develop distinctive yells and chants for home games. During halftime the only diversion was the formation of a block "T" at midfield by the Corps en masse. The student body was known as the "Cadets" or "Farmers," long before the use of the term "Aggie" in the 1920s.[21]

There was very little travel to away games, but this did not dampen the enthusiasm for greeting the returning team. Aggie gridiron star Caesar "Dutch" Hohn '12 recalled, "The Cadet Corps always met the train that was returning the football team from an out-of-town game."[22] One cadet reported in a letter home, "Well we have a wagon out here in front of Ross Hall waiting to go down to meet the football team. We have a long rope tied to the tongue and the 'fish' are going to pull it to the depot and back. Poor fish!"[23] After their escort to the center of campus, the team captain spoke to his fellow cadets. A brief yell practice followed.

Cadet yell leaders appeared at the turn of the century, first as crowd control and gradually as a major force to direct the "old pep" of the students. Visiting former students, friends, and the opposition marveled at near nonstop cheering and cadet yells. Yells were of two types: those done in the stands while the game was in progress, and those done on the field at pregame or half-time with the Corps in the mass block "T" formation. Numerous yells were introduced year after year up until the 1930s, yet only a few survived approval of the cadets. Each year two senior yell leaders and a couple of junior "assistants" were selected by the cadets. They used megaphones, controlled the crowd, and dressed all in white. The top yells (of a dozen tried) prior to World War I were "Rickety Rocky," "Farmers Fight," "Saw Varsity's Horns Off," "Horse Laugh," and "Military"—all of which became incorporated into Aggie lore.[24]

"HULLABALOO"
Hullabaloo! Caneck! Caneck!
Hullabaloo! Caneck! Caneck!
Warhee! Warhee!
Look at the man! Look at the man!
Look at the A. & M. man!

"CHICK-GAR-ROO-GAR-REM"
Chick-gar-roo-gar-rem!
Chick-gar-roo-gar-rem!
Rough! Tough!
Real! Stuff!
Texas A&M!

"SAW VARSITY'S HORNS OFF"
Saw Varsity's Horns Off
Saw Varsity's Horns Off
Saw Varsity's Horns Off
Short!
Varsity's horns are sawed off
Varsity's horns are sawed off
Varsity's horns are sawed off—short!

"FARMERS FIGHT"
Farmers fight!
Farmers fight!
Farmers-fight-fight-fight
Smash 'em up!
Farmers-fight-fight-fight
Farmers, farmers, fight!

"MILITARY"
Squads left! Squads right!
Farmers, Farmers! We're all right!
Load, ready, aim, fire—"Boom!"
Texas A&M, give us room!

"HORSE LAUGH"
Riffety! Riffety! Riff-Raff
Chiffity! Chiffity! Chif-chaf
Riff-raff! Chif-Chaf!
Let's give 'em the horse laugh
Ha-a-a-a-a-a-a-a-a-a!

"SKY ROCKET"
Whistle-e-e-e-e
Boom!
Ah-h-h-h-h-h
Whistle-e-e-e-e
Rah! Boom! Team!

"RICKETY ROCK"
Rickety Rock! Rock! Rock!
Rickety Rock! Rock! Rock!
Rick! Rick!

Hullabaloo!
Zim! Zam!
TAMC!

Campus dance clubs, such as the Kala Kinasis (KKs) and the Swastikas, fostered by cadets with German heritage, flourished to promote social life. Other turn-of-the-century pastimes included periodic campus social events, usually in the spring and hosted by a "sponsor"—a practice begun by Mrs. Ross in the 1890s—in the local community. Many of the sponsors, often pictured in the cadet yearbook alongside the Corps units, were the older daughters of the professors. Many of these ladies, such as the Hutson twins, Mary and Sophie (Dr. Charles W. Hutson), Emma Fountain (Charles B. Fountain—English), Helen Bittle (Dr. Thomas Bittle—chaplain), and the Davis twins (Dr. S. F. Davis—chemistry), routinely attended classes with the cadets. Though A&M was an all-male college, the Hutsons each received a "Certificate of Completion" in Civil Engineering in 1903. Considered "special students," these early coeds were not carried among the alumni rolls until years later. In the fall of 1906 the YMCA was organized on campus and chartered by the national headquarters.[25]

From the earliest time, the cadets developed slang terms for all types of events, professors, the mess hall cuisine, and everyday life. The 1903 *Longhorn* provided a glimpse of the colorful cadet vocabulary, much of which has been passed down to the modern Corps of Cadets:

AXLE-GREASE (Golden)—A substitute for butter used at A. and M., and noted for its vile odor and marked tendency to crawl around the Mess Hall floor; also valued as a lubricant for the campus wagons.

BUGHUNTERS—Cadets majoring in agriculture; members of the "Farmers Club."

BULL—The sonorous title of the warlike head of the military department at A. and M. C.

BULL-TICS—The science of war as expounded to the First and Second Classes by the Bull.

CUSH—A favorite dish of the A. and M. gray-coat—a concoction of soggy pastry and nutritious (?) compounds patented by the Sbisa House.

EXTRA—A delightful (?) stroll of two hours on the pebbles of the parade-ground, with gun on shoulder, while "Old Sol" blazes down pitilessly, exacted of the delinquent cadet as a weekly penalty for his wrong-doing.

FISH—The A. and M. Freshman, a harmless, well-meaning specimen of young America whose delights are in being strapped by upperclassmen and guarding the flag-pole at night during his first few days here, and whose awe-inspired fear of the cadet officer and his ram book amounts almost to reverence.

FOX—A cadet especially brilliant in any certain type of work.

PUNK—The technical name conferred upon the lead-like biscuits served at the Mess Hall.

RAM—A report for any breach of rules and regulations, usually involving the pleasures of an "Extra" on the following Saturday.

REGULATION (Reg)—Sbisa's imitation of extract of sugar-cane.

SAWDUST—Sbisa's refined (?) sugar.[26]

Adding campus activity was crucial given the treatment of the fish by upperclass cadets. Void of any variety of on-campus amusement, cadets were left to create their own fun, and each new year the incoming fish were "fresh meat" (a ready target). Each year after 1900 the freshman class numbered more than 300. The harassment ranged from the general sophomoric tricks and pranks such as selling the fish passes to the natatorium or nonexistent buildings to having them guard the flagpole (with instructions "to stay there all night"), to full-scale hazing. New fish were often dispatched on urgent missions to the Commandant's Office to get a "bucket of reveille" a "box of taps" or a "reveille wrench." It had been hoped that raising the age to enter A&M and further expanding admissions requirements would reduce or at least limit the college's recurring reputation of being nothing more than a reform school.[27] Sensitive to such allegations, the A&M *Catalogue* under the discipline section was blunt to note: "The College is not a reformatory. It encourages the attendance of young men who have a serious purpose."[28]

While rules in the college catalog, the cadet handbook known as the *Blue Book*, and directives from the Commandant's Office stated that hazing would not be permitted, this did not preclude the practice. When did so-called good fun and harassment go over the line to be considered serious hazing? There were no clearly defined guidelines, only a blanket statement against an undefined practice that most entering freshmen had never heard about until arrival on the campus. Hazing as a rite of passage or required initiation was periodically to be the source of much unrest from the turn of the century to the present day. The most common form of agitation

after the turn of the century was related to "strapping"—a term used for the use of either a belt or wooden paddle. Strapping had been a practice since the earliest days of the college.[29]

In the fall of 1903, U.S. Cavalry officer Capt. Herbert H. Sargent was assigned as Commandant of Cadets. His reputation and articulate style as a career soldier had preceded him, due to the fact he was a veteran of the Sioux Wars as well as service in both Cuba and the Philippine Islands during the Spanish-American War. He had authored a number of books on military campaigns, particularly those of Napoleon. Two of his books, *Campaign of Marengo* and *Napoleon Bonaparte's First Campaign,* had long been required reading for all senior cadets.[30] Sargent, although "a strict disciplinarian," quickly became a favorite of the cadets. He made a number of changes to the Corps; one that has remained until today was the naming of a designated Cadet Corps Commander, thus abandoning the previous policy of having the Commander of Company A serve in both capacities. Holding the rank of Cadet Major, the first cadet selected for the top slot was Marion S. Church '05 of Dallas.[31]

The military features and instruction of the college, which Captain Sargent augmented and encouraged, were considered paramount to a cadet's education. The *Battalion,* a monthly publication of the Austin and Calliopean Literary Societies, at the turn of the century was passionate in its advice to the cadets:

The military government of the A. & M. College is exacting, but reasonable in its requirements. A youth to get along easily must develop the idea of duty, realize the importance

Clubs, both official and unofficial, were popular pastimes for the cadets, as is evident by the Company I Cigar Club of 1916. Courtesy Texas A&M University Archives, Cushing Library

of obedience and respect to those instructing. He soon becomes aware of the fact that the most successful, self-reliant man is always under discipline, and the more perfect the discipline, to a greater extent are these qualities attained. A great truth of life is brought clearly before the young man's notice; that is, to be happy and influential it is necessary to control one's words and acts. The cadet must be regular in his habits, punctual at every class and company formation, and systematic in the preparation of his lessons. Neatness of person, tidiness of quarters, and a manly bearing are required of him. These qualities, if made part of the boy, will surely enable him to battle his difficulties in a heroic way. It has been remarked that an A. & M. graduate, after entering the business world, shows remarkable energy, system, and steadiness. It can be truly said that such young men are made by the habits acquired at their alma mater.[32]

Henry Hill Harrington, a professor of chemistry at the college since 1888 and son-in-law of Lawrence Sullivan Ross, was selected by the A&M Board of Directors to be the next president of the college in early September, 1905. The rapid faculty turnover, budget concerns raised by opponents of the college, and renewed criticism of the college's programs once again surfaced. The new president had a number of problems and issues with which to deal both on and off campus. During the first year of the Harrington administration cadet life was rather calm as he focused his attention on off-campus issues. As a gesture of goodwill the cadets dedicated the 1906 *Longhorn* to "a true friend of the student body, our new president."[33]

To the detractors of the college, Harrington addressed the three main external complaints with straightforward solutions and candor that were not always accepted. He believed that the college should be allowed to grow and prosper, and he wanted no opposition to his plans. One of his suggestions was to reduce the overcrowding of three to four cadets to a room by using tents to house cadets. The Harrington strategy was to use tents to focus attention on the overcrowding on campus, thus possibly obtaining more funding from the Texas Legislature for a new dorm.

He also quickly resolved the second issue. Agricultural groups such as the Texas Farmers Congress and the Rice Growers Association had complained that the soil around the college was poor and not suitable for teaching and agricultural experiments that would have validity statewide. To resolve the issue, Harrington began the process to speed up the opening of agricultural experiment farms and substations in key locations around the state, twelve in all by 1909. This system, under the funding and guidelines of the Hatch Act (1887), continues to the present day.

The third concern involved the rural or remote location of the campus in Brazos County. Detractors of the college argued that it was not located near a big city where the students could have closer contact with "practical industry" and big business. Harrington argued that the location was adequate. The "remoteness" of the small rural community of 12,000 was beneficial to limit contacts with "urban temptations."[34] Opponents of Harrington and A&M attempted to criticize the location issue on yet another dimension—its unhealthiness.

He refuted the recurring charge that the location of the college caused a high rate of sickness and malaria as an old issue that had long ago improved with the addition of electricity, better heating for the buildings, and, more importantly, potable water—all of which reduced the incidence of large-scale illness on the campus. However, Harrington's claim would be shaken when the college had to close early on May 25, 1907, short of the June 11 commencement date, due to a typhoid epidemic in the Brazos Valley.[35] Nonetheless, his staunch, often uncompromising approach in deflecting negative questions about the college served the Corps of Cadets and the college well in defending Texas A&M against outside detractors. His actions were well received by the media, yet on campus he faced problems. This headstrong approach failed Harrington in his dealings with the faculty, staff, and the cadets.

CADET LIFE

The spartan campus and military regime clearly set the college apart from institutions throughout the Southwest. Political issues and efforts by the president to place the campus in the best light attracted a great deal of attention in the Corps of Cadets. Leading newspapers in Houston, Dallas, and Galveston were eager to print views of daily cadet life and events on the campus. Reporters were routinely assigned to cover the annual week of commencement activities each June. Competition was keen during graduation to determine the outstanding company. The victorious unit was to have the distinction of "the honor of carrying the flags for the following year—the Texas flag and Old Glory . . . at the head of all battalion formations." One observer noted, "Company locality levels class distinction and tends to put the whole Corps on an equal footing."[36]

However, it was the individual cadet that formed the nucleus and heart of the Corps. New cadets on campus, mostly from rural backgrounds, had had no exposure to military training and discipline. Coverage by the press gave tremendous insight into military life at Aggieland as was reported by the *Houston Post* in mid-1906:

> The attitude of the cadets toward the military feature of their college life here goes to prove that there are as many ways of looking at the subject as there are individuals to do the looking. One man will call it a bore and a vexation, while his roommate may be enthusiastic over tactics and formations; one sleepy-head may grumble over the 6 o'clock reveille, while another fellow is eager and anxious to be "A soldier that's fit for a soldier," and wears the cadet gray with square-shoulder distinction.
>
> Military life means strict attention to duty, promptness, stiff drilling, a demerit system of discipline, absolutely regular hours, and most of the students will acknowledge its advantage and a large percent come sooner or later to enjoy it.[37]

MOSES IN AGGIELAND

After four years as Commandant, the flamboyant and seasoned "thirty-year man" style of Captain Sargent was replaced by a young 1897 graduate of West Point, Capt. Andrew "Bull" Moses.[38] A native Texan raised in Burnet County, Captain Moses received his appointment to the Military Academy from then-Governor Sayers and on active duty was

trained as an artillery officer. He served in a number of stateside assignments, including being one of the first graduates of a special one-year course on antisubmarine defense of coastal artillery. Captain Moses assumed his duties at A&M far from the ocean on September 1, 1907. Though he lacked bravado, he clearly excelled in his ability to organize and train the cadets.

The War Department had plans to increase the level of accountability by way of increased inspections of the military programs at all the nation's fifty-two Morrill land-grant colleges. As a result of training changes made at West Point during his cadet days in the mid-1890s, Moses was prepared to introduce the A&M

Capt. Andrew M. Moses (Commandant, 1907–11) was pivotal in reorganizing the Corps and quelling a major student "strike" in 1908. Photo from 1908 Longhorn

One of the first airplanes to land in Brazos County sits on the 50-yard line of Kyle Field on December 1, 1911. The Wright Flyer flown by Lt. Robert Fowler was an instant marvel for the cadets. In the background are the bleachers built by the cadets, and to the right, faculty homes. Courtesy 1912 Longhorn

cadets to the changes and improvements in weapons and equipment, as well as the changing rigors of the then peacetime "modern army." The small regular Army realized it had to prepare the best citizen-soldiers possible given the possible demands of mobilizing a large force on short notice. In the case of Texas A&M the military training received in the Corps was on par with the instruction at other academies and institutions.[39]

The new Commandant advised the faculty, with the approval of the president, that discipline would be swift, but fair. The fact that there had been poor enforcement of the demerit system prompted Moses to revamp the order of magnitude of both the offenses, which he called "misdemeanors," and the resulting punishment. The revised "Classification of Misdemeanors," considered an addition to the *Blue Book,* ranged from one to ten demerits per offense. A sample is instructive as to the priorities of the time. One demerit was issued to a cadet who was late at any roll call, who misspelled a word in official communications, or who failed to

invert his wash basin. A cadet received two demerits for not keeping step in ranks, sitting in the window of a dormitory, having a dirty gun rack or rifle, or for carrying water during study hours. A cadet found loitering on campus after call-to-quarters, playing a musical instrument during study hours, keeping a light on after taps without permission, or acting boisterously in the mess hall received four demerits. Six demerits was punishment for wearing citizen clothes to Bryan, entering the guard room except on official business, or throwing food in the mess hall. Leadership exacted the stiffest penalty—ten demerits—upon a cadet for using profanity, vulgarity, or exhibiting indecent behavior, for jumping on or off a moving passenger train, or for insulting a sentinel at guard duty.[40]

In the fall of 1907 Captain Moses was introduced to an aspect of A&M that possibly eclipsed his emphasis on training and military custom—football. The 1905, 1906, and 1907 football teams had produced three winning seasons, compiling a three-year record of 19-4-1. The prime out-of-state rival was

Tulane, a series that began in 1902. And in 1907 the cadets maintained their undefeated record against the "Tulaneites" with an 18-0 victory in New Orleans. Few cadets were permitted to attend, but all the campus was on hand for the victorious homecoming. Moses received a first-hand look at cadet spirit:

> The enthusiasm of the A. and M. cadets for their football teams was irrepressible. Last night the team was due to arrive from New Orleans at 12:30 a.m. At 12 reveille blew and the entire student battalion arose from their beds and repaired to the Houston and Texas Central depot, taking as they went along almost everything that would burn. They piled up inflammable material just north of the depot, and had a huge bonfire raging when the train pulled in somewhere about 1 a.m.
>
> The team was greeted with uproarious cheers, and with enlivening music by the college band, which was also out in full force. The conquering heroes were escorted to Foster Hall, . . . being cheered individually and collectively.[41]

Moses was wise enough to know that the esprit de corps and enthusiasm of the cadets for sports was a benefit to the "military features" of his training program. An anonymous commentary on the "Military Department" at the college in the 1908 *Longhorn,* with Moses being the likely author, articulated the evolving purpose and role of military training: "The main object of the military instruction at the A. &. M. College is to thoroughly qualify the student to be company officers of infantry volunteers or militia . . . the daily military routine has been a powerful factor in molding his character and in permanently fixing the habits of neatness,

punctuality, obedience and strict attention to duty." These words and their intent, along with the Commandant's role, in the brewing unrest at the college would prove timely.[42]

THE STRIKE OF 1908

In early 1908 a major schism erupted between the president, the faculty, and the cadets. President Harrington's bold measures to defend the college and to demonstrate that he was in control were viewed as heavy handed when it came to dealing with internal problems on campus. Difficulty began with faculty discontent related to the dismissal of several employees without what was deemed "just cause." The difficulty of this case was made even more sensitive by the fact that a majority of the faculty and staff lived in close proximity to each other on the campus. Furthermore, many faculty and staff, entrenched in old habits, were not agreeable to the efforts to improve the campus and academic standards. By the time of the major confrontation in February, 1908, a petition prepared by the cadets contained complaints that dated back to early 1906.[43]

The boiling point for the faculty was reached over a confrontation between the president and the college physician, Dr. Joe Gilbert '94, concerning the handling and seriousness of an illness on the campus. Fearing for the life of his young child, Harrington (at the insistence of his wife) had wanted the doctor to quarantine the entire family of popular history professor C. W. Hutson in their campus home until their visiting grandchild, sick with whooping cough, was well. Harrington's overreaction led to additional complaints. Other accusations against the president that fueled the feud included charges that he was a "grafter," several times taking food from the mess hall for personal use; that he had failed to allow the mess hall to be opened in early September, 1907, for the football team; and that he had allowed his laundry to be cleaned at no charge by college employees.[44] Cadet A. J. "Niley" Smith '08 was the only student to level charges of a specific nature. Smith, a member of the A&M Livestock Judging Team, protested that the president had blocked the team's request to attend a circus in Bryan for the purpose of viewing "different types of draft animals which were not available here at College," in order for them to be prepared for upcoming judging competition. Harrington responded, "I consider this order [request] of no circumstance" and thus tossed it in the wastebasket.[45] In short order these numerous issues embroiled President Harrington, the entire faculty, the cadets of the Corps and their parents, the governor's office, the A&M Board of Directors, and the A&M alumni. A special hearing by the A&M Board was requested. All parties entangled in the controversy resolved to "fight to the finish."[46] At stake was the future course of both the Corps and the college.

In support of Dr. Gilbert and the Hutson family, and after their petition to the administration was turned down without even being given a credible review, the Cadet Corps went on "strike" by refusing to meet for formation or to attend class.[47] Faculty members who supported the president bolstered his call for all students to return to class or "turn in their arms and be dismissed."[48] The cadets thought the incident would also hamper their dealings with the president. Restless, they wanted a solution. In mid-February, 1908, Captain Moses met numerous times with the leadership of the Corps and each class individually to review their concerns and to appeal to them—especially the senior class—not to abandon their education.

Moses probably had been forewarned by Captain Sargent of Harrington's meddling with the cadets. During April Fool's Day in 1906, the president overruled Captain Sargent publicly in the case of the Commandant administering discipline to a number of rowdy cadets. This action, which Harrington viewed as placing himself in a good light with the cadets, resulted in the public embarrassment of the Commandant and the eventual alienation of the cadets.[49] Repeated efforts by the cadets to have their concerns heard at a higher level by either the president or the board were rejected. During a hastily called meeting, the A&M Board of Directors exonerated Harrington of all charges on February 13.

While members of the senior class of '08 were in favor of a student walkout, they were also concerned with their graduation only weeks away. Realizing they were getting no support, the junior class of '09 met late into the evening of February 18 and voted to leave the campus en masse. By morning a majority of the seniors, sophomores, and freshmen agreed also to strike. Sensing disaster, the A&M former students, primarily those in the Houston chapter of the Alumni Association, tried to intercede to encourage the cadets not to leave and for those who had departed to return to classes. Only a handful of students remained on campus by late February, a far cry from the 580 on the rolls at the beginning of the session.[50]

The alumni statewide converged on Dallas, San Antonio, and Houston to attend meetings in advance of a showdown with the A&M Board on the resolution of the "Great

Trouble." Concerned former students noted that the cadets would be "rattled to a frazzle" if not given proper assistance. In addition to the intervention of the Board of Directors and alumni, parents of cadets statewide took an direct interest in the resolution of the disturbance.[51] Appealing to the junior class to reconsider their blanket stance against returning to campus, the father of cadet William Furneaux sent the following plea to the class of '09:

Now that the smoke of the battle that has been so fiercely raging about dear old A. & M. is lifting, I hope you will pardon me for offering a few suggestions on the situation. . . . That you have made a noble fight every one admits, and while you have not at this moment secured your fullest hopes, you have set to work a movement that will accomplish all you have asked for. To me there seems but one way open now, and that is to return at once and resume your duties like a man. I am afraid in years to come it may be a source of regret if you do not find your name enrolled as a member of the class of '09.[52]

By early March, the alumni and parents of the cadets held additional meetings statewide to appeal for calm at the college. With the assurance of the Alumni Association that they would get to the bottom of the situation, cadets slowly returned to campus, but only about 375 were present and accounted for at the June commencement, at which 49 members of the class of '08 received their diplomas.[53] Shortly thereafter a four-day hearing was held by the A&M Board, which feared that the governor or the legislature might step in to take drastic measures if action was not taken. By now more than 200 parents

contacted the board. The vocal alumni, who had promised the cadets a fair hearing, were again disappointed by the board as they once again concluded that matters would improve in the future and that the president had acted in good faith. In a statement that resembled an unholy compromise of sorts, the board concluded by stating it would "close the school" if needed to gain control of this or any further problems.[54]

Having a field day with the 1908 affair, the media dubbed the affair the "Great Trouble" in newspapers across the state. While detractors of the college tried to distort the facts of the events, hazing apparently played no role in the students' protest. When questioned by the board, Captain Moses indicated, as did all involved, that at all times the cadets were well behaved and orderly. The Commandant failed to mention a series of rowdy gatherings in which the president was burned in effigy. And on April Fool's Day, 1908, a boisterous group of cadets held a mock parade and "cornerstone laying" (a block of old pine) for "Harry Hall 1908."[55] Unwilling to budge and having lost the confidence and esteem of the cadets and faculty, Harrington resigned at the August, 1908, board meeting to be effective September 1. This brought the debacle to an end. And in spite of the "troublous times," once again A&M and the Cadet Corps had been in the spotlight statewide. This incident, like the Crisp affair in the 1880s, only increased interest in the college. Applications for enrollment were at an all-time high in the fall of 1908.[56]

The 1908 incident was a pivotal event in the growth and image of the Cadet Corps. The cadets believed that although they were a part of a quasi-military environment they needed a means to address their concerns.

Most observing the situation as it unfolded were amazed at the unity of the cadets and the loyalty to each other and their college, as well as their willingness to abandon the college and the Corps on what they felt was a matter of principle. Joe Utay '08, among those who departed A&M in the spring of 1908, noted that President Harrington "was a cold blooded administrator interested in only securing his position indefinitely."[57] The cost to those who left will never be known—yet their resolve made a marked impression on the Board of Directors, alumni, and Cadet Corps. Their grievances on particular items did not diminish their closeness to the Corps.

The second major result was the role and adamant concern of the Alumni Association. Not until the active involvement and intervention of the leadership of the former students did the controversy get resolved. The class system, especially in these early days when each group was so small, was close knit and would remain so even after graduation. The alumni understood this and through the active participation of such stalwarts as Charles C. Todd, Frank A. Reichardt, F. Marion Law, L. L. McInnis, James Cravens, E. B. Cushing, and Charles Rogan the situation was resolved. Their intervention forced a hearing before the board in June, 1908, and reinforced alumni intentions and efforts, even in controversy, to protect their alma mater and the Corps of Cadets.[58]

MILNER AND THE TROUBLE OF '13

The Board of Directors quickly selected Col. Robert T. Milner, Texas commissioner of agriculture, as president of the college. An ex-officio member of the A&M Board and knowledgeable about the needs of the college, his link with the agricultural community would prove timely. Though the agricultural community was vocal, it is interesting to note that the A&M graduates between 1880 and 1908 used their college education first as civil engineers, second as businessmen and lawyers, and third in agricultural pursuits. Delighted with the opportunity, Milner quickly worked to place the campus back on somewhat of a normal footing—yet the old problems of overcrowding and the need for legislative funds to provide housing and facilities continued unchecked. If the legislature would not act, then the college felt it must make room for the increased interest in the institution.

Milner had a solution for years of housing shortage. The demand for admissions was so dramatic that the college resorted to the use of 243 canvas tents to house 486 cadets. These tents were constructed of wooden floors and three wooden sides about four feet tall and an upright canvas top, and they acquired the name "Camp Milner" or "Tent City." They provided only a temporary solution and forever changed the college's attitude toward efforts to expand the institution rapidly. Milner proudly boasted in his annual report to the A&M Board, "There are more students in tents than were enrolled in the College in 1906! The student body is the largest under military discipline in the world. There are 600 more cadets in this school than there are in West Point."[59]

The state legislature was well aware of the conditions in "Tent City" yet slow to take speedy action to relieve the situation. The college surgeon's report documented the hazards of "exposure of the students to the heat and cold and the dampness . . . 81 per cent of the la grippe and colds treated by him came

Growth of the Corps required that tents be added to accommodate the overflow of students. This view, circa 1908, shows "Camp Milner" or "Tent City," as the area was called. President Milner boasted that A&M had the world's largest student body under military discipline. Courtesy Texas A&M University Archives, Cushing Library

from the tent section." However in the final report, it was stated that the cadets "informed us" (the special visiting committee from Austin) that the tents were as satisfactory as could be expected, but that they found it practically impossible to keep the tents properly heated and ventilated and clean and free from dust, and that the conditions greatly interfered with their studying.[60]

Concerned with the size of the student body, the board authorized the Commandant to reorganize the Corps of Cadets into a regiment of two battalions each comprised of four companies. In an effort to assist the Commandant, the board authorized the employment of drill sergeants and an administrative assistant. Senior cadet officers were granted "more responsibility" and those who "faithfully discharged" their new duties would be allowed to "carry side arms."[61] A grand gesture—however, side arms were never issued to the cadets.

In the spring of 1909, Colonel Milner and Captain Moses planned a busy semester to keep the cadets occupied. Mother's Day, the forerunner of Parents' Weekend observed annually in late April, was observed with a flower-pinning ceremony of the entire regiment in front of Old Main prior to chapel services. The Corps marched to chapel to the music of "Home Sweet Home" for a lengthy sermon by a visiting chaplain, concluded by an easy presentation titled a "Tribute to Mother" by Caesar "Dutch" Hohn '12 of Yorktown.[62]

The calm on campus in the spring of 1909 attracted a large gathering for the June commencement. The Corps of eight companies, numbering "about 530 cadets," held drill competition exercises.[63] The winner, Company C, commanded by cadet Woodie R. Gilbert '09, received the Howell Award, a Texas flag to be displayed by the winner. Prior to the final commencement ball, Captain Moses addressed the senior class "drawn up in line on the parade ground." Emotions ran high as the Commandant congratulated them on their success at A&M and offered best wishes for the future. He lamented the trouble of '08 yet noted that their loyalty to the college and Corps was important to the future of the college. Moses's leadership throughout his duty at A&M was opportune.[64]

A decade later, the *Alumni Quarterly* paid special tribute to the dynamic young Commandant upon his promotion to brigadier general in mid-1918: "General Moses was the 'oil that stilled the troubled waters' during the strike of 1908 and those who knew him during those trying times will remember him always . . . we rejoice in his promotion."[65]

MARCH TO THE BRAZOS

During the early years the cadets constantly concocted reasons to let off steam. A number of times per year the fish were pitted against the sophomores in a free-for-all melee called the "Cane Rush."[66] The upperclassmen would claim a section of the campus and dare the

fish to take it from them. The Commandant soon found an alternative. With spring visits to Houston and the San Jacinto battlefield limited after 1904, the Corps adopted the practice of marching to the Brazos River. Captain Moses used the occasion for "field training." The "Hike," as it was referred to by the cadets, occurred on April 1, thus reducing disorderly conduct on campus associated with the annual April Fool's Day rowdiness of pranks and mischief—"by taking them out on military duty they don't get a chance for trouble!" Yearly, the Commandant issued general orders for the field exercise. The three-day event was considered a major undertaking.[67]

For the ten-mile trek to the iron bridge on the Brazos, it was mandatory for all able-bodied cadets to attend. The Commandant and the Corps Commander led the hike, followed by the regimental staff, the band and bugle corps, and two artillery field pieces "drawn by Texas mules." The infantry units followed. In 1911, the college hired a retired Army quartermaster sergeant, P. A. Koenig, to handle all logistics. Cadets were issued full rations, carried their own tent and bedroll and water, along with a ten-pound model 1898 Krag .30-caliber rifle with bayonet. Captain Moses allowed the cadets to wear a modified uniform similar to that worn and made famous by Teddy Roosevelt's Rough Riders. The surgeon, Dr. Otto H. Ehlinger, provided an ambulance, and the campus steward, Bernard Sbisa, supplied full commissary stores to serve meals, consisting of spuds, dope (coffee), bacon, bread, and a little "cush." Once in camp, the cadets were kept busy with drills, target practice, and sham battles. The *Longhorn* reported that much time was "given over to fun and frolic, with a dubious eye on drill . . . relieving us of nearly all our surplus 'pep.'"[68] The hike was an annual event until World War I.

Both the former students' association and the Board of Directors kept a watchful eye on the campus. One of the most proactive former cadets during the turn of the century was William A. Trenckmann, Class of 1879. In addition to being one of the first graduates of the college he was also president of the A&M Board of Directors in 1910–11. In order to understand the changes under way on campus as well as to better assess the immediate needs, he spent a number of days each month attending classes, visiting with each department, and talking with the cadets. Trenckmann was impressed with the training and preparation the cadets received. Most graduates entered civilian vocations, due to the continued limited opportunities in the regular Army. Only a few graduates were given the option to take competitive exam for a commission as a second lieutenant. Those who passed and entered the Army had only one active duty option—service with the Nature Troop in the Philippine Islands.[69]

An air of enthusiasm existed on campus going into the 1911 football season. Coach Charlie Moran, very popular with the cadets, had come to A&M in mid-1909 and compiled a 15-1-1 combined record for the 1909 and 1910 seasons—beating the varsity of Texas three times in the two seasons. In the fall of 1911, the Aggies met the varsity at a neutral site in Houston. The already heated rivalry between the two schools became even more heated with an upset victory caused by Texas recovering an Aggie fumble inside the A&M ten-yard line and falling in the end zone. Heated debate followed the 6-0 game both on and off the field—resulting in the two teams canceling all future games. The famed series would take a time-out from 1912 to 1914

The 1910 Longhorn *featured a number of unique groups and clubs. The Beaux-Legged Fraternity was given special recognition along with the Red Head Club. Courtesy 1910* Longhorn

Following the turn of the twentieth century the Corps adopted a uniform complete with Stetson and dark shirt, reminiscent of the style worn by Teddy Roosevelt and the Spanish-American War Rough Riders. Courtesy Texas A&M University Archives, Cushing Library

and not resume play until 1915, with the formation of the Southwest Conference.[70]

The 1915 return to play with Texas was welcomed by the cadets and alumni and heralded as the "greatest football game ever witnessed in Texas." The *Bryan Daily Eagle* highlighted the Aggie 13-0 victory with a banner headline reading, "A. & M. Put The Fixin's on Varsity."[71] The jubilation over the game resulted in the cadets celebrating the great victory Friday night with a bonfire built on West Anderson Street in downtown Bryan. City engineer A. S. Adams claimed that the paving was damaged and that the repairs would cost "about $18." However, Adams reported that the "city will pay this willingly, as the A. and M. team won the game."[72] However, cadets were advised that future bonfires should be restricted to the campus.

Twin tragedies struck the campus with the loss of the mess hall on November 11, 1911, followed by the destruction of the majestic Old Main Building on May 27, 1912, both

by fire.[73] One former cadet in the Class of '12 remembered the magnitude of these events:

1911–1912 was a hard year especially for the 1912 Class. The mess hall burned and the weather turned cold and wet so that Sbisa could not feed us and we were sent home early for Christmas. There was a spinal meningitis epidemic in Texas and many people died. The senior class had to guard the entrances to the campus at night to prevent any one entering and also had to meet the night trains to keep passengers from getting off. Our quarantine did not keep us from having one case of meningitis, but the boy survived with one lame leg. The burning of Old Main just before commencement was another hardship.[74]

While the loss of the mess hall clearly disrupted the daily campus routine for a number of weeks (Bernard Sbisa boasted it was the only time in thirty-seven years a meal was

The Cadet regimental staff of 1910–11. Left to right: *Cadet Colonel of the Corps H. M. Pool, Cadet Lt. Col. C. E. Sanford, Captain and Adjutant H. J. Kelly, Captain and Quartermaster W. S. Moore, and Captain and Ordnance Officer G. W. Robinson. Courtesy Texas A&M University Archives, Cushing Library*

late), the cadets slowly adjusted to standing in line at a make-shift kitchen with mess kits. The greater tragedy, however, was the near total destruction of Old Main. James M. "Cop" Forsyth '12 recalled that from his third-floor window in Leggett Hall he could see that the fire, which started sometime between 2:00 and 3:00 A.M., was out of control on the upper floor of Old Main. The majority of the flames were coming from the room that housed the small library. While the fire was intense, cadets did what they could with buckets of water. Forsyth noted, "One cadet ran to the Commandant's office on the first floor to try to save whatever he could gather in his arms. Upon leaving the burning building the cadet noticed he had saved the ram books [records of misdeeds]. Upon realizing what he had removed and abhorring the ram system he turned and threw the ram books back into the fire!"[75]

The library was lost along with virtually all the records of the college, the president's office, and supplies. The extensive investigation into the fire at Old Main concluded only one finding—that the fire was "of incendiary origin."[76] While there were rumblings to close the college temporarily due to the loss of the two most important facilities on campus, the board obtained speedy approval from Gov. O. B. Colquitt for a special appropriation sufficient to replace the gutted buildings. In appreciation to the governor, the cadets held a yell practice in his honor during his June visit to campus. The governor admonished the college to build new structures that would last "at least" a hundred years. The mess hall was replaced by Sbisa Dining Hall and Old Main by the Academic Building. Both buildings were completed under the supervision of F. E. Giesecke in 1913–14 at a total cost of about $400,000. At Final Review and commencement in May, 1912, the Corps had more than 1,200 cadets.[77]

The growth of the Corps continued into the fall of 1912. The entering freshmen, the Class of '16, totaled more than 225. The 1912 football team, coached by Charlie B. Moran, completed an 8-1 season with a 53-0 defeat over Baylor at Gaston Park in Dallas.[78] The dramatic growth of the Corps was more than even the addition of tents could absorb. The rapid growth along with the loss of two key facilities placed an added strain on the faculty and Commandant. Close monitoring of cadet activities had begun to diminish as early as 1912. In December, 1912, hazing of the fish intensified. The faculty urged the president to act at once to eliminate the practice. Governor Colquitt, citing specific examples from parents, expressed concern to the A&M Board. The college administration did not want a repeat of the 1908 student strike and disruption and instructed the Commandant, Levi Brown, first lieutenant, and his lone assistant, Sgt. Jim Kenny, to concentrate on military training and not to become involved with discipline except as it influenced their ability to carry out their assigned duties. A&M Board Chairman Edward B. Cushing encouraged President Milner to maintain order.

The overcrowding and wanton abuse of the fish was not acceptable. A number of upperclassmen had been issued cadet probations for hazing incidents in the fall of 1912. However, just prior to Christmas the harassment intensified.[79] In response to an incident in which the fish of Company D refused to remove their painting of the class of "1916" from the college water standpipe (tank) or to obtain a Christmas tree for the upperclassmen, all fish were "strapped" prior to or just after the holiday recess. A three-member faculty panel, well aware of a special Texas legislative resolution deploring hazing that had passed in March, 1911, after a University of Texas student being hazed by upperclassmen shot and killed another university student, held a hearing on the A&M incident. On January 28, 1913, the full faculty ordered the immediate dismissal of 27 cadets for hazing underclassmen. Among this group were numerous student leaders.

The Aggie Band leads the evening cadet parade along Military Walk near the turn of the century. Courtesy Texas A&M University Archives, Cushing Library

Though petitioned by a large number of the Corps to review and overturn this order, the administration refused to restore the expelled cadets. A large number of cadets then determined they would not attend class; the faculty responded that failure to comply would be considered a breach of cadet rules and that such insubordination was grounds for dismissal.[80]

Dismissal of the 27 cadets for hazing along with 466 cadets for insubordination was upheld and confirmed to the board in late February, 1913. Victor Barraco '14, having missed most of his senior year, recalled years later "that [the reaction to the insubordination] brought on a strike, which was wrong, of course. Known as the 466 Club. In my estimation that was the 466 fools, of which I was one. I had to pay for that . . . I finally graduated in 1915."[81] An investigation by the Texas Legislature was begun in Austin, but it never resulted in any action. Disgusted with the handling of this situation, cadets prepared to leave en masse. In a dramatic show of unity the Corps held an unscheduled "final" dress parade. One witness to the event noted, "The Seniors arranged themselves along the sidelines as if it were a graduating dress parade, and when the underclassmen passed in review, they saw 140 seniors with bared heads and tear-dimmed eyes watching what they thought was the last military ceremony they would ever witness."[82] After the parade the cadets marched to the armory and turned in their rifles. By now the governor, the lieutenant governor, and the legislature were fully involved in the controversy. Responding to letters from parents and the front-page coverage in the newspapers, the Texas Legislature passed a new bill outlawing the practice of hazing as a misdemeanor

punishable by fines and/or imprisonment. The indecorous struggle was ended when the cadets were given amnesty in exchange for a pledge that they would abstain from all forms of hazing.[83]

Nearly six decades after this turbulent period, E. E. McQuillen '20 addressed the fifty-sixth anniversary reunion of the Class of '13 in the spring of 1969: "This 1913 class was a rugged class. You experienced in full measure both the sweet and the bitter during your college days. You enjoyed great years in football, beating the University of Texas in three of the four games played. . . . you suffered through the sad strike of '13 . . . yours was the largest class to graduate until 1917 . . . those [were] rugged years but you survived them well."[84]

The board convened an extensive inquiry at its August, 1913, meeting in Fort Worth. Two days of testimony were graphically covered by the newspapers statewide and were published in a detailed report issued by the board. Milner, under pressure like Harrington, resigned effective October 1, 1913, and was replaced by acting president Charles Puryear, dean of the college. President Milner had unwittingly played a role in the events that culminated in his sudden departure. Eager to act, his plan to add tents increased the size of the Corps by 400 percent practically overnight. Milner's enthusiasm for growth did not take into account the need for additional staff support, recruitment of new faculty, extracurricular events, and what the cadets termed "amusements," such as a gymnasium. The lack of such care to monitor the growth and the management of the student body engendered an atmosphere in which hazing increased unchecked. The increase in the number of cadet units did not provide for adequate leadership, training, or oversight.[85]

Victor A. Barraco '14 as a yell leader in 1913–14. The duty of leading the "old pep" and yells fell to four cadets, two juniors and two seniors, in the early years. Barraco become one of the first Aggie flag officers in the U.S. Marine Corps. Courtesy Texas A&M University Archives, Cushing Library

Unrest in the Corps cast the A&M College into the political spotlight in Austin. Opponents of the A&M College hoped to use the 1912–13 turmoil as a catalyst to discredit A&M in Austin in order to merge the college with the University of Texas. Five constitutional amendments were voted on by the Texas Legislature, and all were defeated by a two-to-one voter margin. Aggies soon learned that the competition between the two schools ranged far beyond the annual Thanksgiving gridiron clash.[86]

Calm again returned in the fall of 1913. Recognizing the need for more direct supervision, two retired sergeants, George Smart of Maine and John Linder, an ordnance specialist, were "detailed" to Texas A&M to join Sergeant Kenny, thus bringing the total staff in the Commandant's Office to four.[87] Lieutenant Brown increased drill; each dorm was inspected twice a day; roll call was taken at every formation; and additional target practice was scheduled. The cadet leadership in the senior class was provided instruction in how to handle discipline problems properly. The changes and the turmoil had been difficult.

Infighting and turmoil were not new to some cadets. Gradually at the turn of the century A&M began to attract cadets from Mexico and Latin America. These students were routinely accepted in the ranks. And many arrived on campus with unique stories to tell. Three cadets from Mexico City, Miguel Marquez, Enrique Aramburu, and Fernando Iriarte, traveling back to the college were trapped in Monterrey, Mexico, during a pitched battle between Mexican Federal troops and rebels led by Gen. Venustiano Carranza. After being pinned down in the hotel for days, they escaped, losing all their cadet uniforms, and arrived to report on November 18, 1913.[88]

The Class of '14 lamented that they had "been constantly beset by problems," noting that in their four-year class history they seemed to always be looking for "a calm [that] follows a storm"—tents, overcrowding, fires, two meningitis outbreaks, the "Great Trouble," and "the story of the 466." The impact of the turmoil was evident by June, 1914. Of the 477 cadets who enrolled in A&M in the fall of 1910 "only 69 are here, ready to grasp that which stands for success in college—the college diploma."[89]

Given the student unrest and distractions, the irony of the decade preceding 1914 is that military training and standing as well as the preparation of the Corps of Cadets not only to pass but to surpass annual military compliance inspection by the War Department was unprecedented. In the midst of all the controversy, inspectors routinely sent by the War Department gave the military program at Texas A&M its highest rating, a "BA." There were five inspection classifications: A, B, BA, C, and D. Under class "A" was a college whose organization was essentially military in nature—strict on uniform inspections, drill, and "daily conduct according to the principles of military discipline." Class "B" included state land-grant or agricultural colleges that required military tactics in the curriculum. Thus, a "BA" rating were those colleges of the "B" group that attained the military efficiency and excellence required for colleges of the "A" class. Only ten colleges were yearly recognized with a BA "distinguished" rating. Other institutions with a superior rating included The Citadel, VMI, Norwich, New Mexico Military Institute, and Culver Military Academy. The A&M Corps of Cadets, predominantly Texans, proved keen to military regimen, an aspect that would prove critical as war clouds appeared on the horizon.[90]

On the eve of a fully realized world war in Europe, Dr. William B. Bizzell was selected president of the college in August, 1914. The "guns of August" would soon have a dramatic impact on the Corps of Cadets, the college, and its former cadets.

4

A CALL
TO ARMS

*We want men trained so that in case we should ever have a war
there would not be a scarcity of good men for leaders.*
Capt. H. L. Laubach, War Department, after inspection of the A&M
Cadet Corps, April 13, 1914

*It was a military school in a world where military schools were
quite common, its only claim to the uncommon being its attempt
to blend the gentlemanly occupation of the military into the less
gentlemanly avocations of commercial farming and engineering.*
Henry C. Dethloff, *A Centennial History of Texas A&M University, 1876–1976*

THE CAMPUS TURBULENCE of 1908 and 1913, the losses of Old Main and the mess hall to fire, and the seeming inability to obtain adequate funding were grim reminders to all that conditions at the college needed attention if the institution was to survive and grow. The "trouble" of 1913 gave way to a much-welcomed period of calm on campus, yet in Austin debate continued over the possibility of merging the state's two public institutions of higher learning into one university. Yet no Aggie could envision such an event.

Under the Texas Constitution of 1876, Texas A&M had been designated a "branch" of the yet unfounded University of Texas.

Thus, those who supported and promoted consolidation, the "One University Plan," felt that such action would eliminate duplication in such areas as engineering and science as well as the redundant need for libraries, classrooms, laboratories, and faculty. While cost savings were touted, the true objective was to increase funding for the University of Texas. To this end the Texas Legislature passed a joint resolution that outlined the terms to amend the constitution and unify the institutions. Such action would require voter approval. The A&M Alumni Association organized quickly, launching a statewide campaign in opposition to any merger. In a mid-July, 1913, election, shortly after Milner submitted his resignation to the A&M Board, the amendment was defeated. The arrival of a new president helped put the issue to rest.[1]

Dr. William B. Bizzell officially assumed the office of president on August 25, 1914. The new president, faculty, and Corps of Cadets had little warning of the dramatic changes in store for all involved with Texas A&M. As an estimated 375 (new students would come and go without notice) new fish arrived for orientation and class and the upperclassmen jostled for room assignments and new uniforms, word arrived that the large advancing German army was at the gates of Paris—close enough to hear the peal of the church bells at Notre Dame cathedral. The bloody onslaught had already cost 300,000 casualties. The college, the Corps of Cadets, and the former students had little hint of the lasting impact such faraway events were to have on their lives, their alma mater, and the nation.[2]

Those who experienced the changes and those who have assessed the period define World War I as a true "watershed" in the history of Texas A&M. The college and the

Corps of Cadets would never again have an enrollment of fewer than 1,000 and would begin a growth spurt that would average in excess of 15 percent per year. Between 1915 and 1920 the size of the Corps of Cadets more than doubled. The extensive building coordinated by Professor Giesecke with the addition of dorms, new classrooms, new academic building, and mess hall seemed destined to create an excess of space as well as a buffer for future growth. Such impressions were dashed with the coming Great War. Furthermore, the extensive role of the college to provide training for the war effort as well as the contribution of Texas A&M former students were to polish the image and increase recognition of the small rural college on the national scene.[3]

Bizzell—born on October 14, 1876, at Independence, Texas, on the Brazos River, a few miles south of the college—was an 1898 graduate of Baylor University in Waco. Fully aware of the troubles of his predecessors, he worked to bring a higher standard of morality to the faculty, staff, and Cadet Corps. Bizzell stressed improved scholarship, expanded religious activities, an emphasis on athletics, the outlawing of all hazing, and the continued expansion and development of the facilities and equipment needed to foster a first-class college.[4]

Very keen to "find out" or test newcomers to the Texas A&M campus, the cadets at first expressed grave reservations and concerns about the resolve of their new president who "parted his hair in the middle." His most recent job had been as president (1910–14) of the "girls" school at Denton, Texas—the College of Industrial Arts (CIA), known to Aggies as the "College of Innocent Angels"—which later became Texas Woman's University in 1957.[5] Bizzell prevailed to become very popular with

The train depot at "College Station" provided A&M's primary north-south transportation link for more than a half-century. This picture was taken in 1911. Courtesy Texas A&M University Archives, Cushing Library

the cadets. In addition to being a good day-to-day administrator, he strongly encouraged intramural athletics (especially football) and launched the first systematic effort to develop a comprehensive facilities plan for the college. These efforts would largely set the course for the college throughout the roaring twenties and into the early thirties.[6]

The era of good feeling among everyone on campus was a welcome improvement. Discipline now was more strongly enforced with a larger staff in the Commandant's Office. And all went well even when the Class of '17 painted their class year on the standpipe (water tank). The Commandant called a class meeting to determine how to remove the unauthorized painting, whereupon a "committee of one was appointed who saw that it was removed."[7]

By focusing attention on averting the same types of problems experienced in the past, Bizzell's guidelines concentrated on the building of a more solid foundation so the college would indeed grow and prosper. In addition to pushing forward the completion of Guion Hall and the YMCA (with an indoor swimming pool and billiards hall)—to give the cadets a place of wholesome recreation and mentoring away from the dorms and classes—Bizzell raised admission requirements in 1914–15, employed new faculty, and strongly supported the expansion of both intercollegiate sports and facilities at the college.[8] President Bizzell's call to arms in the daily affairs and mentoring of the faculty and cadets was but gentle ground work for the graver challenges ahead far from the rural remote campus.

Football quickly became a central focus of both cadet and alumni support of the college. The gridiron success of the Farmers recorded a 75 percent winning record (73-18-4) from 1902 to 1910. Yearly, the Corps became more involved with the support of the team. Yell leaders led new yells, and the band was expanded to eighty-one members. Travel to away games became more routine and was much enjoyed by the cadets. The catalyst for winning was supplied by Coaches Charles B. Moran (1909–14), E. H. W. "Jigger" Harlan (1915–18), and Dana X. Bible (1919–28). During the 1914 and 1915 seasons, the Farmers lost only one game per season, defeating the University of Texas on Thanksgiving, 1915, the first-ever meeting of the two rivals at Kyle Field, by a score of 13-0, prompting one cadet to proclaim, the "maroon and white is wild with joy!"[9]

The growth of football statewide and the demands for organized intercollegiate sports resulted in the formation of the Southwest Intercollegiate Athletic Conference composed of Texas A&M, Baylor, Arkansas, Rice, Oklahoma A&M, University of Oklahoma, Southwestern University, and Texas—in time for the 1915 football season. Competition in the new conference was spirited. In the fall of 1917 cadets branded the new University of Texas mascot, a longhorn steer, with the 13-0 score of the 1915 game in hopes of firing up the team and fans. By now the term "Aggie" had begun to appear to denote students and alumni of the college. The Corps numbered 1,242 cadets in 1917.[10]

WAR CLOUDS: NATIONAL DEFENSE ACT OF 1916

President Woodrow Wilson had, from the onset of the growing European conflict, expressed somber concerns about American involvement "over there." Shortly after winning the 1916 presidential campaign, his slogan being "He Kept Us Out of War," the War Department was instructed to begin quietly and quickly to assess plans and manpower requirements in the event a mobilization was needed. The generals found the country ill equipped and undermanned to field and support an army. The U.S. Army in 1915 numbered a regular force of about 25,000 with little or no "prepared" reserves. Military training at the land-grant colleges had indeed introduced many citizen-soldier-cadets to the basics of soldiering, military procedure, drills, and target practice. Unfortunately, most of the students who were a product of such two- to four-year training were far removed from their student days and in civilian jobs and occupations that had little relevance to the demands of an army in the field. Notwithstanding, it was exactly this diverse core of men— trained in land-grant colleges between 1880 and 1917—that would fill the ranks of U.S. Army units in 1917–19, most in leadership positions.[11] In an address to the annual meeting of the Alumni Association, Bizzell noted with pride that "after forty years the military instruction of the College has received ample vindication and justified [sic] by those who succeeded in including military science as a required subject in the land-grant college."[12]

By the winter of 1915–16, as the debate over raising a "citizen army" intensified nationwide, it was apparent that the United States would be involved with the European conflict in some fashion.[13] German U-boat activity in the Atlantic in early and mid-1915 intensified to break a British naval blockade, completely ignoring the shipping of neutral nations. The sinking of the British liner *Lusitania* in early May, 1915, shocked the world—killing 1,198 passengers in eighteen minutes, including 128 Americans. The war effort would demand a broad military mobilization nationwide. The concept of military training under the limited provisions of the Morrill Act of 1862 needed quick attention. To create a broader bridge between college-trained officers and their transition to regular Army status in time of war, Congress passed the National Defense Act of June 3, 1916. The major component of the new act was the enhancement of officer training in what was to be known as the Reserve Officers' Training Program [Corps] or ROTC.[14] The land-grant colleges would supply a majority of college-trained officers in this ROTC program. The National Defense Act of 1916 dramatically expanded the role of the federal government in securing college-educated men for the officer corps of the regular Army, and it created state national guard units to replace the variety of unregulated state militia programs. Such new reserve units would be equipped and compensated by the federal government as well as required to meet all federal standards.[15]

The concept of a broad citizen-soldier army dates back to the American colonial period, buttressed by a requirement in 1792 that each state establish a "militia system." Under the initial guidelines all free, white, male citizens between eighteen and forty-five years of age were eligible for military service. This state-oriented militia base of volunteer citizen-soldiers would have major significance for the nation. Noting that "the obligation to

military service is universal," Maj. Gen. Leonard Wood spearheaded a nationwide prewar debate on military preparedness.[16] The first suggestion for the ROTC concept dated back a century prior to the 1916 National Defense Act. In 1815 Capt. Alden Partridge, superintendent at West Point, noted that the military academy alone could not supply adequate officers in time of war, thus "the diffusion of military science throughout our country" into other academies and institutions would be "requisite for officering any additional force that might be necessary."[17]

Partridge established the first nonfederal military college, outside the traditional federal training offered at West Point or Annapolis, in 1819 in Vermont. The primary function of the new institution, called the American Literary, Science, and Military Academy was to identify and train future militia officers. In 1834 this fledgling academy changed its name to Norwich University. Shortly thereafter, numerous military colleges were founded in the South, the most prominent being Virginia Military Institute in 1839 and The Citadel in 1842. Due to a strong military tradition in the South, in the years prior to the Civil War, military academies could be found in each state below the Mason-Dixon line. Though most were purely ceremonial, these academies, in fact, provided a valuable source of graduates for state militia organizations. When the Civil War broke out in 1861 the South was able to field units much more rapidly with officers who had military training. It was this southern military tradition that the Corps of Cadets at Texas A&M aspired to after the college was opened in 1876.[18]

It quickly became apparent that the citizen-soldier would play a critical role in the North-South conflict. In April, 1861, there were only 684 West Point graduates in the regular U.S. Army, of a total officer strength of 1,098. Within weeks of the beginning of hostilities, more than 300 officers resigned their U.S. Army and Naval commissions to serve with the Confederacy. In the North, governors attempting to raise units during the mobilization of the state militia for the Union Army appointed officers at will, and few had prior military experience or regard for military expertise or qualification. Many of the new state units held elections to fill key officer slots. There was a noticeable shortage of trained men to provide leadership for the army.[19]

This shortage of officers in the North was the primary reason for the swift completion and passage of the Morrill Land Grant Act—signed by President Lincoln on July 2, 1862, only hours after the Army of the Potomac had abandoned hopes of capturing the Confederate capital at Richmond and narrowly escaped to the James River for evacuation at the conclusion of the "Seven Days" battle. The end of the Peninsula Campaign in late June, 1862, marked the largest land engagement and concentration of armed forces ever assembled on American soil. The enabling legislation, championed by Justin Morrill for more than a decade, basically provided for a certain amount of "military schooling" as a central part of the newly created "land-grant" institutions in each state. Morrill thought the potential of a broad base of officer procurement would be in the best spirit of the democratic ideals to control the size and maintenance of a large standing army in peacetime but would also provide a source of officers in time of national emergencies. The act provided large tracts of public lands to fund such new colleges. The negative of the Morrill Act was that the extent of the military training was not well defined or uniform from college to college; moreover, some colleges questioned the need for such military training to be compulsory. Neither the Congress nor the War Department clarified the meaning or scope of the land-grant institutions' mandate regarding instruction in "military tactics."[20]

Furthermore, the Morrill Act initially provided no provision for financial support of on-campus military programs to provide Army officers, equipment, and course materials. At first it was assumed that a faculty member with prior military training would fill the requirement of the college, which is exactly why "Major" Morris was hastily named as the first Commandant at Texas A&M in 1876. He was the only member of the new faculty with prior (although limited) military experience at VMI. The Congress began to correct this funding and training oversight with a number of supplemental pieces of legislation in 1866, 1888, and 1891. These new measures allowed the War Department to pay and assign regular U.S. Army officers to each land-grant campus as the professor of military science and tactics, PMS&T. Furthermore, the role of the officer on duty varied with the institution. Some assigned officers served as "the Commandant," others only as the chief military instructor of tactics. By 1898, only forty-two institutions nationwide offered some form of military training.[21]

After the Spanish-American War, a better system of military preparedness was deemed necessary. However, after much debate no sweeping changes were made until the eve of World War I. In the interim, one practice that was adopted after the Spanish-American War was the increase in the number of inspections and evaluations of existing campus programs. A rating system was established to rank all colleges offering military instruction. Modest amounts of training manuals, rifles, and equipment were provided to each institution. The top ten military programs would receive recognition as a "distinguished institution." Those, like Texas A&M, which repeatedly placed in the top ten received additional military arms and equipment. Furthermore, each college was to identify yearly one outstanding cadet for a commission in the regular Army. With no developed state militia or reserve system, graduation marked the end of most semiformal training and indoctrination in any aspect of the military. One key to the college program during this period was the level of organization, enthusiasm, and expertise of the Commandant.[22]

Ironically, the ROTC program mandated by the National Defense Act of 1916 produced no cadet officers trained from Texas A&M or any other college for World War I. The first A&M officer commissioned under the ROTC program was in June, 1920. The new ROTC program did not go into effect until late 1917 or early 1918 and then only on a limited basis due to the fact that most available active duty Army officers had been recalled for duty in Europe. The War Department reported that "military efficiency was hampered by the supply of officers constantly lagging behind the active duty demand."[23] In the case of A&M, an interim retired Army officer, Col. Charles J. Crane, functioned as Commandant of cadets. By the time the application and approvals were obtained by the Bizzell administration from the War Department on October 19, 1916, and implemented by college officials, the war had ended.

The slow training cycle and supply of fresh new officers constantly lagged behind the demands of the Army, thus hampering military efficiency throughout the war. However, this did not preclude an all-out effort by the administration, the A&M Cadet Corps, and its alumni. When the war in Europe was over, the War Department reconstituted and expanded the new ROTC program, and by 1919 there were Army ROTC detachments on the campuses of 191 colleges and universities (four times the number operating under the old Morrill Act when the nation entered the war). In 1920, the U.S. Army Air Corps (then the Air Service) established the first four ROTC detachments at the University of Illinois, University of California, MIT, and Texas A&M. And the first Naval ROTC units were established in six institutions in 1926, not including Texas A&M. In the meantime, Texas A&M would prepare for war.[24]

FARMERS FIGHT

On April 6, 1917, President Wilson presented a formal declaration of war on Germany to the U.S. Congress. By mid-June, the U.S. commander in Europe, Gen. John J. "Blackjack" Pershing, had landed in France. The first call to man the American effort required 250,000 troops. The response by the A&M faculty, alumni and former students, and cadets was swift and overwhelming. Texas A&M became the first college in America to offer the War Department its full cooperation, facilities, and personnel for military training. President Bizzell went directly to Washington, D.C., to lobby for supplies, equipment, and additional Army officers in order to be prepared to offer whatever was needed. By mid-April, 1917, mobilization began—the campus was on a total war footing, a state of emergency that would last until early 1919.[25]

The Commandant, Capt. C. H. Muller, and his assistant, Lt. W. H. Morris, were recalled to active Army units and replaced by Charles J. Crane, former Texas A&M Commandant (1881–84) and retired Army colonel, for the duration of the war.[26] The college also hired Lt. A. F. McMann as Assistant to the Commandant and Capt. William Martin of the Seventy-ninth Cameron Highlanders Royal Canadian Army as chief instructor of tactics. Martin, an early veteran of the war in France, whom President Bizzell made a personal trip to Ottawa to recruit, gave detailed instruction on battlefield conditions and trench warfare in France.[27]

In order to provide a speedy source of specialty military training and production of officers, three primary methods were employed by the War Department. The first method was a call to arms that began in early 1917 prior to Wilson's formal declaration. Under provisions of the National Defense Act of 1916, the Army issued guidelines for "affording students and graduates an opportunity to take competitive exams for provisional appointments" as second lieutenants in the Officer Reserve Corps (ORC). Starting pay was $1,700 yearly. Candidates had to be twenty-one to twenty-seven years of age. Graduates of a recognized college could be exempt from exams and given direct commissions. Due to the increase in the size of the Army, new recruits were told "promotion promises to be unusually rapid." Commissioning and rank for those with special skills in engineering, medicine, and chemistry would be determined upon induction. And the age limit could and would be waived.[28] Alumni and former students, due to prior exposure to military training in the Corps of Cadets, were quickly recommended and many were directly commissioned as officers at the time of enlistment.[29] A prime example of this on-site rapid promotion was Col. Edward B. Cushing '80, Seventeenth U.S. Engineers AEF. As soon as Wilson declared war, Cushing (one of the first graduates of Texas A&M, past president of the Association of Former Students, 1900–1901, and chairman of the Texas A&M Board of Directors, 1912–14) took a leave of absence as general manager of the Southern Pacific Railroad to enlist for two years of duty in France. By war's end he had served as a logistics and transportation specialist (rail, ports, and distribution) on General Pershing's staff. Hundreds of other A&M men followed his example.[30]

However, some Aggies were already in France fighting against the Kaiser's forces long before the formal U.S. declaration of war. Lt. John Ashton '06 had graduated and departed for England to train as a veterinarian (the School of Veterinary Medicine was not established at A&M until 1916, so those seeking training had to go elsewhere). When the war erupted in late 1914, he joined the Army Veterinary Corps, British Expeditionary Force, for duty in France.[31] The Great War would depend heavily on the horse and mule—the last full-scale use of the beast of burden in warfare. One of the most decorated Aggies of the war was Lt. Georges P. F. Jouine '07. At the outbreak of war, he was working as a surveyor on levee construction on the Mississippi River for the Corps of Engineers, but he left to join the French army as a private in late August, 1914. He was soon commissioned a lieutenant, and for thirty-four months he commanded a trench mortar company on the front lines. In early 1916, after being wounded

Pvt. J. V. "Pinky" Wilson '21, author of the "Aggie War Hymn," pictured on active duty during World War I. Courtesy Sam Houston Sanders Corps Center

four times in action and awarded the French Legion of Honor, he became one of the first Americans (though in a foreign army) to command in combat a company with a revolutionary new war machine—referred to by H. G. Wells in 1903 as the "Land Ironsides"—the tank. He was wounded twice more while in the tank corps. By war's end, he had received a dozen military combat decorations from three nations. Shortly after his death in 1958, the Houston A&M Mothers Club established and yearly sponsored the "Jouine Award" to recognize annually the most outstanding scholastic unit in the Corps of Cadets.[32]

The second method for training and commissioning officers rapidly was more specific. U.S. Army training camps in various regions of the country were established to train the first wave of "90-day wonders." In late April, 1917, the cadets followed the alumni lead, when most members of the Class of '17 were dispatched to Officer Training School (OTS) at Camp Funston (under the command of Colonel Scott, the former A&M Commandant in the 1880s), near Leon Springs, Texas. Others enlisted directly in a branch of the armed forces. Only weeks short of graduation, their diplomas were approved by the board and presented in a brief, rare, off-campus ceremony under a big oak tree near the training camp. Others also joined. All the sport's captains for the 1917–18 season joined the Army, and popular head football coach Dana Xenophon Bible enlisted and served as a pursuit pilot, Twenty-second Aero Squadron of the First Army in France.[33] Bible's last prewar team captured the Southwest Conference championship in an unprecedented showing. The 1917 Aggie football team was undefeated, untied, and unscored upon—amassing an 8-0 record and scoring

270 points to 0 by their opponents. (Only two teams succeeded in penetrating inside the twenty-yard line.)[34] With the departure of so many upperclassmen, juniors assumed leadership positions until they either enlisted or were called to duty. Juniors of draft age in the Class of '18 were given academic credit by the college for courses completed and dismissed by President Bizzell to attend training camp.[35]

The third means of preparing and mobilizing men for the war was on-campus training. With no commencement in June, 1917, the Cadet Corps dwindled to a few hundred cadets. Those cadets under the legal age to be drafted continued training, drill, and intensive target practice. The staff and faculty were quick to endorse and support the all-out war effort.[36] Cadets and faculty remained focused on preparation throughout the summer. The open-door invitation of Bizzell to the government was acknowledged in August, 1917, when the first of a wave of wartime training programs was assigned to the College Station campus. Facilities, barracks, and an expansion to the mess hall and officers' quarters were hastily constructed. To support the troops, Bernard Sbisa prepared to feed more than twice as many meals. While the Corps of Cadets went about its somewhat routine daily schedule, hundreds of new Army and Navy recruits in various specialties or training groups were established on campus. In early 1918, the first A&M man to give his life, Pvt. Norman G. Crocker '18, died shortly after enlisting in the Army as a result of a German torpedo attack in the North Atlantic on the *Tuscania* on February 5.[37]

The U.S. Army Signal Corps set up the radio communications, radio electricians, and repair course taught by the Department of

World War I training on campus was very extensive. Here cadets work to overhaul automotive frames. Courtesy Texas A&M University Archives, Cushing Library

Electrical Engineering. Cadets were permitted to enroll in these courses followed by OTS, commissioning, and immediate assignment overseas. The success of the signal corps training program was followed by an intensive radio mechanics, aircraft engine repair, and technicians program that later expanded to an automotive and motor-truck mechanics course conducted by the Department of Agriculture Engineering. By the fall, 1918, the auto mechanics course produced 1,731 mechanics. Intensive training followed in horseshoeing, surveying, meteorology, blacksmithing, carpentry, and machine tool operation.[38]

There was no question that the war required a cooperative effort and that the Aggies were going to contribute everything possible. On February 2, 1918, representatives of the Dallas A. and M. Club—Joe Utay '08, Thomas H. Barton '01, Marion Church '05, and Francis K. McGinnis '00—were appointed to call on Governor Hobby for the purpose of obtaining authorization to form a regiment of cavalry in the state national guard.[39] Hobby agreed, but he could not give the final authorization until guidelines were received from Washington. In Dallas detailed plans were drafted and equipment assembled. In July, 1918, authority was received designating the unit as the "Third Regiment of the First Brigade of Texas Infantry" (later renamed the Tenth Regiment, Texas National Guard), but to all it "will be known as the A. and M. College Regiment."[40] The regiment was fully organized and manned in September and began intensive training awaiting "federalization" for overseas duty. Joe Utay, captain of the 1907 football team, member of the Ross Volunteers, and a feisty Southwest Conference football referee (1912–36), spent an entire month in Washington lobbying the War Department to activate the A&M regiment. Potential recruits were told, "It should be borne in mind that this regiment is not an 'air bubble' or 'play thing' . . . but a fighting unit of loyal A&M men."[41] In late October, the "A. and M. Regiment" was at full strength, had passed all federal inspections, and awaited mobilization by the War Department. Commanding the regiment was Col. Abe Gross '96 of Waco; deputy commander was Lt. Col. Marion S. Church '05 of Dallas; and the three battalion commanders were Maj. Joe Utay '08 of Dallas, Maj. John D. McCall of Austin,

and Maj. Richard H. Standifer '08 of Fort Worth.[42] Only the November, 1918, armistice kept these Aggie citizen-soldiers from active duty.

Back on campus, the cadets marveled at the intensity of activity, as the campus became, in many respects, a full-fledged Army-Navy training camp with little regard for the normal course of college protocol and academics.[43] Nearly the entire senior Class of '18 had been drafted or joined the Army to attend officers' training camp. Cadet Corps Commander Harry C. Knickerbocker '18 and the top cadet leadership joined the Army, each recommended for active duty by Colonel Crane. Commencement exercises in June, 1918, were canceled.[44] Cadets, the traditional Corps organizational structure, and the regular academic studies had become a low priority. This was further accentuated by the fact that nearly all of the Class of 1919 went in the summer of 1918 to summer camp training at Fort Sheridan, Illinois, trained, and went directly to Army duty.[45]

While eager to assist the Army, Bizzell and the faculty by late summer 1918 were concerned at the plight of the college curriculum—the campus was virtually "federalized" for the duration of the war. One challenge to an already crowded campus was the creation (by presidential directive) of the Student Army Training Corps (SATC) to utilize students and the campus facilities nationwide for the selection and training of officer candidates.[46] In the fall of 1918, SATC was opened to all eighteen-year-olds who planned to attend college and had been cleared by their hometown draft boards. They were inducted as privates in the Army with full pay and assigned to an SATC unit. The SATC program was the last full measure to mobilize all able-bodied reserves and to identify potential officers. A series of intensive twelve-week courses in engineering, chemistry, and military studies was geared to prepare these college trainees in less than two years to be ready for immediate duty in the active Army. By October, enrollment in the SATC program at A&M exceeded 1,100 cadets.[47] Colonel Crane by early 1918 performed only a ceremonial role as Commandant of Cadets. The regular Army officers were in total charge, and discipline remained very strict. Army-style KP (kitchen police) and work details for punishment took the place of the Cadet Corps demerit system. Cadets and SATC students that "showed no ability" were subject to be ordered to active service "for the duration of the war." Vernor G. Woolsey '22 recalled a lighter moment during his fish year in what was a serious wartime setting:

> The only instance that I ever saw of a trainee's getting the best of an officer in charge was always a source of great amusement to me. While in formation and marching to the mess hall, strict discipline was enforced. No talking or "monkey business" of any kind was tolerated. In this instance one of the privates in formation began to whistle. The lieutenant in charge of the platoon heard and saw the one who was whistling, and promptly pulled him out of formation. He made him march to the mess hall alone. When the entire Corps was in and seated, the lieutenant ordered the offender to stand up on the table and finish his whistling so that everyone could hear and enjoy his solo. The young man obeyed the command and stepped up on the table in front of him. He began whistling the "Star Spangled Banner," and everyone in the mess hall, including all

of the commissioned officers present, had to stand at attention until the whistling of the National Anthem was completed. No penalty was imposed on the whistler. Truly, he had the makings of a general![48]

The SATC program in effect took over where the Cadet Corps left off—except it was administered by the U.S. Army under contract with Texas A&M. Colonel Crane resigned to take an assignment at Rice University in Houston, and Maj. Fred W. Zeller of the SATC program became Commandant of both the Army training and the Corps of Cadets on September 1, 1918. Zeller, at A&M for nearly a year in command of the Army's special training programs, moved swiftly to minimize chaos and reorganize the cadet trainees, who jokingly referred to SATC as "Stick Around 'Till Christmas" or the "Sunday Afternoon Tea Club."[49] Though lampooned by both cadets and trainees, SATC in fact by the fall of 1918 "nationalized" all able-bodied men over eighteen years of age for direct induction by the War Department into the armed forces at a moment's notice. Four special training units commanded by commissioned U.S. officers took over: Company One was housed in Milner Hall; Company Two in Leggett Hall; Company Three in Mitchell and Austin Halls; and Company Four, including the first Naval training unit at Texas A&M and the college band, in Foster Hall. Those existing cadets under eighteen were technically not eligible for the SATC program yet were blended into the units and received "the same intensive military drill."[50]

The war effort was very demanding. At A&M, Major Zeller proved very effective in maximizing the ten- to twelve-hour days for training and class. The War Department placed a high priority and rigorous demand on the college-based SATC program, due to what was seen as a tremendous pending need for officers and men by mid-1919 to man either a second front via the Balkans or an extensive counteroffensive through France and into the heartland of Germany.[51] Former A&M Commandant Herbert H. Sargent was recalled to active duty in 1917 as a prime strategist in the war plans division of the general staff. As dozens his former A&M cadets, among them George Moore, G. P. F. Jouine, John Warden, and Douglas Netherwood, filled the active ranks in the field, Major Sargent helped draft the grand scheme to win the war with the minimum American casualties. The challenge was compelling. French and British forces, after nearly four years of war, were depleted. Thus General Pershing and the AEF would have been expected to take the lead in the offensive. U.S. forces in Europe exceeded two million men by the fall of 1918.[52]

Much to the relief of all, hints of a cease-fire in war-torn Europe became reality on November 11, 1918, as the Allies and Germans agreed to an armistice. The silence of the guns of war were cause for jubilation on the campus and across the nation. Only seventy days after the full-scale implementation of the SATC program the war was over—over there and over here. Class of '20 historian Thomas A. Cheeves expressed considerable relief in the 1920 *Longhorn*: "From September 15–December 20 [1918] inclusive is well called the darkest period in the history of the school . . . the only thing that saved A.& M. as well as many other things was the signing of the Armistice." Concerned about any misinterpretation of the college's commitment to the SATC, President Bizzell stressed that the "college will continue its military and academic work without interruption—regardless of the armistice with Germany."[53]

Orders from the War Department to unit commanders at colleges and institutions nationwide strongly supported Bizzell's actions and comments. Commanders were subsequently ordered to work closely with the college administrators to phase out the SATC and all campus military training programs and "commanding officers are further directed to reduce to a minimum such detail of individual soldiers as interferes with their academic studies . . . . studies get first place."[54]

While President Bizzell and the A&M administration were ready to begin to get back to normal, there was concern with the postwar transition or demobilization. On November 14, 1918, Maj. Fred Zeller was reassigned as commandant at St. Louis University and replaced by Maj. Grant M. Miles.[55] One of the first items Miles, formerly with a training unit in Chicago, and Bizzell agreed on was that all members of the SATC and the Cadet Corps be granted permission to attend the annual Thanksgiving Day football game in Austin.[56] On November 27, Bizzell took the lead in the transition process and held a mandatory mass meeting at Guion Hall for all college personnel, cadets, and trainees to review the demobilization procedure. There were four key components:

1. technical students—in the accelerated two-year SATC program will be processed out and allowed to consider a regular college course in January, 1919.
2. academic students—primarily current cadets and special students (veterans)

would need to reconfirm their enrollment in the college and be prepared to start the new term or depart the college.

3. SATC cadets—who left jobs voluntarily to come to the college to prepare themselves for military duty would be given "preference" in processing out to return home to work.

4. demobilization—the War Department ordered each college and institution to be complete with the demobilization and transition by December 21, 1918.[57]

President Bizzell realized that this was a turning point for the trainees and cadets he was addressing in Guion, as well as for the college. He urged that the transition be orderly, that strict military discipline would be observed, that any cadets or trainees who wanted advice on "mapping out your future" should see Professor Frank C. Bolton. The SATC would be fully phased out and all that would remain was the Corps of Cadets and ROTC. The campus would began a new era on January 1, 1919. President Bizzell urged the cadets to focus the return to normalcy on the core values and traditions of the college: "We have been compelled to sacrifice the College *spirit* which we have cherished for years by taking in much larger numbers than we could do justice to, but we did it to try and aid in winning the war and have no apologies to offer for it. We hope to see normal conditions restored soon, however, and with the restoration of these there will come the restoration of our College *spirit*."[58]

AGGIE WAR LEGACY

The joy over the sudden end to the war and the demobilization that followed was a welcome challenge. Never before had the college been called on to alter fully its course and purpose. Without a doubt, President Bizzell provided timely leadership, enthusiasm, and heartfelt restraint in his management of the wartime campus.

Yet it was time to bring back the old-time Texas A&M College "spirit." The energetic president planned and executed the same level of hands-on leadership used during the war to transform the college and the Corps of Cadets back to as normal as possible operations beginning in early 1919. Seniors from the classes of 1918 and 1919, most of whom had received wartime "certificates" of attendance and not degrees, were encouraged to return for a special four-week mini-semester that was to begin on January 3. However, there were some changes that could not be turned back.[59]

The record of accomplishment by the cadets, former students, faculty assigned military officers, and alumni during the war was overwhelming. Both on the home front and in France, the Aggie citizen-soldiers, due primarily to their training at Texas A&M, made an immediate impact. Four former A&M Commandants—Moses, Bartlett, Scott, and Carleton—became general officers during the war. These generals, along with Col. Levi Brown, Col. C. H. Muller, and future Commandant Col. Ike Ashburn, commanded A&M men in all sectors of the conflict. Colonel Ashburn noted with great admiration, "I followed the records of the A. and M. men in the military service closely and I have known none of them but was a credit to this

TABLE 4-1
World War I Officer Production

School	Officers	Other Ranks	Total	General Officers	Killed or Died
Texas A&M	1,233	984	2,217	(2)*	53
VMI	1,176	231	1,407	5	57
VPI	638	1657	2,297	2	24
Clemson	A	A	1,549	0	25
Norwich	345	147	495	2	16
N. Georgia	A	A	371	0	10
Penn Military	208	142	350	4	12
Citadel	276	40	316	0	16

Notes:

A = Unknown or unavailable

*The *Annual Review* in 1919 lists the service of two Aggie general officers but not by name. Records are inconclusive as to who held this rank.

Source: Texas A&M *Alumni Quarterly;* Kraus, *Civilian Military Colleges,* p. 487; Couper, *One Hundred Years at V.M.I.,* pp. 226–27; and VMI Museum.

institution . . . one thing that impressed me in France was the fact that spirit counted for even more than efficiency."[60] At home, General Pershing's praise of American farmers reflected positively on A&M agriculture graduates, the agricultural experiment and extension stations, and production research advancements that contributed to the massive increase in wartime agricultural yield.[61] However, it was the contribution of Aggie officers and soldiers that most brought nationwide notice to the college.

No college or university in the United States—not even the military service academies—had made a larger contribution than Texas A&M "over there."[62] The Alumni Association realized this dramatic contribution early on and had initially begun to collect data on the whereabouts of Aggies as part of a project to complete a huge fifteen-by-twenty-six-foot flag in red, blue, white, and gold. The "Service Flag" would recognize the effort and sacrifice of the "A. and M. men" to the nation. The statistics were breathtaking. More than 50 percent of all living Aggie graduates served on active duty during the war. No other college could make this claim. Furthermore, hundreds of ex-cadets served on active duty. However, exact numbers are difficult to confirm. Also, many Texans who were not former cadets of Texas A&M claimed to have attended the college. One of the most puzzling cases is that of Daniel R. Edwards, a recipient of the Congressional Medal of Honor near Soissons, France, on July 18, 1918. For decades in printed stories, personal memoirs, and *Who's Who,* Edwards claimed to have been "for a year and half" a cadet and member of the football team at A&M. With the destruction of all the college records in the Old Main fire of May, 1912, his claim has remained unverifiable to this day. He would be neither the first nor the last to spuriously claim attendance at Texas A&M.[63]

Notwithstanding, Texas A&M furnished 1,233 officers and 984 noncommissioned former students and cadets, representing nearly every A&M class from the beginning (1879) of the college to 1921. Of this number, 53 lost their lives during the conflict: 67 percent killed in action, 20 percent in war-related training accidents (primarily airplane crashes), and 13 percent from illness. Among those memorialized for their contributions were Lt. Luke Witt Loftus '17 from Dolores, Texas, who was killed in a skirmish with Mexican Federal troops at Nogales, Arizona, on August 27, 1918, and the popular Jesse L. Easterwood '11, a Naval aviator killed in a training accident in Panama on May 19, 1919, shortly after returning home from the war. The Class of '18 made the largest contribution in number of volunteers and also had the highest KIA casualties (14), followed by the classes of 1913 (9), 1919 (5), and 1911 and 1917 (3 each). Most A&M casualties occurred in an eight-month period prior to armistice. Thus, concerns about the war dragging on into 1919 were well founded; it could have taken a dramatically larger toll.[64]

To recognize the accomplishments of the wartime Aggies, Bizzell led the way in ensuring their contribution would not be forgotten. The World War I Service Flag, honoring more than 2,200 Aggies who served (an estimated 702 graduates and 1,515 former cadets or former students), was first hung in Guion Hall and later moved to hang from the fourth floor in the rotunda of the Academic Building until the early 1940s.[65] Of the 702 A&M

Corps of Cadets fall review in 1919: Gen. T. M. Scott, reviewing officer, along with Cadet Corps Commander Waller T. Burns '20. Buildings in the background are, left to right, *Ross Hall, Foster Hall, Bizzell Hall, and Guion Hall. Courtesy Texas A&M University Archives, Cushing Library*

graduates who joined the military, 668, or slightly more than 95 percent, were commissioned officers. Many of the former cadets won commissions via OTS; however, the exact number is unknown. In November, 1918, Bizzell envisioned "a more dignified, elaborate and permanent memorial that shall be made to perform some useful function on the campus at the same time that it perpetuates the names of the men who gave up their lives for freedom."[66]

Bizzell invited suggestions from the alumni and former students on a proper type of "memorial building." A number of ideas were considered—a building for student activities (which predates 1945 formulation of plans for the Memorial Student Center completed in 1951 to honor the Aggies of World Wars I and II), a memorial library, or a new "Memorial Football Stadium." The Memorial Student Building concept (student activities were at the time allotted only cramped space in the YMCA) and an expanded library were far ahead of their time in 1918. Thus, the idea to raise funds for a new "concrete" football stadium began to gain momentum in late 1919, shortly after the annual Thanksgiving Day meeting of A&M and Texas, but it never was fully realized. In the meantime, the classes of 1923–26 collected funds for a nine-ton, granite, flag-draped World War I memorial that was first located next to Guion Hall and then moved to the West Gate entrance of campus. On April 21, 1924, dedication of the monument was keynoted by Gov. Pat M. Neff. The Cadet Colonel of the Corps, Herman L. Roberts '24, read the Roll Call of Aggie dead. Additionally, at the suggestion of Laurence J. Hart, president of the Board of Directors, live oak trees were planted on March 2, 1920, around the main drill field in honor of each

Aggie killed in the war, and American flags were added at Kyle Field in the 1920s.[67]

THE POSTWAR CORPS OF CADETS

Texas A&M and the Corps of Cadets that mustered in early 1919 were even stronger than their prewar models. In 1918, the "cadet gray" was worn for the last time as cadets took on the olive drab and khaki uniform of the regular U.S. Army. There were other changes. The march or "hike" to the Brazos had been discontinued, along with mandatory Sunday chapel. The college newspaper, the *Battalion*, was shut down (temporarily replaced by the *Daily Bulletin*) for much of 1918, and the *Longhorn* was delayed. Sports activities were curtailed and the Corps trips rescheduled. Eager to attract attention in the days following the war, the college opened a zoo filled with exotic animals—lions, tigers, an elephant, and ostrich. The 1919 *Longhorn* lamented that "it seemed as if the pep had died out altogether." There had been no campus commencement ceremonies since June, 1916.[68]

Bizzell, thanks largely to his wartime experience, realized that to train and mentor the largest Cadet Corps in the history of the college would require a Commandant and additional military staff of the highest caliber. Comfortable in dealing with the War Department, he took an active role in the selection of these key individuals. The large Corps, mixed with hundreds of returning veterans, known then as "casuals" or "cits" (citizens), would be a challenge. To fill the post of Commandant, Bizzell prevailed in having Col. Charles H. Muller, Commandant at the college in 1916–17, reassigned as the professor of military science and tactics in January, 1919.

In the post–World War I years during the first introduction of ROTC, Texas A&M became one of the leading campuses with training in all branches of the army. Here cadets train on the main drill field in caisson formation facing east toward Bizzell Hall in the center and Guion Hall on the right. Photo from 1920 Longhorn

A native Texan and West Pointer, he had distinguished himself in Europe and upon return became familiar with the new role of ROTC in the postwar period. The selection of Muller proved critical to launching a new era of military training at A&M during the 1920s.[69]

Thus, the decade ended with Texas A&M and the Corps of Cadets adjusting to a postwar era the nation came to enjoy. Demobilization of the large standing regular Army and the realization of the role and impact of citizen-soldiers on the outcome of the war only heightened the resolve of postwar planners to institutionalize the training of potential officers through ROTC on college campuses nationwide. At no place were such programs more welcomed than at Texas A&M. The Corps of Cadets gladly accepted the challenge.

5

HAIL AGGIES!
HAIL MAROON
AND WHITE!

The College is entering upon a new period in its history and the people of Texas will look to it more and more for leadership and guidance in developing the wonderful resources of the State.

Charles Edwin Friley '19, College Alumni Secretary

Few institutions are better fitted for the training of F.A. units than is A. & M. of Texas.

Field Artillery Journal, November, 1919

THE IMPLEMENTATION of the Reserve Officers' Training Corps (ROTC) program at Texas A&M began in earnest during early 1919. While the provisions of the National Defense Act of 1916 had been in force for nearly two years, the war had precluded an organized activation of the new officer training structure for college-level instruc-

tion. The delay in implementation caused by the war actually assisted in the assessment and planning by the War Department of how best to staff and train college students to become future officers. The manpower needs and experience gleaned during the war, as well as the postwar reorganization of the Army, reaffirmed that a broad base of military education was needed to provide a reserve of citizen-soldiers for times of national emergency. Thus, it was deemed imperative to encourage the "technical colleges" such as Texas A&M to expand general military instruction beyond basic infantry drills and into more detailed instruction in special areas or "branches" such as field artillery, chemical warfare, signal corps, ordnance, air corps, and military engineering. To help accomplish this, the ROTC Act of 1916 was amended with the National Defense Act of June 4, 1920, to broaden the scope of officer training as well as expand the role of federal oversight over military units on college campuses nationwide.[1]

Under the new campus military program courses were to be designed by the Office of the Secretary of War. In addition to increased allotments of arms, uniforms, and equipment, the War Department was authorized to establish pay rates, called "commutation of subsistence," for all cadets enrolled in training. Furthermore, to emphasize the importance of the new program, the president was authorized "to detail on such duty as many officers up to the grade of colonel as he deemed necessary to act as professors and assistant professors of military science and tactics." With militia affairs thus "federalized," the traditional state role in controlling the militia, dating from the founding of the nation, was effectively eliminated.[2]

A postwar meeting of the American Association of Land Grant Colleges and Experiment Stations in Baltimore agreed to "correlate" technical instruction with campus military courses. While in no way was it meant to be mandatory or binding on the member colleges, the general agreement on how to implement ROTC was significant. Unlike some colleges, such as the University of Texas, where the faculty made "strong statements opposing the organization of an ROTC unit on campus," Texas A&M was very comfortable in pursuing the mandate and challenge of expanded military training under ROTC.[3] Dr. Frank C. Bolton, head of engineering and most recently the 1917–18 chief of wartime training at Texas A&M, explained: "It is not the plan of the College to magnify, unduly, the military features, but it is believed that the revised plans of the War Department will not only prepare Reserve Officers for the Army, but also greatly strengthen the technical courses at the College."[4]

The Corps, in total disruption during much of 1918, prepared in January to re-emerge and grow. The 1917–18 school year had been decimated by the voluntary exodus of most of the junior (1919) and senior (1918) classes. A two-regiment organizational structure of four battalions and sixteen companies of cadets was reduced to one regiment of three battalions and eight companies. In September—due in large part to the growing SATC program—the lack of upperclassman leadership (seniors of '19) and the constant drive to recruit and train officers at the OTS camps all but halted normal Corps operations on campus.[5] The readjustment in January, 1919, was swift. The administration was surprised at the large number of returning seniors. The postwar adjustments proved a welcome relief. For example, 1918–19 Cadet Corps Commander Douglas W. Howell '19 had left the college in the fall of 1918 and was commissioned a lieutenant in the regular Army. Following the armistice, the *Reveille* campus paper reported on January 4, 1919, that Howell, a native of Bryan, would return to A&M to "assume his position as Colonel of the Cadet

Corps." In the spring of 1919, an auxiliary battalion was formed to accommodate two signal corps units and a battery of field artillery. Benefits of the new ROTC program served as an inducement to many new and returning cadets as they were given a uniform allowance (valued at $41.83 per year), received pay for summer camp attendance, and were given unlimited use of issued Army equipment "in the sum of $50."[6]

An expanded role of instruction in ROTC course work meant more resources and personnel for each college program. Courses in the different Army branches were to include a balance of both the theoretical and the practical—leadership fundamentals, planning and tactics, as well as live fire and field exercises. President Bizzell and Professor Bolton quickly realized the benefits to the college's instruction programs. For example, as a result of actively lobbying the War Department, training materials from the war and signal corps equipment were to remain at the college in the electrical engineering department. Thus, Texas A&M would be among the first colleges to receive approval for new branches of instruction as well as additional training equipment, uniforms, arms, and instructors. The first significant installment would be horses and a "large number of guns" for instruction and maneuvers in the field artillery unit.[7] Commandant and PMS&T head, Colonel Muller, an infantry officer, added additional staff to match each new branch of instruction. Three new field artillery officers, Lt. Col. Louis R. Dougherty, Captain Dunkum, and Capt. Winthrop W. Leach, were added to the staff in February, 1919. In addition, as infantry officer, Lt. J. W. McKeena and six sergeants rounded out the largest Commandant's staff in the history of the college. Within the Corps

the first field artillery unit, "Battery 'A' Field Artillery, Texas A&M College ROTC," was formally authorized and organized on April 2, 1919. Reporting favorably on the changes at the college, the *Battalion* ran a masthead that championed the new image of the Aggies, "The Men behind the Big Guns Always Win." The postwar Corps of Cadets had begun a new era of military instruction at Texas A&M.[8]

The campus was ready to return to a more normal schedule during the spring semester of 1919. More than eight hundred students registered, a fact that was considered "remarkable" given the "disorganization and interruption caused by the SATC" program in late 1918. More than sixty seniors, presented only "war certificates" for partial completion of course work, returned to complete their degrees. Athletics came to the forefront with the return of Coach Dana Bible from France. As the returning cadets prepared for basketball and spring baseball, Bible's priority was to begin to assemble a team for the fall, 1919, football season.[9]

However, Bible's efforts were one of many activities and changes under way across the campus. A life-size bronze statue of Gen. Sul Ross was dedicated in front of the Academic Building, the weekly *Battalion* resumed publication on March 27, the *Longhorn* staff was reorganized, daily cadet guard mounting and sentinels were discontinued, the zoo west of the railroad tracks was expanded with the addition of more exotic animals, spring social events became more important than the intense wartime drilling, and cadets marveled at the variety of new Army uniforms, given the limited use of the old-style cadet "grays."[10] The Army-issue green and tan uniforms could be mixed for wear, plus for the first time boot trousers and lace-up boots were available for daily use. Military training for returning veterans in the junior and senior class was made optional. A number of alumni and students expressed concern that these "non-military" or "casual" students were allowed to wear "cits" or civilian clothes on campus. And early planning was under way for a grand 1919 commencement and Final Review, the first on-campus ceremonies since June, 1916. The old A&M College "spirit" was beginning to return.[11]

Nationwide, ROTC units more than doubled to include detachments at 191 colleges and universities. While most institutions were eager to dissolve the SATC military program, Texas A&M viewed the experience as a means to transfer military training on campus to the ROTC.[12] The financial enticements from the War Department were also very attractive to colleges that otherwise would have possibly opted not to participate. Federal funding to

Two views of the dedication of the statue of General Lawrence Sullivan Ross in 1919. It is most unusual to see the old Confederate warrior's likeness draped in the Stars and Stripes. Courtesy Texas A&M University Archives, Cushing Library

campus ROTC and military training programs in the 1920s was second only to the appropriation made to the U.S. Army Corps of Engineering projects nationwide. Nonetheless, within two years the number of participating campuses fell to 124. The bulk of the ongoing ROTC programs were in the land-grant colleges in the southern and midwestern sections of the United States and in military institutions such as Norwich, VMI, and The Citadel that had a long military tradition and spirit of public service.[13] By 1919, Texas A&M and the Corps of Cadets was rated one of the top five distinguished military college programs in the United States.[14]

In an era before colorful recruiting brochures and large advertising budgets, a War Department memorandum to all college campuses highlighted the benefits of ROTC to the student, the college, and the nation. Students in ROTC would secure, among other things, discipline; physical and leadership training; scholarships of about $125 a year during the last two years of the course; the assurance of service as officers in a period of emergency; technical training helpful in their civilian careers; and the opportunity to attend free ROTC summer camps. Participating colleges would gain a close connection with the national government, the War Department, the U.S. armed services, and other ROTC colleges; an enriched curriculum; funds for scholarships; scientific and military equipment and instructors; and, interestingly, improved college spirit and loyalty. "Each agency that brings together college men for a concrete purpose and common interest assists materially in linking them to the college." (Texas A&M and the Corps of Cadets already well understood this particular benefit.) Last, included in ROTC's many benefits to the na-

tion were a "large group of well-trained reserve officers qualified as teachers and leaders for emergency service; the practical application of science to warfare and a stimulation of scientific interest in things military"; available training facilities in the event of a national emergency; civilians with a general appreciation of military training; and opportunities to develop training methods.[15]

To prepare for the expansion of military training at Texas A&M, a 172-acre tract of land in the John Tauber homestead northwest of Old Main was purchased by the college. Barns, feed bins, fencing, and tack rooms were constructed. In June, 105 horses for use by the field artillery units arrived at the campus. For the next two decades A&M cadets would train in the classical art of warfare—based on infantry drill, movement, and execution of horse-drawn artillery, and mounted cavalry maneuvers. Only a very few of the nation's campus ROTC programs would become so involved in military training in the 1920s and 1930s.[16] An additional new military feature added by ROTC was the requirement for summer training camp, usually at a military post in the branch of the cadet's specialty. The first such practical field training began in the summer of 1919 and has continued for the last eight decades. For most cadets in the 1920s and 1930s, summer camp was their first time to travel outside of the state of Texas. The first one hundred Aggie summer camp trainees were sent to Camp Zachary Taylor, near Louisville, Kentucky. At Camp Taylor, the cadets excelled in all aspects of field training and demonstration of leadership ability, thus setting the high standard for all future Aggies.[17]

In June, 1919, President Bizzell made a significant change in the Office of the Commandant with the appointment of the first

Ike S. Ashburn
Lt. Col. O.R.C. 306th Inf.
Commandant

World War I hero Lt. Col. Ike S. Ashburn returned to Texas A&M as Commandant from 1921 to 1923. He routinely rode with the Cavalry units of the Corps of Cadets. Courtesy 1923 Longhorn

permanent "civilian" Commandant of Cadets, retired Lt. Col. Isaac S. "Ike" Ashburn. The popular Ashburn, a recent war hero, was a close friend of the president and well known in alumni circles. Ashburn, a native of Gainesville, first came to Texas A&M in the wake of the 1913 cadet unrest as the first director of public relations for the college and secretary to the Board of Directors.[18] In 1916, realizing the country could soon be at war, he took a leave of absence to attend an introductory Army summer training camp at Plattsburg, New York. In August, 1917, he completed the first Officers' Training Camp at Leon Springs and was given a direct Army commission as a captain and made regimental adjutant in the 358th Infantry stationed at Camp Travis in San Antonio. In January, 1918, he was promoted to major and selected as a battalion commander, and in June, 1918, he was shipped to France. After a brief period of training in France, the 358th went into action at the St. Mihiel salient offensive in September. Though temporarily paralyzed by a wound to the neck on September 12, Ashburn "refused to leave his men and continued in active command" of his battalion until two days later, when he was shot in the thigh by a sniper's bullet. He remained in the hospital until January, 1919, was awarded the Distinguished Service Cross along with numerous French medals, and returned to active duty as the executive officer of the 90th Division during the occupation. He returned to Texas after his March discharge.[19]

Colonel Muller remained as the PMS&T, with Colonel Ashburn overseeing issues involved with preparing for the largest Corps of Cadets in the college's history. Ashburn had a broad appeal to the cadets, faculty, and alumni and proved to be a keen recruiter of

future cadets. He helped organize the "Texas A. and M. College Post" of the American Legion and was elected the first president of the Ninetieth Division Association, composed entirely of veterans of World War I.[20] Applications for new freshmen and returning cadets flooded the campus throughout the summer of 1919. The Corps of Cadets in September was organized into fifteen units—two artillery batteries, nine infantry companies, three signal corps units, and the band. Daily the college *Bulletin* carried a special section stating the growing enrollment figures: October 7—1,346; October 14—1,386; October 17—1,405. Cadets continued to arrive weekly until Thanksgiving, pushing the enrollment over 1,500.[21]

In the fall of 1919, Colonel Muller posted a notice for applications to the new cavalry unit. Cadets from all Corps units could apply. To prepare for the cavalry unit, $25,000 was provided to build five additional buildings—a gun shed to hold at least twenty-five pieces of field artillery and tack, a two-story barracks, a cavalry barn large enough to stable sixty horses, a mess hall for the Army troops assigned to the campus, and a storage barn. The campus of thirty-three buildings had begun a growth phase that would last four years. By January, 122 applications had been received, and Troop A Cavalry was formally organized in late February.

New cadets continued to arrive on campus after the Christmas break, driving enrollment to 1,745 by mid-February, 1920. To accommodate the growth, the Corps was reorganized and two new units added. Colonel Ashburn scheduled a Saturday dress parade during the spring on any weekend that had "good weather." Ashburn was popular with the cadets. His hands-on daily contact and example of leadership proved timely at a period when other college military programs were having second thoughts about the level of commitment to ROTC. He personified the image of a citizen-soldier and worked to bring recognition to the benefits of membership in the Corps and college to citizens across the state. Cadets dedicated the 1920 *Longhorn* to Colonel Ashburn.

Able cadet leadership in the Class of 1920 proved timely. E. E. McQuillen '20—class president, infantry regiment commander, and valedictorian—would have an impact on A&M long after graduation, serving as the executive secretary of the Association of Former Students from 1926 to 1947. Except for a new wave of charges of hazing, morale was high during the spring banquet events, hops, and graduation. At Final Review the Corps of Cadets numbered 1,800. Cadets and alumni alike boasted that the old "pep" was back on campus.[22]

SAW VARSITY'S HORNS OFF

While the ROTC program came full force to Texas A&M, it is important to note that the college and the Corps of Cadets had by this time a solid military identity and had formed the base of Aggie traditions. Class distinction by the 1920s was well established, the various sports teams were widely acclaimed, and the image and appeal of the Cadet Corps had survived the "troubles" of 1908 and 1913. Postwar alumni and former students reentering the civilian work force nationwide carried the Aggie message and lore far and wide. Traditionalists lamented the changing of the cadet uniform from the gray to the olive drab, yet the new military dress did little to dampen an

enthusiastic student body and vocal former students. The spectacular A&M war record of service focused attention on the institution. Returning veterans brought back more than just memories of France—letters home from the front and reunions reaffirmed that the Aggie traditions and esprit de corps transcended far beyond the campus. The A&M Cadet Corps continued its unbroken string of recognition by the War Department, begun in 1910, as a distinguished military college. Already with a solid base, the traditions of the Corps and college were to get a major reaffirmation and augmentation in the early 1920s. The roaring twenties was a time of enthusiasm, postwar euphoria, and all-around good times. The Aggies were no strangers to this national mood.

In the fall of 1919, Coach Bible once again assembled a powerful championship football team, going undefeated, untied, and unscored upon—amassing 275 points to the opponents' 0. The Aggies once again claimed the trophy as the champions of the Southwest. Bible stressed the fundamentals of football to the cadets and based victory on a solid defense. During the final game of the 1919 season against Texas, the Aggies made no substitutions and took no time-outs during the 7-0 victory. Coach Bible would become one of the legends of early college football. He led the Aggies to five championships in eleven seasons between 1919 to 1928. His A&M success attracted national attention. Bible coached most of the early A&M football stars—Hershal Burgess, Sam Sanders, R. G. Higginbotham, Puny Wilson, and Joel Hunt. When Knute Rockne of Notre Dame was offered the head coaching job at Nebraska in late 1928 and turned it down, he recommended Bible as "the best young coach in America." Jerome Rektorik '28, who played on the A&M 1926, 1927, and 1928 teams, recalled, "Bible may not have been the most brilliant coach, but his ability to motivate the players to put out and *do* their best was tremendous."[23]

One of Bible and Texas A&M's greatest athletes was Jack Mahan '21, World War I veteran, all-around sportsman, football star, and team captain. Mahan was the first Aggie to compete in the Summer Olympic Games, finishing sixth in the javelin at the August, 1920, world gathering in Antwerp, Belgium.[24] Back on campus the games were less dramatic, yet no less important. With the help of cadet labor, the sixty-row wooden bleachers at Kyle Field were expanded, but delays in raising adequate donations postponed the vision of

an expanded concrete stadium. Construction of a "Memorial Stadium" at Texas A&M never materialized and was indefinitely delayed in 1920; thus, the official name of Kyle Field was retained.[25]

The 1919 championship was followed in January, 1920, with a very unusual event. President Bizzell, the athletic council chaired by Ike Ashburn, a few faculty members, and a number of former students were invited to Austin for a barbecue of the University of Texas mascot—Bevo I. The individuals who had branded the "orange and white" longhorn (which was to have been a Texas alumni gift to the University of Texas Athletic Council) with the foot-high numerals "13-0" (score of the 1915 game) were also invited.[26] Alfred Bull '16, one of the seven Aggie cadets to brand

Kickoff of the A&M-Texas game on November 26, 1925. Note the size of the stands in the north end zone. In the background is DeWare Field House. Courtesy Texas A&M University Archives, Cushing Library

You are cordially invited to attend the Barbecue of the Longhorn Steer, Tuesday evening at 6:30, January 20th, at the Men's Gymnasium. By the time this invitation reaches you, Bevo will have chewed his last cud and his juicy steaks will be awaiting your appetite at the pre-arranged time. The T girls of the University will do the serving. Besides the barbecue, there will be short speeches, good music and informal dancing. We are anticipating your coming.

Athletic Council
By Chas. W. Ramsdell, Chairman.

The invitation to the January, 1920, barbecue of Bevo, the longhorn mascot of the University of Texas, held in Austin. Courtesy Texas A&M University Archives, Cushing Library

the steer shortly after the mascot was given to the University of Texas on Thanksgiving, 1916, gave an account of the late-night escapade. The hide of the original Bevo, "except that part bearing the box car size brand 13-0," was to be placed in the university trophy case in Austin. The hide with the brand was given to A&M for display in College Station. Initial plans were to hang the hide from the wall in the YMCA, prior to being placed in a permanent display case.[27] However, no trace of Bevo's branded hide remains today.

The Corps numbered more than two thousand cadets for the first time in the fall of 1920, and spirits remained high. The 1920s proved to be a good time for the Aggies. Football games and Corps trips marked the key events in the fall of each year. The Corps followed the team in alternating years to an away game either in Waco, Houston, Austin, Fort Worth, or Dallas. The Corps was transported en masse by as many as four specially chartered trains to the out-of-town events, paraded through the streets of each city, and treated to a "hurried" pre-game meal. During the 1920s, half-time entertainment was provided by the entire Corps (minus the band) forming a block "T" on the field; the band usually remained in the stands. After the game there was usually a dance in the evening. For example, in Waco the event was held at the Coliseum or Cotton Palace; in Dallas, at the Adolphus Hotel; and in Houston, in the ballroom of the Rice Hotel. At midnight the cadets loaded the trains for the return journey to the campus.[28]

The *Longhorn* yearly highlighted each season's Corps trips: "The one thing that made a hit was the 'T' formation between halves. Another source of favorable comment was the yelling, no, gentle reader, we don't cheer, we yelled during the entire game. Go where you may and there is nothing that will compare with the pep of the Texas A and M College, win or lose. Are we proud of it? I'll say we are. All the pep was rewarded by a 10-0 victory over the Baylor Bears."[29]

Corps trips continued for two decades until the outbreak of World War II. The khaki-clad Aggies wearing campaign hats were a big attraction and much appreciated by veterans' groups and the former students of the college with each visit. A town's economy boomed

Corps trips were an important part of the fall out-of-town gridiron schedule. This 1921 picture was taken on Main Street in Waco, Texas. Courtesy Texas A&M University Archives, Cushing Library

when the Aggies visited. The streets were often lined twenty to twenty-five people deep to see the cadets in review. The excitement the day of the game and the rivalry were intense. The enthusiasm, however, got out of control and tragedy struck at half-time of the 1926 Baylor game in Waco. A fight broke out and resulted in numerous injuries. One Aggie, cadet Charles M. Sessums, was killed in the fight after being hit with a chair or club. A lengthy investigation followed, but it was inconclusive. This tragic event was followed by the suspension of all games between the two colleges for five years.[30]

In 1921, the Corps introduced a new unit of cadets, composed of only senior cadets. Unlike the Ross Volunteers, which made only ceremonial appearances, this was apparently the first actual drill team in the Corps. The unit was called the Shock Troop. In addition to wearing distinctive uniforms of a khaki suit, pleated blouse, and Sam Browne belt ("such as has been specified by General Pershing for wear by regular army officers"), on the left shoulder they "wore the 'A. and M. Fourragere'—a decoration with an elaborate coiled cord of red and white with a gold tip." Marching just behind the band in the parade, in a "column of squads they changed to extended order, merged to form a hollow 'T' and passed the reviewing stand in platoon front."[31] Future Texas A&M President and Chancellor Marion T. Harrington '22 was a member of this unit and recalled:

The Corps uniform during the post–World War I era was a mix of after-war styles and colors. Boot leggings in this 1925 photo of senior cadet officers were an issue item and very popular. The only common item is the campaign hat. Courtesy Texas A&M University Archives, Cushing Library

The Commanding Officer of this troop was Cadet Major Charles Wright Thomas, later A&M's first Rhodes Scholar, the 1st Lieutenant was Chester H. Chambers, 2nd Lieutenant was William W. Lynch, and 1st Sergeant was Captain Robert E. L. Pattillo. The other 94 seniors were commissioned [cadet] captain and lieutenants and were members of the troop. The Shock Troop members were assigned to the 12 companies for drill training but did not live with that unit in the dormitories.

When commencement time came, the Military Science Department told Shock Troop members that they would not march in the final review as a unit but instead march with the units they had worked with in drill. The Shock Troop members decided that they would not participate in the final review except as a unit, and a plea was made to the commandant, Ike Ashburn, who granted permission. After that challenge, the famous Shock Troop of A&M was disbanded in 1922.[32]

By the early 1920s the college colors of "maroon and white" and the terms "Aggie" and "Gig 'em Aggie" had become a fixed part of Texas A&M lore, gradually replacing the earlier labels of "cadets" and "farmers."[33] There was no official mascot yet, but the "wildcat" was strongly considered by the cadets and then rejected.[34] However, it was the drafting, introduction, and acceptance of a fight song that over time brought a great deal of identity and attention to Texas A&M, its cadets, and former students. Until the 1920s the college did not have an "official" fight song for sports events or a standard alma mater song. Numerous songs had been introduced, most fashioned on the melodies of popular songs of the day, but none became popular enough

to be accepted as the school standard. A wave of possible songs, ballads, and verses were introduced and considered. Then, in late 1920, a returning former cadet and Marine veteran, James Vernon "Pinky" Wilson '20, introduced a new one-verse song he had composed while standing guard duty on the Rhine River in occupied Germany. Drafted and sung as a quartet ballad by soldiers in Europe, it first appeared under a number of different titles, "The Battle Hymn of A. & M." and "Good-Bye to Texas University." Upon his return to A&M, the original thirteen-line fight song was sung numerous times by Wilson's newly assembled Aggie quartet at intermission of the daily movie at the Palace Theater in downtown

Sam Houston "Sammy" Sanders and Coach Frank Anderson in 1921. Anderson would serve as Commandant of Cadets 1935–37. Courtesy Sam Houston Sanders Corps Center

Bryan. The *Battalion* first printed the lyrics in October, 1920.[35]

During the spring of 1921, working around the words drafted by Wilson, Aggie Band Director George Fairleigh "put into music the irresistible swing of the march song that can be easily transposed into 'jazz' time, when the occasion demands, that makes it appeal to the martial spirit at A. and M. College."[36] Wilson's song, comprised of various time-honored portions of old Aggie yells, was given an original melody by Fairleigh that set it apart from any previously considered songs (see yells in chapter 3). You could sing it, march to it, and without a doubt understand its words and clear meaning directed toward Texas A&M's primary rival—the varsity at Texas University, or t.u. In time, the popular title became the "Aggie War Hymn," which was printed as a subtitle on the original 1921 copyrighted version—entitled "Good-Bye to Texas University" and comprised of only *one* verse. According to Wilson, "Before the war, A&M had one of the greatest yell leaders of all time, Wrathall King 'Runt' Hanson '16 of San Antonio. It was Runt who came up with the idea of having the Cadets come marching in and form that famed Texas Aggie 'T.' When the Cadet Corps marched in and formed that 'T' in 1921 before the first football game, then came 'The Aggie War Hymn.'"[37] The timing could not have been better, because Texas A&M football and Coach Bible exploded into the national championship spotlight with a New Year's Day 1922 victory at the Dixie Classic (which later became the Cotton Bowl) in Dallas over the highly favored "Praying Colonels" from Centre College. Cadet Sam Houston "Sammy" Sanders, Jr., '22, the primary benefactor of the Sanders Corps of Cadets Center, returned the opening kickoff fifty-five yards, "which started the ball rolling for A&M."[38]

Years later at the urging of the cadet yell leaders and the former students Wilson drafted a second verse. A brief dispute over the copyright to the new two-verse version of the "Aggie War Hymn," which temporarily prohibited the Aggie Band's playing the fight song, was resolved in 1938. Though the new second verse was introduced to the cadets to sing when playing an opponent other than "t.u.," the students for decades have steadfastly adhered to singing the one and only original rendition.[39]

Hullabaloo, Caneck! Caneck!
Hullabaloo, Caneck! Caneck!
Good-bye to Texas University,
So long to the orange and the white
Good luck to the dear old Texas Aggies
They are the boys that show the real old
*   fight*
"The Eyes of Texas Are Upon You"
That is the song they sing so well,—
So Good-bye to Texas University,
We're goin' to beat you all to—
Chig-gar-roo-gar-rem! Chig-gar-roo-
*   gar-rem!*
Rough! Tough! Real Stuff! Texas A&M.

In 1925, a senior cadet from Marlin, Texas, Marvin H. Mimms '26, along with Band Director Maj. Richard C. Dunn developed a new song titled the "Spirit of Aggieland." With the encouragement of Col. Ike Ashburn and the yell leaders, it was introduced by A&M's one-hundred-piece band at a yell practice—"first loud and then softly, with a soloist [John D. Langford, Jr., '26] singing"—in front of the YMCA in mid-October, 1925.[40] Yell Leader Ervin O. Buck '26 extolled the merits of having a second song. "The Spirit of Aggieland" became an instant hit with the Corps.

Some may boast of prowess bold,
Of the school they think so grand,
But there's a spirit can ne'er be told,
It's the spirit of Aggieland.
We are the Aggies—the Aggies are we,
True to each other as Aggies can be,
We've got to Fight, boys,
We've got to FIGHT!
We've got to fight for maroon and white.
After they've boosted all the rest,
They will come and join the best,
For we are the Aggies—the Aggies
*   are we,*
We're from Texas A.M.C.

The third of the big three A&M songs is the "Twelfth Man," written by Mrs. Ford Munnerlyn, whose husband was also in the Class of 1926. The song was introduced at half-time of the 1941 A&M-Texas Thanksgiving game in College Station. Other songs about Texas A&M and the Aggies would follow throughout the 1930s, such as "There Shall Be No Regrets," revised by Richard Dunn and Yell Leaders Tom Dooley and "Peewee" Burks in 1934; "The Men of Aggieland," by R. E. Bayless '28; "Give Your Heart to A. & M." by Ernest Ford; "Texas A.M.C. War Song," by George E. Perfect; and the popular World War II swing era tune "I'd Rather Be a Texas Aggie," words and music by Jack H. Littlejohn '39. Yet none were as popular and as universally accepted as the "Aggie War Hymn" and the "Spirit of Aggieland."[41]

THE "HOLLYWOOD" SHACKS

The intense esprit de corps gradually attracted more cadets, placing a strain on the college facilities and housing. The Corps of Cadets in 1922 numbered 2,100—from twenty-three states and twelve foreign countries. The single composite cadet regiment structure, in use from September, 1920, to June, 1923, was expanded to include two mounted units of the cavalry—A Troop and B Troop. For the first time an Army aviation officer was assigned to the Commandant's Office to organize an Air Service Squadron that comprised two flights by the fall of 1922. In spite of the historic dominance of the Army training and commissioning of officers from A&M, Otto P. "Opie" Weyland '23, Aggie Band Commander in 1922–23 and product of the Army Air Service Squadron, became the first A&M graduate to obtain the rank of four-star general in the U.S. Air Force.[42]

During the 1921–22 school year Colonel Ashburn added the distinguished student award, a distinctive recognition for academic excellence. The Class of '26, which entered the college on September 16, 1922, was the largest group, to date, at the college—850 fish. Such rapid growth was to precipitate problems. Housing once again became critical, as more than 300 cadets were accommodated in tents on the current-day site of Law and Puryear Hall. A harsh 1922–23 winter further aggravated the situation. Some funding was forthcoming to renovate fifty-year-old Gathright Hall from its use as headquarters of the extension service back into a dormitory. This proved only a token effort, prompting the Association of Former Students to demand the Texas Senate Finance Committee take proper action. The *Texas Aggie* noted that "prospects point to a greater attendance next year, and if there is, Texas youth again will be living in tents amid rain and sleet and snow because the 'Empire State of the Union' of which we are so proud to boast is too poor to provide proper accommodations for them."[43]

Efforts to replace the tents—initial cost of $60 each—became more imperative as the monthly cost of maintenance and serious risk of fire increased due to the oil soaked on the canvas for waterproofing. As an interim solution, in early 1923 the tents were replaced, on the same plank flooring or foundation, with twenty-foot-square wooden buildings equipped with camp stoves. These shacks became known as "Hollywood." Although they did not increase the bedding capacity, the makeshift housing was considered by some to be safer and warmer than the tents. In January, 1923, Colonel Ashburn left the Commandant's Office to become the executive secretary of the Association of Former Students (1923–26) and later the general manager of the Houston Chamber of Commerce. The popular Ashburn was replaced by one of the few individuals who could follow in his footsteps as Commandant—Col. Charles C. Todd '97.[44]

The former Cadet Corps Commander (1896–97), valedictorian of the Class of 1897, and Spanish-American War veteran found the Corps in good shape and growing. To solve problems associated with the growth, Todd conducted a major reorganization. The original First Regiment, with eighteen units, was expanded to two regiments with twenty units. In the first regiment, there were three battalions of three companies each; in the second or "composite" regiment there were four battalions—signal corps, cavalry, field artillery, and air service—and the band. This two-regiment organization of twenty to twenty-two units would continue for six years until Final Review in June, 1929. The entering freshman class in September, 1923, totaled more than 825. This, coupled with increased retention, placed overall Corps

strength at over 2,300 cadets. Ever-growing freshman classes, excellent retention, as well as the growing number of cadets who enrolled in advanced ROTC in their junior and senior years reflected the full impact and significance of the Texas A&M Corps of Cadets when it was reported that A&M had the largest Corps in the country by early 1924.[45]

Cadet life and training in the 1920s was very regimented. Flush with a new energetic cadre of ROTC officers and equipment, the Corps maintained a brisk regimen. Every effort was taken to run the campus in a "strictly

Col. Charles C. Todd '97 was one of the first career regular Army officers from Texas A&M. Cadet Corps Commander in 1896–97, he served two terms as Commandant of Cadets, April–July, 1898, and 1924–1925. During his retirement years he was active in the fight to allow women to attend Texas A&M. Courtesy 1923 Longhorn

TABLE 5-1
Organization of the Corps of Cadets from September, 1923, to June, 1929

Corps Staff

Infantry Regiment

First Batt	Second Batt	Third Batt
Company A Inf	Company D Inf	Company G Inf
Company B Inf	Company E Inf	Company H Inf
Company C Inf	Company F Inf	Company I Inf

Composite Regiment

Field Artillery Batt	Cavalry Squadron	Signal Corps	Air Service*	Band
Battery A	Troop A	Company A	First Flight	
Battery B	Troop B	Company B	Second Flight	
Battery C	Troop C	Company C		
Battery D	Troop D			

* Note: The air service flights were temporarily phased out in mid-1927, and an engineering battalion of two companies was added to the Composite Regiment for 1928–29.

military fashion." Rising before dawn, the cadets in the new field artillery and cavalry units made pre-breakfast preparation of all the equipment. All cadets were required to know not only the manual of arms, but also how to disassemble, clean, and assemble every weapon assigned to their unit. Cadets marched to breakfast, lunch, and evening supper. Bennie Zinn '26 of Company B, field artillery, recalled that "ROTC in those days was designed to train the cadets for every possible aspect of a field officer's responsibilities." Zinn recalled that cadet life was very regimented:

At the table we ate just like we did at home. We didn't have people asking us questions and making us sit up straight and stiff. For announcements, we sat up with our hands in our laps. We were restricted on certain kinds of cush. For a year I didn't have any pineapple or apricot pie. That was no great loss. I liked other kinds better anyway.

Immediately after evening chow there was a brief yell practice, held on the steps of the YMCA. The cadets went directly from yell practice to their quarters, where they hit the books. Each fish had an upperclassman assigned to help him with problem courses.

We did go home for Thanksgiving and Christmas, but we were required to wear our uniform at all times. It was good advertising for the College.[46]

The days of only basic drill and training for the cadets were dramatically changed in the late 1920s and early 1930s. The School of Military Science and Tactics became a well-staffed and well-supplied operation. Many regular Army officers assigned to Texas A&M during this period had seen duty in France. The emphasis on hands-on training was paramount. The five primary branches—infantry, field artillery, signal corps, cavalry, and air service—reflected an effort to provide training that at least closely resembled the active Army. The infantry was "equipped with every piece of equipment that a regular army regiment of the United States Army has . . . rifles, pistols, machine guns, hand and rifle grenades, gallery rifles, ammunition and field engineering tools." Training was conducted weekly. There was an indoor "gallery range" and a thousand-yard outdoor rifle range.[47]

The A&M field artillery unit was equipped with eight caissons, ten limbers, two battery and store wagons, a battery reel cart, ninety horses, four mules, and all accessories. In addition to three 75mm howitzers, they were provided a 155mm gun, two five-ton caterpillar tractors and ammunition trucks, as well as "trench periscopes, prismatic compasses and sitogoniometers." The cavalry, with sixty horses, two pack mules, and four draft mules, introduced cadets to such military aspects as saber drill, small unit tactics, and field operations on a larger scale. The signal corps continued much of the same training from World War I but introduced new advances in telegraphic and "heliograph" field communications. Signal corps operations were supported by horse-drawn wire-carts, motorcycles, and cable-splicing equipment. A single-engine

Family-style dining and waiters in Sbisa Dining Hall were a hallmark of Corps life and training. Courtesy Sam Houston Sanders Corps Center

Liberty airplane—fitted with a machine gun and bomb sights along with "dummy drop bombs"—was assigned to the college for use by the air service. In addition to learning the basics of flight, cadets were given courses in engine repair and tactics. Cadets were encouraged to excel in all phases of training and given professional instruction by the expanded staffing of the Commandant's Office. This training on campus was continued at the yearly summer camp in each of the branches of service.[48]

THE BAND MURDER AND FISH CORPORALS

Texas A&M in the 1920s attracted cadets from all across the state and nation, but most students were Texans. As each new fish class continued to grow in size, so too did the magnitude of pranks and high jinks directed at the new cadets. The YMCA provided a somewhat neutral haven to which fish could escape, yet even this was not foolproof. Housing remained crowded, yet the enthusiasm among the Cadet Corps remained high. Each new entering class vied to make sure its class number was painted high atop the water tower; cush (dessert) was off limits to fish; and hitchhiking was the primary means to get to Bryan to see a movie or to go home for the holidays. The Corps used the train for Corps trips, and cadets loved to attempt to stow away on trains going to Houston or north to Waco and Dallas. Some cadets hid in the overhead luggage rack covered with coats, others between the folding seats, and some hung on the outside of the train to avoid detection. One account noted: "Once a train official looked out as we were going around

a curve and saw three boys hanging out the side of one car. The conductor came to the car, opened a window and called to a student, 'What are you doing out there? Have you got a ticket?' and the kid responded, 'I just like to ride this way. Sure, I've got a ticket.'" The conductor eventually was able to get the cadets safely back in the train.[49]

Every spring the class balls and class dinners occupied time and planning. The annual junior banquet attracted the most attention and enormous class rivalry. The main objective of the sophomore class was to keep the junior class toastmaster away from his own banquet in Sbisa Hall, and it was up to the juniors to protect the toastmaster at all costs and deliver him safely to the event. Elaborate plans and schemes were devised by both sides in this yearly tug-of-war.[50] Tremendous ingenuity and effort was displayed in a number of such endeavors. Such energy resulted in a number of firsts. Cadets, for example, formed a radio station in 1920 and broadcast the first known live play-by-play of a football game.[51] The Aggie Band had a number of small orchestras and jazz combos that were always eager to play. While hazing occasionally cropped up, it seemed to be conducted in what historian Henry Dethloff called "a spirit of mischief or to perpetuate a college tradition" and not one of abuse of the fish.[52]

With the advent of different branches of the Army now identified for cadet training, an intense rivalry began to grow among the cadet units. Company-level awards and yearly recognition began to be more frequent in the 1920s. The increased size of the staff in the Commandant's Office and the demands placed to comply with the ROTC standards set forth by the War Department gradually caused the

Corps of Cadets and its officers to be more accountable for their actions and the actions of those in their units.[53] Discipline became more regimented. However, no defined guidelines or fast rules applied to the pranks and charades the upperclass cadets inflicted on new "greenhorn" cadets. With ample free time and a desire to outdo the previous class or prior year, different variations on a number of well-planned pranks unfolded during the early weeks of each new school year. Two such activities—the "band murder" and the selection of the "fish corporal"—became elaborate, time-honored rituals.[54]

One of the best accounts of these yearly happenings, briefly portrayed in the movie *We've Never Been Licked,* is provided by Alanzo C. Taylor '24, a member of Battery A, field artillery:

The band murder usually took place early in the first term of the school, probably about the second or third Saturday evening of the term. It was executed solely by the band members. It began by two juniors, who occupied one of the large corner rooms on the third or fourth stoop of Milner Hall openly and profanely quarreling as they went to and fro in the building during the day, attracting doubtful glances from the freshmen who overheard them. This continued for several days and when the fish began to ask questions of the sophomores they were told this had been going on since the two were freshmen and not to pay attention to it. When the curious fish asked why they continued to room together, they were told that both were so obnoxious that no one else would room with either of them and they stayed together because both were so obstinate neither would give up their large and commodious room

to live with someone else, even if they could find another roommate.

Things came to a head just before taps on the chosen evening. The fish who roomed close to them heard unusually loud voices coming from their room, accompanied by oaths and threats. Then they heard loud cries, overturned furniture and finally a scream. Some of them rushed to the door but found it locked. When a large crowd had gathered, including a few seniors and other upperclassmen, the door was opened from the inside, one of the occupants pushed through the crowd and disappeared down the nearby stairway. The room was dark but the crowd punched the light switch and entered the room. Near the center lay the body of the second occupant, half covered with an army blanket and blood, or something that looked like blood, scattered around the body. A bayonet appeared to have been thrust through the man's breast. Actually it was driven into the floor between the man's left arm and his body, but it appeared to be in his left breast.

Quickly the seniors pushed the fish from the room and herded them into an adjacent or nearby room which was unoccupied. A junior noncom was left to guard them and keep them together. The other upperclassmen rushed to the Infirmary, about two blocks away, obtained a stretcher, and loaded the maimed student on it. They apparently carried the man to the Infirmary, and upon returning advised the group that he was still alive and the doctor thought it would be best to send him to a Bryan hospital where surgery could be performed, if needed.

The incident was apparently reported from the Infirmary to the Sheriff's Department in Bryan and in a half hour or so two deputies appeared to question the witnesses. The fish were closely questioned by these lawmen and after they left, the seniors lectured the alarmed fish on the need for silence. They were not to discuss the incident with anyone not in the band and already acquainted with the matter—they were not to write home about it. Hopefully, the young man would recover and the good name of A&M could be protected. They were advised that an all points notice of the affair and a description of the prospective "murderer" was being sent to nearby law agencies. It was anticipated he would be found very quickly. Actually both boys never left the campus. They went to one of the more distant dormitories where they had friends, and occupied an empty room, to which they had previously transferred the clothing and books they would need for a few days. They attended their usual classes, being careful not to be seen by any members of the band who were fish. Bulletins were issued by the officers of the band from time to time on the recovery of the wounded man, which were encouraging, and also to the effect that nothing had been learned of the fugitive.

This went on for nearly a week. Finally both boys came back to their room with no explanation for the freshmen. After a little more time and thought they realized they had been had. All of the "lawmen" and other "outsiders" involved were actually students and mostly casuals who were older than the regular students in the Corps. I don't know when the joke began but it had been going on for several years when I was a fish in the 1920s.[55]

Freshmen cadets at Texas A&M have generally not held any cadet rank or leadership position. However, new Aggie fish began to realize that rank—even the lowest cadet rank—carried with it more privileges and prestige. Upperclassmen were keen to identify the most eager as well as the most gullible fish. To determine such eagerness outside of formal ROTC training and Corps promotion channels, a yearly ritual to select the most deserving freshman to be the "fish corporal" (or sometimes referred to as "fish sergeant") was conducted.[56] This account by Taylor of the 1920 episode tells the story:

About three weeks after we began the school term, there appeared on the bulletin board in Leggett Hall, where all the Artillery unit was quartered, a notice to the effect that freshmen candidates for the rank of corporal would be considered. The notice explained that in the battery there were not enough sophomores who were qualified academically for the grade of corporal and therefore one freshman would have to be selected for this duty to fill out the complement of the battery. This announcement was repeated at the next drill formation by the Captain of the Battery, with the injunction that freshmen should be particularly careful during the drill session, as their performance would be evaluated by the Seniors and Juniors and would be given consideration by the promotion board in the selection of the successful candidate. It was reiterated that those wishing to be considered should file a formal application by a specified date.

This brought on some anxiety on the part of the more gullible fish. They immediately went to their sophomore friends for advice on how the application should be phrased and what qualities should be stressed. They got all kinds of conflicting advice on this point. Some told them to be sure and mention they had R.O.T.C. training in high school, if that were a fact, and if they had not had this experience perhaps Boy Scout training would help, particularly if they had won merit badges in this activity. Some of the sophomores even volunteered to write their applications for them. After the drill session in which they were to be tested, a list of the freshmen and the grades they obtained in the drill were posted on the bulletin board. My roommates and I had had R.O.T.C. training in the Ft. Worth school system and we noticed we were graded near the bottom of the list, so we smelled a rat. The ones who were near the top of the list were, for the most part, unsophisticated country boys who had never worn a uniform until they came to A&M. As the years passed, other elements were added to the preliminary phases of the selection. For example, the year we were Juniors, the three of us conceived the idea of holding a Board examination on the following Saturday evening, in which three Judges, all Seniors, would assemble the candidates in our room, with all the Senior officers seated in chairs around the room (the furniture had all been removed) dressed in the No. 1 uniforms with sabers to listen to the examination. One by one, the candidates, who had been briefed in an adjoining room about conduct and attitude, were brought in and questioned by the Board. A sergeant, at an adjoining table with a typewriter, made a record of the proceedings. The candidates were told some of the questions might appear trivial, but they were intended to reveal psychological attitudes and thinking processes and should

not be taken lightly. The candidate was then asked a well chosen series of silly questions. I remember one which went as follows: "Have you ever seen a one-armed, naked man floating down a stream on a slab of marble, pick up a shot gun and shoot a duck flying by, pick it up and put it in his pocket?" Then the candidate, with sweat running down his face, answered, "no sir," and one of the Judges remarked to the sergeant, "Put him down as poor in observation, sergeant." Strange as it may seem, we learned a good deal about the thinking qualities of the candidates by this means. Some soon caught on that it was a huge joke and gave us about as good as we were sending without cracking a smile. Others were completely befuddled. It was from the latter group that we selected the successful candidate.

When the decision was reached it was announced at the formation for our midday meal and the successful candidate was conducted to the table in the mess hall where the [Cadet] Colonel Commanding, and his staff, were seated and introduced as the *only* freshman in the history of A&M who had been selected for promotion to corporal. The Colonel and staff members congratulated him, introduced him to the assembled Corps, and asked him to have lunch with them. We left him in their care. Later in the day he had corporal's stripes sewn on his shirt and took his place in our Battery formation as such for several days until smarter fish friends convinced him it was a joke.[57]

These good-natured activities were constantly under the watchful eyes of both the unit commander and first sergeant. The ability to think on his feet, understand that even as a fish a cadet was accountable for his activities, and accept that the Corps and college endeavored to develop the man was central to the rites of passage at A&M. An Aggie company commander, Louis A. Hartung '29, posted the following guidelines for his fish:

FISH DON'TS

1. Don't forget to knock and remove your hat on entering an old man's room.
2. Don't be caught hanging around the Y.M.C.A.
3. Don't, on going into any room, state your business until you are sure that you have met every one in the room.
4. Don't ever tag a senior at a dance.
5. Don't forget that everything that every one of you fish say or do comes back to me, so act at all times and all places as you do here in your own troop and own hall.
6. Don't forget to salute all officers.
7. Don't forget to say "sir" to all old men and call them all mister.
8. Don't ever talk back to nor argue with an upperclassman.
9. Don't let a sophomore make you carry out a senior nor a junior. Don't let a junior make you carry out a senior [refers to kidnapping the class toastmaster].
10. Don't forget to come to attention when an upperclassman enters your room.
11. Don't forget to speak to everyone you meet, whether you know him or not. Everybody here is your friend.
12. Always see that the upperclassman seated next to you at the table has been helped before you help yourself.[58]

END OF THE BIZZELL ERA

In June, 1925, President Bizzell announced his departure from Texas A&M to accept the presidency of the University of Oklahoma. His decade of service at A&M came at a time when progressive leadership was needed. In addition to the growth and reorganization of the Corps of Cadets, the college slowly began to emerge as a leading academic institution in the fields of engineering and agriculture. In addition to campus programs, Bizzell enhanced extension service operations, alumni relations, and political contacts in Austin, all of which had a tremendous impact on the image and role of Texas A&M statewide. However, the popular president voiced a word of caution. In a parting message to the cadets, faculty, and alumni, Bizzell stated that attention was needed in a number of key areas if the college was to grow and gain prominence: first, he encouraged the expansion of the fledgling research programs and graduate studies; second, he urged all involved with the college to give full-time diligence to increasing legislative funding and political support in Austin; and last, he questioned the role of compulsory student membership in military training, declaring that a closer study was needed to chart the future of the Corps of Cadets. On the Corps he noted, "the housing plan at the College is archaic and inadequate"; this never-ending problem needed solving.[59]

Colonel Todd also departed, retiring to practice law in Bryan, and was replaced by Col. F. H. Turner. A veteran of two wars, Turner had joined the Army during the Spanish-American War as an enlisted man and then earned a lieutenant's commission in 1901. In addition to running the officers' special warfare school in France during the war, he had been Assistant Commandant under Todd in 1924. He had a staff of sixteen officers and six sergeants and made few changes in the cadet training program.[60] The A&M Board of Directors selected Thomas O. Walton (1925–43) to be the next president of the college and scheduled his formal inauguration to coincide with semi-centennial celebration of the college in mid-October, 1926.[61]

The Corps continued to grow gradually in size throughout the late 1920s. In addition to the ever more common practice of cadets having "water fights" on warm days, the cadets began to build a bonfire to be burned the night before the Thanksgiving football game with the University of Texas. These early bonfires were not much more than a pile of old wood, outhouses, and trees hastily dragged into place a few days before burning—a far cry from the mammoth fires that followed in the 1960s. One aspect of cadet life that began to attract concern was the underground activities of secret organizations. The *Battalion* termed their members "those who have withdrawn into the Invisible Empire of secret fraternities." When the Corps was small in number such organizations had been rumored to exist, but they seldom were sustained from year to year. As the student body grew larger these organizations resurfaced. The Class of '32, which enrolled in September, 1928, totaled 1,100 fish; total enrollment for the 1928–29 session was 2,770. The best known and most often cited of the secret societies that grew with the student population were Kala Kinasis or "KKs," the Swastikas or "Stickers," and the True Texans or "TTs." Some cadets and alumni assumed that such secret organizations were allowed on campus with the approval of the administration. In fact, they were not sanctioned. These quasi-fraternal groups hoped to sway Corps policies and appointments to key leadership positions, but the actual influence of such organizations would vary.[62]

Col. Charles J. Nelson became Commandant in 1928. As the ROTC program was expanded and better staffed nationwide, competition among the cadets and various detachments both in their annual sessions and at summer camp training began to increase. Complaints of hazing surfaced in letters to President Walton in 1928–29 and in published reports in the *Battalion* concerning upperclassmen using freshmen in prohibited manners to run personal errands, clean rooms, and conduct late-night details known as "fish calls."[63] The A&M Board of Directors ordered an investigation that resulted in the cadets inviting, by a vote of the senior class, members of the state legislature to visit the campus personally in order "to quell rumors which are said to be grossly exaggerated."[64] The issue soon died and no action was taken. Notwithstanding, Texas A&M yearly maintained its "distinguished military rating," uninterrupted since 1910.

TABLE 5-2
Corps of Cadets Enrollment, 1919–39

1919–20	1,383	1929–30	2,734
1920–21	1,393	1930–31	2,433
1921–22	1,525	1931–32	2,194
1922–23	1,879	1932–33	2,001
1923–24	2,091	1933–34	2,158
1924–25	2,256	1934–35	2,998
1925–26	2,170	1935–36	3,430
1926–27	2,395	1936–37	4,130
1927–28	2,543	1937–38	4,926
1928–29	2,770	1938–39	5,582

Source: Office of the Registrar, Texas A&M University

Colonel Nelson and his staff began to reorganize the Corps to accommodate both the growth and the addition of new branches of instruction. The Commandant's Office worked to standardize the cadet uniform fully by requiring leather leggings and matching uniforms—secondhand light-colored serge shirts (primarily World War I surplus) would not be acceptable—only regulation olive drab. Campus housing, including the Hollywood shacks, remained overcrowded. However, some relief was at hand with the completion of Hart Hall (1930) and Walton Hall (1931).[65]

THE AGGIE RING

In 1929 there was a movement to change the design of the Aggie ring. Debate over the changes was heated; the class ring, first worn in the early 1880s, seemed to some too plain and not "collegiate enough." The debate began when the junior cadets in the Class of '30 questioned the design of the ring that bore on its crest the letters "AMC," an eagle, and the class year. This design of the 1894 ring had been accepted as the official ring by the Alumni Association in 1911. The classes of both '30 and '31 wanted the ring updated to include either the word "Texas" or the full name of the college. Others questioned the pricing of the ring and the quality that was provided by a local Bryan jewelry store. After a number of class meetings and votes it was decided to seek a redesign of the ring. However, no major changes occurred, and pricing would be by competitive bid. In 1933, President Walton appointed a permanent committee to oversee both the qualifications for granting of the ring and any proposed

design changes. Concerns voiced by members of the classes of 1930 and 1931 were well intended, and slight modifications to the ring have been made since the 1930s. Yet in no way have such changes diminished the significance and importance of the ring to Aggies.[66]

In the heat of the 1929–30 ring debate the editor of the *Battalion* captured the essence of the Aggie ring: "This ring has grown to mean A. and M. wherever A. and M. men are wont to go; they have carried it to the four corners of the earth on their business ventures; they have worn it gallantly on a hundred battlefields; and they have come to look for it on the hand of the men they meet as a symbol of comradeship."[67]

Thus, the ring has remained virtually unchanged for the past century. Efforts to add a large stone on the crown of the ring, or to designate a special ring for coeds or veterinary students were all rejected in favor of one traditional ring for all Aggies.

THE DEPRESSION ERA

The decade of the 1920s ended with the Great Crash of the stock market in October, 1929. The era of jazz, cars, electricity, bootleg whisky, and the good life abruptly came to a halt as the national economy spiraled out of control, ballooning unemployment from 1.5 million in 1929 to more than 12 million in 1932. However, the impact on the state of Texas and Texas A&M was generally moderate. Although cadet enrollment dropped for a couple of years from 1930 to 1932, the revenues from the permanent university fund based on oil royalties smoothed over the reverberations of the downturn in the national economy. By late 1930, Texas A&M had produced 4,272 graduates. The Class of '30 with

Yearly prior to 1940, the outstanding company in the Corps was honored by the designation as the color guard unit. Note the World War I style leggings, a standard part of the dress uniform in the 1920s and 1930s. Courtesy Texas A&M University Archives, Cushing Library

367 grads was the largest to date.[68] However, the depression era did limit those who could pay for a college education, which at A&M cost about $140 per semester in 1933. Those bold enough to enroll without enough money scrambled for one of the hundreds of part-time campus jobs that were created for cadets. Cooperative housing, called the "Project Houses," was set up in the local community to help defray costs. These houses were first old homes in the nearby community, with a housemother who provided room and board for a very low fee. This concept was later expanded to include additional student housing both on and off campus.[69] At the lowest point of the depression when many colleges were adding new students, the enrollment at A&M dropped about 500 students, faculty salaries were cut, and warrants were issued in place of money for pay.

Notwithstanding, the cadets still found time and money to take the bus (for $0.08) into Bryan for the Saturday movie, which in early 1931 featured Joan Crawford in *Dance Fools Dance*. Colonel Nelson was given a one-year extension of his tour at A&M, the first time a Commandant would serve for five consecutive years. In the fall, a coastal artillery unit was added to the Corps, the band under the direction of Col. Richard J. Dunn modified their uniforms to include a white Sam Browne belt, and the Corps of Cadets conducted the first out-of-state Corps trip to Shreveport, Louisiana, for the opening football game of the 1931 season against the Gents of Centenary College.[70] On campus "Prexy's Moon," a light on the top of the Academic Building that for decades had been fair game for unauthorized target practice, was removed forever. And the Aggies outscored Texas 7-6 on Thanksgiving Day.

In September, 1932, Col. Ambrose Robert Emery replaced Nelson as PMS&T. Military training at the college continued under the federal guidelines and annual inspections. Although many A&M cadets trained four years and attended a branch-specific Army summer camp, only a few remained on active duty once they received their reserve officers' commissions. To keep the cadets busy Emery devised a series of "sham" battles or maneuvers in which the cadets would practice repelling an attack on College Station by an imaginary foe invading from the south.[71]

In spite of a January, 1933, campuswide flu epidemic that placed more than 400 cadets in the hospital, the Ross Volunteers escorted newly elected Texas governor Miriam A. "Ma" Ferguson to her Austin inauguration and evening ball. While Colonel Emery was charged to run cadet training, Col. J. E. Mitchell was named Commandant and senior Army officer on campus. Maintaining discipline and uniformity among the cadets occupied much of his time. The Corps began to conform to subtle changes under way in the regular Army. Uniforms that were not functional and a burden, such as the lace-up leggings, were under revision. The solution was to issue straight-leg slacks and riding breeches. One concern was to have the 230 cadets in the junior class ('34) well prepared for the rigors of summer camp. A&M cadets continued to excel at camp and returned to A&M in the fall in time for an uproar over allowing women to attend class on campus. While the cadets voiced displeasure over the need for coeducational changes, the fight over women at A&M was primarily carried on by attorneys, the administration, and alumni off campus. In November, 1933, the local court ruled that A&M would remain all male.[72] In a strange event—the only form of

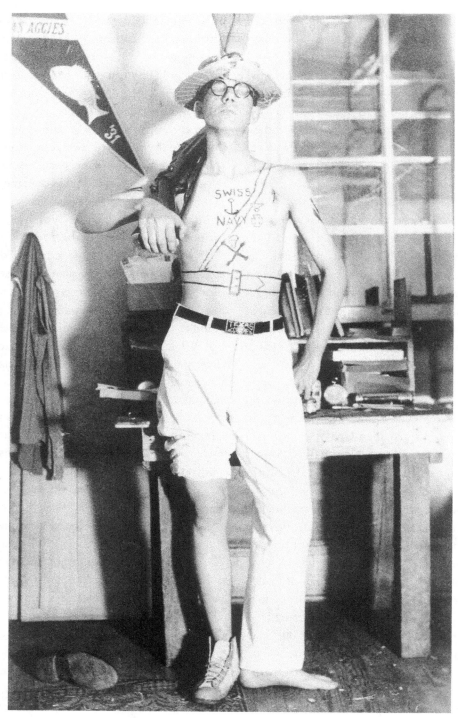

For decades, killing time became a priority for fish such as freshman T. M. Sewell '31 in the fall of 1927. Courtesy Texas A&M University Archives, Cushing Library

indirect protest on the coed issue by cadets—a group of "radicals" in early November attempted to burn the bonfire prior to Thanksgiving Eve. This unidentified group failed in their efforts. The campus settled down in time for the traditional senior class Thanksgiving rite of passage known as "Elephant Walk."[73] The 1933 *Battalion* highlighted the background of the tradition that continues to this day:

> On the eve of the Thanksgiving game, the seniors of the Cadet Corps gathered in a solemn group behind the "Y" Chapel to take part in one of the oldest and most sacred of A and M traditions, the "Elephant Walk." The ceremony itself carries out the legend that the old elephant, upon seeing the end drawing near, leaves the rest of the herd and seeks a secluded spot for his death-bed. This rite represents the last act of the senior class as a body; their going opens the way for those who follow in their footsteps. Thursday the seniors will see their last conference football game as students of the institution.[74]

Campuswide enrollment dipped one more time, from 3,000 to 2,600 cadets in 1934–35, and has never again shown a decrease except for the last two years of World War II. By 1935, in spite of depression concerns, Texas A&M charted a new path and aggressive program to weather the economy. This course of action came in various forms that were not all linked to one another. However, growth and enrollment were the two primary benchmarks. These efforts and plans were partly the result of reactions to the Griffenhagen Report, prepared by a commission retained by the state legislature to study recommendations to streamline the higher education system in Texas as well as save taxpayer funds. The legislature-directed study

proved very controversial at Texas A&M on a vast array of academic and administrative topics. The report included a recommendation that "the military system should be relaxed."[75]

The military legacy by the early 1930s was an integral part of the college. Thus, there would not be any major changes. A sign of the times, the Aggie Band began to play a new favorite of the cadets, "There Shall Be No Regrets." The 1930s ballad penned by Curtis Vinson faded from use during the war years.[76]

THERE SHALL BE NO REGRETS

Warm beat the hearts of all thy sons,
Oh Aggieland of Texas,
None but the true thy banner bear,
None but the staunch thy emblems wear,
In war, in peace, where duty leads,
High flames thy spirit in their deeds.

Chorus:
In war or peace, staunch hearts we
  plight,
Staunch hearts unwav'ring in the right,
Thousands before us,
Thousands to follow,
Saber or plow, to the fight!
To win or to lose, whatever besets,
May there be none who forgets,
Courage our ensign,
Valor our guidon,
And there shall be no regrets.

Strong is the tie that makes as one
Ole Army's Texas Aggies,
In fellowship, in pride of corps,
In glories won by hosts of yore,
Unbroken ranks to face and fight,
Hail Aggies! Hail Maroon and White!
(Repeat chorus)

To put a new face on growth and the future at Texas A&M, the campus turned the main entrance toward the east with the construction of the System Administration Building, later (in 1998) named for Dr. Jack K. Williams, to face the opening of a new north-south state road—Highway 6. Additional monies were spent to improve support buildings and dorms, and the seven branches of the ROTC training program added a chemical warfare unit in the fall of 1935.[77] However, the hallmark of the viability of the Corps and Texas A&M would be the attraction of new students. In the spring of 1935 an all-out effort was begun to tell the A&M story. To tout the new changes at the college, the rich heritage, and opportunities on the campus, the *Longhorn* was sent annually to every Texas high school library, and at least once yearly more than 7,000 special-issue *Battalions* were mailed to high school juniors and seniors statewide. On campus, the senior class held the first spring Ring Dance—complete with an eight-foot-tall replica of the Aggie ring.[78] The first full-scale awards ceremony to recognize cadet achievement and military excellence was begun at Final Review in May, 1935. The awards and their sponsors brought added attention to the Corps of Cadets and Texas A&M.[79]

These efforts and events were to pay off. Fall enrollment in 1935 totaled 3,400 cadets, followed by a steady pattern of growth—3,500 cadets in January, 1936; 4,075 in September, 1936; and more than 4,400 by January, 1937.[80] The growth of A&M surprised even detractors of the institution as well as proponents of a strict implementation of changes to higher education in Texas. Many did not realize that Texas A&M had gained a broad public appeal. The alumni remained supportive and

vocal, reflecting the esprit de corps of their campus days. And in 1939 the Aggie football team defeated Tulane to win the national championship. The Corps was alive and well. A sense of the 1930s at A&M is best captured by the *Centennial History:* "Despite, or perhaps because of, the hard times, the student body at Texas A&M developed a particularly intensive spirit of cooperation and loyalty to one another, and to the school."[81] This view is further confirmed by Ormond R. Simpson '36 regarding the remote nature of campus: "There were only 7 students who owned automobiles on campus and the keys to those were kept in the Commandant's Office. Obviously, there were no traffic or parking problems on campus. Travel was by train for those who could afford it . . . or follow the strict protocol for the lineup at the Main gate to catch a ride."[82]

Nationwide growth in military training at the college or university level was not as smooth. An anti-ROTC, anti-military element caused many campuses to reduce or eliminate the ROTC training instead of fighting for its place on campus. Vigorous debate ensued on the merits of military training in peacetime as the number of ROTC dropped from nearly 200 to 122 detachments on college and university campuses nationwide. A feature article in *Infantry Journal* summarized opponents' principal objections to college-level military training: "first, that it [ROTC] is not required by the Act of 1862; second, that such education fosters militarism that is not in harmony with American ideals; third, that the spirit of compulsion is out of harmony with the spirit of modern education; fourth, that the practice assumes the probability of future wars and is out of harmony with the present trend of thought in American life."[83] A fifth objection often stated was that "its [ROTC] system is conducive to hazing."[84] In spite of a U.S. Supreme Court ruling in late 1934 that upheld the right of a university to require students to enroll in compulsory military training, by the mid-1930s, pacifist agitation resulted in 17 colleges dropping reserve officer training altogether and 7 others changing from compulsory to elective.[85]

In marked contrast, Texas A&M clamored at all levels to be able to educate and commission more officers. The armed services gradually expanded the officer corps to nonservice academy graduates. The U.S. Marine Corps in the spring of 1935 approved the direct commissioning of 100 new lieutenants to be recruited from 50 colleges and universities across the nation. The Commandant at A&M was advised that the Corps of Cadets had 2 slots and 2 alternates. However, all 4 of the A&M selectees in the Class of '35 were given regular Marine Corps commissions—Raymond L. Murray, Bruno A. Hochmuth, Odell M. "Dog Eye" Conoley, and the 1934–35 Cadet Corps Commander, Joe C. McHaney. All except McHaney were among the first Aggie Marine general officers.[86] Furthermore, the Congress, not dissuaded by those against military training, passed the Thompson Act, which by 1937 provided for as many as 1,000 reserve officers to go on active duty for one year and further stipulated that 50 could be offered regular Army commissions each year. Additionally, FDR's New Deal established further opportunity for reserve officers to gain valuable leadership skills in the Civilian Conservation Corps (CCC).[87]

Gen. Bernard A. Schriever '31 recalled that in 1935–36 as a second lieutenant and CCC camp officer in New Mexico, the "CCC was

TABLE 5-3
Corps of Cadets Awards and Recognition, 1935

Howell Trophy (best infantry company):
  Company A
Waldrop Trophy (best field artillery battery):
  Battery A
Gulf States Utilities Award (best engineering company): Company C
Brandon Lawrence Trophy (best cavalry troop):
  Troop A
General Jacob Wolters Cup (best proficiency):
  Troop B
Southwestern Bell Telephone Award (best signal corps company): Company B
William Randolph Hearst Trophy (rifle team award)

Source: *Battalion*, May 29, 1935, p. 1.

a great experience . . . it was one of the best learning periods I ever had . . . you had to lead by brains . . . the camp program developed many future leaders."[88] The Air Force considered Schriever America's prototype general of the future. A "three-letter man"—pilot, scientist, and engineer—he fostered "ideas, systems techniques and management principles that changed most of the regulations in the book."[89] General Schriever was the primary architect of the 1950s intercontinental ballistic missile program. He could also be found at the military roots of the manned space program that followed in the early 1960s. He became Texas A&M's second former cadet to be promoted to the flag rank of four-star general. Like Schriever, hundreds of Aggie officers languished in inactive reserve units and the stagnant peacetime armed forces. Thus, the rapid expansion of the Army during the 1942 emergency mobilization would not have been possible without this trained nationwide reserve force.[90]

With prosperity once again at Texas A&M, a number of problems on campus resurfaced: overcrowding and the increasing demand for additional on-campus student facilities such as a student "union building." The increase in enrollment warranted A&M's claiming the largest agricultural enrollment "in the world."[91] In 1937, the Corps of Cadets continued as the largest in the nation, and Texas A&M was one of only three institutions that offered military instruction and commissions in all seven branches of the Army— infantry, cavalry, field artillery, coastal artillery, engineers, signal corps, and chemical corps. There is little doubt that President Franklin Roosevelt's visit to the campus in mid-May, 1937, to review the Corps was more than just a casual sojourn. The winds of war were already being fanned in both Europe and Asia.[92] Escorted by Cadet Corps Commander Louis E. Lee '37, Commandant Col. Frank Anderson, and President Walton, Roosevelt received a twenty-one-gun salute, a campus tour, and a gift of a Hereford calf for his farm in Warm Springs, Georgia. FDR knew that the young, eager Cadet Corps, whose alumni mentors had been battle-hardened in World War I,

would, if needed, prove critical in any military mobilization efforts by the United States. The 20,000-plus visitors—many on the A&M campus for the first time—who filled Kyle Field to get a glimpse of the president also were equally impressed by the military bearing and esprit de corps of the Aggie Corps of Cadets.[93]

What would be learned in the course of a few months was that Texas A&M had already begun to prepare for the possibility of war. By 1940, the A&M Corps of Cadets since the end of World War I and the inception of ROTC had trained more than 16,000 cadets. While most graduated with little or no active military duty, due in part to the shortage of active duty slots in the Army— most stood ready to answer the call to duty in the event of national crisis. Detractors of military training and the citizen-soldier would soon come to appreciate the tremendous contribution made by the cadets, faculty, staff, alumni, and resources of Texas A&M. One who had a direct and timely role in directing the World War II destiny of the Corps of Cadets was a new Commandant who arrived on campus in the fall of 1937—Col. George F. Moore, Class of '08.

6

WE'VE NEVER BEEN LICKED

The important part that A&M trained men will play in the next war cannot be exaggerated. In event of hostilities they would be in a position to go into action as instructors or as officers in actual combat.

L. E. Thompson '40, *Battalion*, October 21, 1938

Texas A&M is writing its own history in the blood of its graduates.

Gen. Douglas MacArthur, 1942

Seek, strike, destroy—then get the hell out of there!

Maj. Gen. A. D. "Andy" Bruce '16, Tank Destroyer Corps, 1943

THE INVOLVEMENT and contribution of Texas A&M to the Second World War began, unbeknownst to most, in the late 1930s. President Franklin Roosevelt knew war was near but did not know when and where it would involve the United States. In May, 1937, in an informal address at Texas A&M to the cadets and guests Roosevelt noted,

"Some think of military training in terms of pacifism. You and I do not. We think of it in terms of the preservation of the nation. Coming here today is a great inspiration to me."[1] Articles gradually filled the media on America's potential role in the face of the rise of "Hitlerism" and aggression in Europe as well as the designs Japan had for Asia.[2] Though preoccupied with the depression, classes, Corps trips, and campus activities, there was a growing awareness of these unfolding events among the cadets, faculty, and staff. Yet none envisioned the level of American involvement in the years ahead.

The debate on the role of military training on college campuses often set A&M apart from those institutions reluctant to allow a significant level of on-campus military training.[3] Whenever there was a dispute on ROTC or military training at another campus, Texas A&M was held up as an example of either military excellence or military excess. The citizen-soldier theme that was ingrained in all aspects of Texas A&M and the ongoing support of the administration, board, and alumni for the Corps made Texas A&M distinctive. For example, an early 1939 editorial in the *Fort Worth Star-Telegram* on the opposition to military training "in any form" at the University of Texas noted, "Texas A&M should be appreciative of the publicity-by-contrast."[4] Thus, Roosevelt and his staff saw no excess, only excellence and opportunities at Texas A&M. Shortly after the President's campus visit in 1937, new emphasis was quietly placed on expanding the military training program of the Corps of Cadets and the housing facilities at Texas A&M.[5] This was the beginning of a decade-long period of dramatic growth and change for the Corps of Cadets from 1937 until the end of World War II.

In the fall of 1937, Col. George Fleming Moore '08 became Commandant of Cadets and professor of military science and tactics (PMS&T). These two duties, which had been separated under Colonels Anderson and Emery, were recombined under Moore. Colonel Moore grew up in Fort Worth and entered A&M in the fall of 1904. While a cadet, he majored in civil engineering and lettered (as a guard and tackle) on the football team, served as a cadet officer in Company B, and was a member of the Ross Volunteers and the "T" Association. He earned the nickname of "Maude" for being able to kick a football like a mule—a reference to a popular cartoon strip at the turn of the century.[6] In the three decades prior to returning to Texas A&M, he had seen service with the coastal artillery, the field artillery, and the ordnance department. A graduate of Command and General Staff School, he was a veteran of World War I and had twice been assigned to assist with the defense and fortification of the Philippines. The Corps of Cadets in the fall of 1937 numbered 4,933 and the combined military staff had been expanded to 17 officers and 40 enlisted personnel.[7] Colonel Moore was faced with many of the same challenges of his predecessors, most notably the need for additional housing and training facilities so badly needed for the growing student body.

The age-old element that impeded the growth of the Corps of Cadets was adequate dormitories. While the Project House concept had eased short-term housing demands, A&M still remained in need of more on-campus dorm rooms.[8] The building of support facilities and classrooms gradually continued on campus in the mid-1930s with money from the university endowment fund, yet no new dorms were built. However, in June, 1937, an application (clearly encouraged by President Roosevelt) was submitted to the Reconstruction Finance Corporation (RFC) for a $2 million loan to build only dorms. The RFC was administered by Jesse H. Jones of Houston, who had been awarded an honorary doctor of law degree by Texas A&M in 1936. And the membership (1936–38) of Elliott Roosevelt, the president's youngest son, on the Texas A&M Board of Directors would also prove timely. Elliott was a member of the board's subcommittee on buildings, along with Robert W. Briggs '17 of Pharr and Edwin J. Kiest, editor and publisher of the *Dallas Times Herald.* Thus the team of Jones, Briggs, Kiest, and young Roosevelt prepared an application for one of the largest single projects since the inception of the college.

Review of the application by the RFC was completed in late 1937, and the contract was let in early 1938. The plans called for twelve new dorms and a mess hall close by to feed the new cadets housed in "at least 1,250 rooms." With this action, the "New Corps Dorm Area" or "Quadrangle" would be the future site of growth and expansion of the Corps Roosevelt and others envisioned. Netting out design cost and the mess hall, the twelve Corps dorms cost $124,500 each when completed in 1939. The loan to the RFC was to be repaid with semiannual payments of $55,250 from room fees charged the cadets, about $25 per semester. Colonel Moore provided the A&M Board with timely information for the final site location on each dorm. Original plans had called for Dorms 9 and 11, and 10 and 12 to be located side by side. In the final plans, Dorms 11 and 12 were moved to either side of Duncan Dining Hall. To provide adequate space for the new mess hall both the college cemetery and the local Consolidated Grammar School had to be relocated off campus. To maximize the number of new rooms within the slim budget, the four-story, all brick and concrete dorms were designed to be very spartan, with few or no amenities. Final room count: 1,315. Duncan Dining Hall with a seating capacity of 3,000—in reality two dining halls separated by a central kitchen—

cost $361,181. Additional funds, both federal and state, soon followed and were directed to the campus for new offices, a new mule barn, and renovation of Guion Hall and the Aggieland Inn. Given the impending threat of war, the expansion of Easterwood Airport adjacent to the west side of the campus was given high priority with a $229,970 grant from the Civil Aeronautics Authority and the WPA.[9]

During his nearly three years as Commandant, Colonel Moore and his staff presided over the largest peacetime growth of the Corps in the college's history. All aspects of campus life and growth were magnified— housing, support services such as meals, and even logistical demands in transporting the Corps on out-of-town Corps trips. Prior to the completion of the new Corps dorms, housing and overcrowding became so critical that more than a hundred cadets were housed in the Hoyle Hotel in downtown Navasota and bused daily twenty-three miles to campus for classes and training.[10] Col. Ike Ashburn, who had returned to Texas A&M in September, 1937, to become the executive assistant to the president, was charged with handling public relations with the media to limit negative publicity surrounding the campus overcrowding.[11] Furthermore, Ashburn's prime assignments were maintaining political ties with Austin and overseeing the housing and feeding of the growing Corps. Keeping the cadets fed was a twenty-four-hour, seven-days-a-week process involving the college staff and student labor. The ability to seat the entire Corps for meals in Sbisa and Duncan greatly impressed visitors to the campus. Cadet J. Wayne Stark '39 during the late 1930s wrote a weekly column in the *Battalion* titled "Aggie Scrapbook" in which he covered facts and events on the campus. The mess hall caught his attention in October, 1938: "At some meals the Aggies will consume close to 1,000 pounds of meat, 1,200 pounds of potatoes, 6,000 rolls, 5,000 cream puffs, 3,000 (1/2 pints) bottles of milk, 70 pounds of coffee, besides the remainder of the food served during the meal. Three meals daily, the Aggies will probably eat 50,000 pounds of meat in one month."[12]

The robust appetite of the cadets was second only to their enthusiasm for sports. Corps trips became major operations. Colonel Moore's first-ever experience with moving the Corps was for the mid-October, 1937, Texas Christian University football game and parade in Fort Worth. His dealing with more than 4,000 enthusiastic cadets as well as coaches, yell leaders, the railroads, the Fort Worth A&M Mothers Club, alumni, and college administrators proved insightful. Realizing the need to ensure full communication, he increased contacts with the cadet officers and required his staff to take a broader role in mentoring the cadets. The Corps took a second trip to the Rice game in Houston. Both the TCU and Rice games ended in a tie. Returning to campus, the Aggie team, led by All-American Joe Routt, prepared for their last home game against the University of Texas. It would be a unique Thanksgiving. In the forty-eight hours prior to the annual game, more than six inches of snow fell on the campus and Kyle Field. Cadets rolled and played in the snow, and classes and drill were temporarily suspended. Though the field was partially covered with a tarp, there was concern about game day conditions at Kyle Field. This annual event became even more dramatic with the return of popular former Aggie head coach Dana X. Bible, who had left Nebraska for Texas—lured away by a twenty-year, $15,000-per-year

contract—a sum greater than the pay of the president of the university. Bible envisioned no greater homecoming than to defeat the Aggies at Kyle Field. By game time the field was clear of snow and a crowd of 33,000 witnessed the Ags beat the Longhorns, 7-0.[13]

Aggie gridiron enthusiasm, along with the traditions and esprit de corps, was one of the main reasons A&M attracted more students. As enrollment grew so did the varying number of classifications of students. The types of A&M students began to be reflected in student interest in programs other than the Corps of Cadets. As the growth in the ROTC branches offered to the cadets expanded so too did the number of students in the graduate studies program. Furthermore, there was an influx of foreign students as well as a growing group of "day dodgers" (students living off campus called "day ducks" in the 1960s), and numerous special students who participated in military training. Colonel Moore worked with President Walton to better define the makeup and organization of the student body. To ensure all were aware of the requirements as cadets and the very small "nonmilitary" element on campus, new guidelines were printed in the *Battalion*. Students were grouped into different categories depending on academic status, physical "unfitness," or level of completion of military science. An executive committee was established to review special circumstances. Notwithstanding these efforts to better qualify the students, mild attempts were made to have the coeducational issue reviewed. The issue of coeds in the late 1930s was not going to dramatically change the direction of the college. Texas A&M throughout the balance of the 1930s and into the 1940s would remain essentially an all-male military college.

Enrollment in the fall of 1938 totaled nearly 5,200, with 1,973 new fish—members of the Class of 1942. Little did the Class of '42 realize what was in store for them.[14]

BONFIRE

By 1938, the annual Bonfire before Thanksgiving Day was a major fall event. Originally built off campus *after* a victory over the University of Texas, as early as November, 1907 (prior to the game's becoming a yearly Texas A&M–Texas Thanksgiving match) the event was moved to campus beginning in the early 1920s. Consisting of a pile of old scrap wood and junk, the annual Bonfire constructed on the main drill field began to consume hundreds of hours of freshman labor to build

TABLE 6-1
Organization of the A&M Student Body, September, 1938

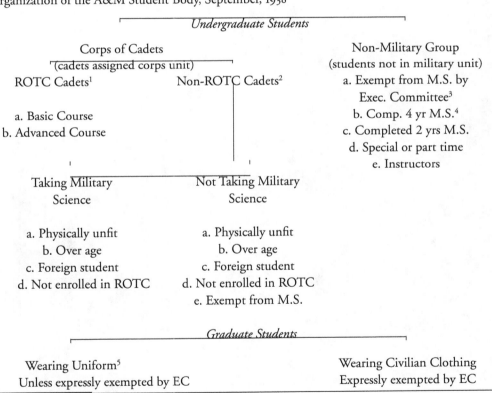

[1]Wears "AMC," insignia of branch and rank (if any), ROTC "patch."

[2]Same as note 1 above, substituting "Aggie Patch" for ROTC "patch."

[3]Students exempt from M.S. may be "non-ROTC cadets" or "nonMilitary students."

[4]Students who have completed the basic or advanced ROTC course may be "non-ROTC cadets" or be assigned to nonmilitary group at their option.

[5]Wears "AMC" and "NM," "Aggie Patch," and purple hat cord. No insignia of rank or branch. Officer's hat cord, boots, Sam Browne belt not authorized.

Year after year, Aggies built the Bonfire bigger. The 1946 Aggie Bonfire was one of the first to use a "center pole" to allow for a more uniform stack of about 45 feet. Photo from author's collection

a bigger and bigger fire. In the late 1930s, the late-night, pregame Yell Practice on the steps of the YMCA became a ritual. As the crowds grew, Yell Practice was moved to the north side of Goodwin Hall. Cadets and fans would meet in front of the Goodwin for a pep talk from the yell leaders and music by the Aggie Band, followed by brief remarks from the coach and various football players. Noted former students, such as "Dutch" Hohn, of Brenham, would for years be called from the crowd to recount the tradition and significance of the traditional "turkey day" clash with Texas.[15]

Campus activities such as Bonfire and statewide news of Aggie athletic teams were pivotal to recruiting efforts. By 1939–40, Texas A&M successfully completed and filled the new Corps dorms to capacity. Each new entering freshman class was growing, and there was an annual increase in the number of graduates. With 623 seniors, the Class of 1939 was the largest class to date, and they received degrees at the June commencement in Guion Hall.[16] While the depression era hobbled the national economy, there was no shortage of students at Texas A&M. The total nine-month cost for tuition, room, laundry, medical fee, and board was $360. Fees were payable monthly or by the semester.

To assist cadets in paying fees, a variety of programs were developed to locate on-campus student jobs. The A&M Student Labor Committee chaired by ex-cadet Ormond R. Simpson '36, director of student employment, helped distribute funds provided by a grant from the National Youth Administration (NYA).

Simpson helped obtain financial assistance for 647 students in the spring of 1939.[17]

TABLE 6-2

Texas A&M Corps of Cadets, 1939–40

Corps Staff							
Infantry Regiment			*Field Artillery Regiment*			*Composite Regiment*	
1BN	2BN	3BN	1BN	2BN	3BN	Signal Corps	Chemical
A	E	I	HQ	HQ	HQ	HQ	A
B	F	K	A	C	E	A	B
C	G	L	B	D	F	B	C
D	H	M					

Cavalry Regiment			*Engineering Regiment*		*Coast Artillery Regiment*	
HQ	1sq	2sq	1BN	2BN	1BN	2BN
Machine	A	C	A	D	A	E
Gun	B	D	B	E	B	F
Troop	C	F	C	F	C	G
					D	H

Band	
Infantry	Field Artillery

Cadets looked not only for alternative means to finance their education but also for ways to have a good time inexpensively. In addition to saving a few dollars for the annual Corps trips, cadets formed new clubs and informal groups such as the "Clickety Clack Club"—with no formal charter or function. Others challenged class members in an unofficial contest to see who could hitchhike farthest over a single weekend and return in time for Monday morning formation. To assist Aggie weekend travelers, the YMCA printed 20,000 "hitch-hiking thank you cards" to be given by cadets to the driver as a token of appreciation. Hitchhiking "stands" were informally set up on all major roads out of the campus. The unpretentious rules of the road stated that an Aggie would not "upstream" or cut in front of cadets already in line. On campus, cadets mounted a Corps-wide challenge to determine who could eat the most live goldfish or frogs. The record set by Cadet Pat Rose '41—a cavalry sophomore who swallowed seventeen frogs to win a $1 bet—stands to this day.[18]

COL. GEORGE MOORE

In order to fill the new dorms in the fall of 1939, Colonel Moore strongly supported the new corps housing rules that effectively required all cadets to live on campus. Day student outfits were disbanded and only the most severe cases of "lack of sufficient funds" would constitute an acceptable reason to live off campus. Moore further looked to reorganize cadet training to better align with active duty Army standards and procedures. As the Army modified and changed drill procedures, so too did training officers at A&M adjust the cadet training. Often the cadets were faster at

adapting to such modifications. One change was in the organization of cadet units. Each cadet company in the Corps would have a minimum of 74 cadet members—34 freshmen, 20 sophomores, 12 juniors, and 8 seniors. However, in special cases unit strength could total as many as 100 due to assignments of fifth-year seniors and day students. The Aggie Band of approximately 200 members was organized into two sections: the artillery band and the infantry band. Requirements for training within each specific Army branch would be closely monitored to comply with Army guidelines and thus better prepare cadets for summer camp and active duty service after graduation. For example, in the signal corps only electrical engineers were eligible; in the cavalry units no cadet could weigh more than 170 pounds; and in the coastal artillery branch (anti-aircraft), cadets were required to have an excellent knowledge of plane trigonometry by the end of their sophomore year.[19]

Due largely to his three decades of active Army experience and expertise, Colonel Moore was the first Commandant to stress an urgency toward full implementation of training and accountability within the ROTC branch system. "Old Maude" became tremendously demanding of both his staff and the cadets. This demand for excellence reinforced the respect cadets held for him. The story of his kicking a football over Ross Hall in 1906 had become near legend. While war seemed distant, Moore, as a result of his lengthy tours in the Philippines, sensed a conflict in the Pacific was just as likely as the nation's being drawn into the European conflict. The prospect of war was driven home to Colonel Moore and all the Aggie Corps on September 3, 1939, when news that the Moores' only daughter was presumed lost at sea, a victim

The U.S. Army Cavalry branch was introduced to Texas A&M in the early 1920s and was an active part of the Corps until the early years of World War II. Hughes "Buddy" Seewald '42 is pictured in a cadet riding uniform with mounts. Courtesy Office of the Commandant

of a late-night German submarine attack and sinking of the British liner *Athenia.* At 6 P.M. on the same day, both France and England had declared war on Germany, thus escalating German justification for U-boats to sink British shipping. However, Anne Moore survived the attack and was rescued after being afloat in the cold and rough North Atlantic for more than eight hours. She returned to Aggieland in early October to a near hero's welcome. Thus, America's path toward war had begun in earnest in the fall of 1939 as FDR proclaimed the United States the "arsenal of democracy" and reluctantly initiated aid for both France and England as well as Russia.[20]

On campus, far from the growing war, cadets focused on a history-making football season leading to the 1939 National Championship, capped off by the defeat of Tulane, 14-13, in the Sugar Bowl. And Colonel Moore gradually intensified cadet training until notified in early 1940 to prepare for his third Pacific assignment to Corregidor Island in the Philippines.[21]

In the spring of 1940, events further unfolded that were to put Texas A&M and the Corps of Cadets on a heightened level of preparedness for the war. Colonel Moore continued his insistence on the details of cadet training. Furthermore, the War Department expanded the scope of cadet training when the Army announced the availability of aviation cadet slots in the Flying Cadets Training Program of the U.S. Army Air Corps. The air corps training centers at Kelly and Randolph Fields in San Antonio were eager to pre-qualify the Aggie cadets for the nine-month course. Speculating on the role of Texas A&M in a new conflict, the *Battalion* in 1940 ran numerous editorials long before Pearl Harbor, such as "Texas

Aggies and the World War—No. 2."[22] By April, the War Department began to reassign key regular Army officers as well as call on ROTC training nationwide to accelerate officer output. Colonel Moore was notified of his pending transfer to the Philippines via Hawaii—an age-old staging area for Pacific area operations. Ormond Simpson, director of student employment and a reserve U.S. Marine officer, became the recruiting liaison on campus for cadets and staff members interested in a commission in the Marine Corps. And in the midst of what amounted to a pre-mobilization phase of the war, John O. Pasco, a graduate assistant in mechanical engineering, released a light-hearted novel in May, *Fish Sergeant,* on cadet life and antics in the Corps. A fictional account of life and times in the Corps at A&M, the popular book has remained in print for five decades.

However, the greatest campuswide attention at the time was drawn to the A&M response to the National Defense Program, under which college administrators in mid-1940 "offered all A&M facilities for the national defense to use."[23] Mindful of the sluggish national mobilization response in 1917 to World War I, it was hoped that an early all-out response from Texas A&M would increase the number of commissions allocated to the classes of 1941 and 1942. President Roosevelt's 1937–38 investment in expanding facilities for the Corps of Cadets would soon prove both timely and critical to the national war effort.[24]

At final review in June, 1940, the Corps numbered in excess of 6,000 cadets. Rumors of potential new national draft legislation and the nationalization of selected Army National Guard units intensified cadet and administration focus on obtaining more commissions for A&M graduates. In the case of the Class of

1940, while the number of graduates totaled 677, only 358 under the Army quota system were able to receive an Army officer's commission. The Aggies wasted little time in lobbying the War Department, reminding Washington that the Aggie Corps of Cadets was the largest military college in the world. Texas A&M officials were not convinced or calmed by War Department response that there are "too many ROTC officers!" The *Battalion* noted, "It's impossible to tell what an officer is worth, but it's easy to know how much one costs the government—$20,000 at West Point . . . $400 at Texas A&M . . . all that Texas A&M College wants is the word to 'Go'!"[25]

Shortly after Colonel Moore departed, Lt. Col. James A. Watson became Commandant in August, 1940. Originally from Ohio, Watson was a career infantry officer. An avid horseman, he had held Army assignments that included a tour of duty as PMS&T at the University of Wyoming. Although Colonel Watson inherited a program in excellent condition and high morale, international events were to have an increasingly tremendous impact on the Corps of Cadets. Shortly after the collapse of France, American neutrality evaporated and military preparedness intensified. In September, rumors of a draft became reality as the U.S. Congress enacted the Selective Training and Service Act, a "precautionary measure" to establish the authority for a peacetime conscription. Furthermore, $18 billion was authorized by Congress for the "national defense."[26] Such enhanced preparation for an even larger call-up of new Army recruits resulted in many high school seniors opting to enroll in a college ROTC program as a means of deferring induction into the armed forces. In the wake of a growing national concern over the gradual

war mobilization, enrollment at A&M swelled to more than 6,600. The enrollment increase of 8 percent caused a reversal in the ban on off-campus housing and the limited day student status. Cadets were permitted to seek room and board accommodations either in the Project Houses or in "college approved" residences in Bryan. William A. Becker '41, the 1940–41 Cadet Corps Commander and a retired Army general, recalled, "the project houses were a brilliant concept and they really worked to handle the overcrowding."[27]

In order to obtain more office space, Colonel Watson received college approval to move the staff of the Commandant's Office from the Academic Building to the first floor of Ross Hall.[28] A young Army lieutenant, Joe Davis '29, served as assistant to the Commandant for administrative matters. Following guidelines stipulated by the Selective Service Act (the first peacetime conscription law in the nation's history), the Brazos County clerk worked with Texas A&M registrar Eugene J. Howell to comply with the mandate to register all men between twenty-one and thirty-six inclusive who were not "specifically exempt" from the draft. During the first fourteen hours of registration on October 16, 1940, 1,445 Aggies (cadets, faculty, and staff) registered. This action, "whether or not a state of war existed," was being conducted nationwide (more than 16 million registered) in advance of an expected "calling-to-the-colors" and induction of an estimated 800,000 draftees by late November. Within months the draft age would be lowered to eighteen.[29]

Colonel Watson assured those cadets in good academic standing and enrolled in ROTC that they would not be affected. Campus rumors on the full meaning and impact of prewar preparations gradually intensified, yet did not preclude more than 1,500 cadets from obtaining weekend passes for the so called "Second Battle of the Alamo" for the second football game of the 1940 season against Tulsa at Alamo Memorial Stadium in San Antonio. The *Batt* reported that "Corps trip fever combined with the 'Aggie Spirit' . . . will give them [cadets] the opportunity to resume any unfinished business they might have" following the summer camp training at military installations in the San Antonio area in mid-1940. Freshman cadets were excluded from travel, and only the 150 upperclass members of the Aggie Band were permitted to attend the season's initial away-from-college trip. The Aggies ended the 1940 season in a tie for the Southwest Conference Championship.[30]

By early 1941, it became very apparent that the nation would in some manner be involved in war. Former Cadet Corps Commander William Becker recalled that "it [the war] was the main subject of our bull sessions as seniors." The renewed pledge of Texas A&M facilities and manpower was reaffirmed to Washington. The seven branches of Army training on the campus, staffed with 31 officers and 55 enlisted personnel, proved very effective in training the 6,500-man Corps. Of the 767 graduates in the Class of '41, 535 received commissions upon graduation. The *Battalion* boasted that the "$18,000,000 A&M Plant" was the national leader in training officers. Of the more than 12,000 living A&M alumni in 1941, it was estimated that 5,000 held reserve commissions and a vast majority of the rest were prepared to be sworn in when additional officer slots were made available. Further courses and facilities modifications, under the direction of Army personnel, were under way to teach flight training, mechanics, and machinist courses

and expanded classes in the fundamentals of the signal corps. To keep the leadership of the Corps informed, Cadet Colonel Becker and his staff held a dinner with all the unit commanders once a month in Sbisa, "where we could do the Corps'[s] business." To address the hectic schedule, the day-to-day cadet uniform was modified to include lightweight Bombay khaki slacks, worn with the campaign hat. Nonetheless, the Aggies eagerly awaited any solid news to dispel the gossip of war that periodically circulated on campus or appeared in the press.[31]

Concern over world events became more evident daily. Ike Ashburn, popular former Commandant and assistant to the president, resigned his A&M post to become director of the Industrial Peace Board in Houston, a clearinghouse to ease management-labor issues in the event large-scale production was needed for a war effort. Campus rumors focused on the impact of the new draft law. A Gallup poll determined that 69 percent of college students nationwide felt they should be able to complete their college work prior to being inducted in the military due to a draft. Administrators at A&M—as well as their counterparts nationwide—mirrored student concerns and were fearful of declining enrollments in the event of a full-scale war. Nonetheless, training of the cadets began to take on a more serious tone. Thus, preparation for the 1941 cadet summer camps was intense, as most of the over 400 members of the Class of '42 bound for training sensed they would soon be on active duty employing the skills and tactics learned in their branch specialization.[32] In addition to receiving transportation to and from their assigned branch specific training camp scattered across the country, each cadet was paid $0.70 per day.

However, more important considerations concerned the future course for each cadet. A solemn editorial in the June 1, 1941, *Battalion* entitled "After the War" took a bold look at the existing concerns and a very perceptive (almost prophetic) focus on the future of the American nation in world affairs:

Now another war worse than before. . . . Purely aside from any supposition that we may be drawn directly into the European conflict, the very magnitude of our defense program is so great that it leaves many floundering in confusion as to how we are going to pull through. . . . A considerable portion of what we are now undertaking as defense will be long continued. This suggests the constantly enlarging role of the United States—not necessarily a voluntary role—in world affairs. It requires long-continued and constantly broadened efforts to match that role in our commercial life, to say nothing of the necessary naval and other defense roles.[33]

SOS—SAVE OUR SPOONS

In the fall of 1941, the Corps numbered nearly 6,600 cadets. Advance fall freshman registration due in some part to draft deferments, recruiting by individual branches within the Corps such as the coastal artillery and engineers, as well as overbooking in the dorms sent a clear signal to the A&M administration that the 1941–42 academic year would be a record for the college.[34] Although four new dorms (Moses, Moore, Crocker, and Davis-Gary Halls) were under construction on the north side of the campus, a shortage of bricklayers delayed the construction projects. The fish

Class of 1945, numbering 2,413, would be dispersed throughout the Corps units in the following manner: 25 percent to the infantry, 25 percent to field artillery, 17 percent to coastal artillery, 11 percent to the engineers, 6 percent each to the signal corps and chemical corps. The dominating two cadet regiments were the infantry and the field artillery, and they controlled the Quad. While each dorm had long ago been formally named, cadets routinely referred to them by number. The infantry occupied Dorms 3, 5, 7, and 9; the field artillery were located in Dorms 4, 6, 8, and 10. Dorm 1 housed the signal corps and Dorm 2 the chemical warfare branch, while the Aggie Band occupied Dorm 11. The athletes or "jocks" took over Dorm 12. The balance of the Corps was spread around the campus—coastal artillery in Bizzell, Mitchell, and Walton; engineers in Walton, the cavalry in Law and Puryear. Corps headquarters was in Leggett.[35]

The Aggie training and preparation paid off as they dominated each 1941 branch specialty summer camp they attended. Well prepared and trained at A&M, they were eager to lead, and with strength in numbers they cut a wide swath.[36] Being off the A&M campus did not preclude the cadets from imparting a bit of their Aggie way of cadet life to strangers. Marveling at the conduct of the cadets, a student from the University of Tennessee wrote the following in the camp newspaper, the *Lachrimator,* while attending the chemical warfare branch session with the Aggie cadets at the Edgewood Arsenal in Maryland:

Then came those Texas Aggies. It's legend now, the first one to enter the mess hall calmly spat tobacco juice through the rafters and scored a direct bull's-eye in the finger bowl, twenty feet away. It was clammy calamity, when the vanguard of the passel of 35 gave way to the main body. They came, like the black plague in winter, to darken our fair mess hall, to create confusion everlasting.

"Shoot the bullneck," boomed one Texan, meanwhile preparing his greasy gun hand for action. The meat was passed. "Pass the stud," roared his colleague; and we of the lighter blood, quivered with fear. He got his ice tea quick-like.

Another Aggie yearned for some sugar. We could see it in his eye. "Sawdust!"

That was his only command.

"Sand and," hissed another. We appreciated the undertone, even though it was a hiss. One shaking campmate finally shook enough salt and pepper in the plate to appease his master.

"Deal One," snarled the previously mentioned bread-eater, having devoured a mere snack of eight pieces. The bread sailed gently throughout the foul air and was snatched, in flight, by an expert meat-hook.

"The blood!" . . . (pause) . . . "the B-l-o-o-o-d!!!" Only white corpuscles circulated in our veins at that moment; we later discovered, however, the vibrating catsup bottle gave one man the clue which saved our skins momentarily. . . . What once catered to the rocking cradle, gave way to the battle's roar. That was before the advent of the Texas Aggies, gentlemen. They sure played Hell with our mess hall.[37]

And these were the fightin' Texas Aggies that would soon play a major role in winning the next war.

Prewar enthusiasm ran high on the campus, and change was constantly under way.

Just prior to the start of classes in 1941, Lieutenant Colonel Watson was hastily reassigned to San Antonio for duty with the Civilian Conservation Corps. The CCC, one cornerstone to FDR's New Deal, was a little-known holding area for newly commissioned officers due to the shortage of regular Army active duty slots. Called on to organize CCC activities and projects, the new young Army reserve lieutenants were expected to "gain valuable experience in command techniques."[38] In the interim, Lt. Joe Davis '30, an administrative officer in the Commandant's Office at the college for nearly a decade, served as temporary Commandant. Cadet leadership in the fall of 1941 was in the hands of Cadet Corps Commander Tom Gillis '42 and the senior class president, Dick Hervey '42.

Once again gridiron fever proved a major diversion for the cadets. The football team again opened the season in San Antonio, this time against Texas A&I. It was not an official Corps trip, so only the band and a large contingent of upperclassmen attended the game. The band, 210 members strong, and "those students who do not have to attend Saturday classes" paraded through the city. Following a fifteen-minute live concert by the band over radio station KTSA, they marched to the Gunter Hotel for Yell Practice. College regulations allowed only one "official" Corps trip, although a second trip was usually granted to the Corps by way of a petition from the senior class to the president. The official trip in 1941 was to TCU on October 18, and the second road trip agreed to was to Houston for the Rice game on November 18.[39]

The new permanent Commandant, Col. Maurice D. Welty, was selected in September, but due to his remote assignment in Newfoundland, he did not arrive on campus until

Military Walk in the 1940s, looking north toward Sbisa Mess Hall. Following the destruction of the original mess hall by fire on November 11, 1911, Sbisa Hall, was completed in 1912. Courtesy Texas A&M University Archives, Cushing Library

early November. In the meantime, in spite of increasing indications of war the cadets planned an eventful schedule of campus events. The 1941–42 cadet calendar included more than thirty dances and balls—ranging from a dance for each class, to the Cattlemen's Ball, to the senior ring dance. In September the Singing Cadets introduced a new song, "The Twelfth Man," to the cadets at the Tuesday night Yell Practice on the steps of Goodwin Hall.[40] Acceptance was quick and it was decided to make the formal introduction of the new fight song at the Thanksgiving Day clash in late November during half-time dedication ceremonies to those Aggies "now in the service of Uncle Sam." Plans for Bonfire

on the main drill field were coordinated by the yell leaders, and aging campus mascot Reveille celebrated her tenth anniversary in Aggieland. Charlie Babcock '43, a feature writer for the *Batt,* noted, "Rev eats where and when she feels like it." Another highlight of the *Battalion* was the cartoon strip depicting day-to-day cadet life—"Fish Blotto"—drawn by Pete Tumlinson '42.[41]

However, the potential for war reported in the *Batt* and the notion that "the movement of the country toward the European conflict is unmistakable" served as constant reminders to all at Texas A&M.[42] Civilian defense training and drill became a regular part of the daily routine. At every turn cadets were reminded of the likelihood of war. Shortages of basic materials for the chemistry labs, construction materials for the new dorms, gasoline, and paper napkins in the mess hall became more noticeable.[43] And the price of the Aggie senior ring jumped $20 due to a new "defense tax." Constant reminders of the "shortage of materials for the national defense" became common, as the mess hall requested all Aggies to return hundreds of missing spoons taken from the mess hall. "Save our Spoons," Aggies were warned, or you will be "forced to eat without certain utensils!" The spoons, most of which were never returned, would soon be a minor issue.[44]

BEAT THE HELL OUT'A JAPAN

Early on Sunday morning, December 7, 1941, a Japanese air attack on the U.S. Naval fleet docked at Pearl Harbor ended months of rumor and speculation—America and the Aggies were at war. Ever mindful of the impact created by rapid mobilization during the First World War, Texas A&M administrators had done as much as possible to prepare the college for another all-out effort. Additional technical and science courses were added to the curriculum. A&M contacted the War Department to add more contracts for juniors and seniors. In 1939–40, Texas A&M produced 919 new first lieutenants. Yet no one was really prepared for the intense impact it would have on the cadets, faculty, staff, and alumni. A&M in early January, 1942, would once again be transformed into an extensive military training camp. The normal academic structure, class schedules, graduations, and requirements—even final exams—would all change or be eliminated. Including the 535

members commissioned out of the Class of 1941, more than 6,000 former students "in the commissioned reserve Army" were already on active duty; 4,842 Aggie cadets were already prequalified and enrolled in military courses. Once again on December 8, the A&M administration and faculty pledged their full support of the national war effort; this time their message to Washington was accompanied by a special resolution from the Corps of Cadets signed by Corps Commander Tom Gillis and directed to President Roosevelt.[45]

Knowing full well that war meant a complete reorganization in the priorities of the college, President Walton quickly addressed the repercussions of mobilization. Walton prepared a personal letter to the parents of every cadet on January 30, 1942, expressing his deep concern and noting in part:

Since your boy resumed his studies in September momentous events have transpired which vitally affect this institution. The Congress of the United States has lowered the selective service age limits, and unfriendly nations have launched a most violent and treacherous attack upon our country.

Both actions have deeply stirred the emotions of our boys, and the writer knows it has caused you inconsolable worry. Fortunately these events have solidified the people in their determination to give their utmost to the defense of our Nation and to the ultimate victory of the allied cause. . . . In view of the passing need, the College has completely changed its schedule and is now on a year-round basis of educational endeavor, with a stricter military discipline in effect than ever before. . . . programs are formulated which will be best for the boy, and will best serve the Nation.[46]

Colonel Welty, on campus for less than a month, proved very effective in dealing with the cadets and administrators at this time of uncertainty and crisis. Knowing that the college would be swiftly called upon to provide as many officers as possible, both President Walton and Colonel Welty urged the cadets to continue their studies and stay in school as long as possible. They vowed to work to ensure that any cadet who was drafted would receive credit for work completed, but no such credit would be offered if a cadet quit without an official notice of induction.[47] Concluding his meeting with the Corps in Sbisa shortly after the U.S. Congress declared war, the Commandant urged the seniors to help maintain order and calm within the ranks with these remarks: "The U.S. was struck an unfair blow, but we will gather up steam and retaliate in such a fashion that everything will be all right in the end, and we will emerge victorious."[48] Cadet sentiment concurred in a December 9 *Battalion* editorial—"Army, Stand Ready!"

1942—FULL WAR FOOTING

The Corps of Cadets returned after a brief Christmas break to a vastly different campus. The normal two-semester schedule was scrapped for an accelerated year-round program "for the duration of the war emergency."[49] The Army wanted more new lieutenants, so graduation of the Class of '42 was moved up to May 16. Once graduation and commissioning were completed in Guion Hall, most of the new lieutenants said good-bye and went directly to the College Station train depot—en route to their first Army assignment. For the next three years,

as during the World War I experience, A&M abandoned most routine college practices. The A&M Board of Directors quickly approved a schedule for the first year of three sixteen-week semesters (I. January 26–May 16, 1942; II. June 1–September 19, 1942; III. September 28, 1942–January 23, 1943), six days a week with no holidays, and only a week at Christmas and between sessions. Under this accelerated wartime program it was possible for a new entering freshman to complete a degree in two years and eight months.[50]

President Walton stressed that the new plan ensured that there would be no "idle" time.[51] Once again new time schedules, stricter discipline, closely monitored call-to-quarters, and the intensified military training were viewed by some to be an attack on certain so-called traditions such as the sophomore role of disciplining the fish and having them perform "fish services" or personal errands on call. Use of the "board"—the paddling of underclassmen—was declared off limits. These new wartime rules, termed the "New Order" by the cadets, required juniors and seniors to pick up their own laundry and clean their rooms without the help of the fish. The dissatisfaction with the changes was quickly overshadowed by even more sweeping schedule changes on campus. The *Battalion* stressed that the Aggies were about to start a "new and unknown life" and in so doing "the Aggies are making every possible attempt to get into the best physical and mental shape to 'beat the hell out of the Japs.'"[52]

In February, the War Department announced that a broad number of service programs to teach enlisted personnel technical skills would begin in March. Even though the college lost more than 300 faculty and staff to the initial wave of inductions and volunteer enlistments, the regular A&M faculty taught most of the courses. Both the U.S. Navy or "Bluejackets," as the cadets called them, and the U.S. Army established programs, followed by the U.S. Marines. The Army Air Corps conducted a civil pilot training program at Easterwood Airport. Many of the new arrivals were perplexed by the Aggies and their customs. Technical courses, generally one semester long, were taught in radio operations and repair, mechanics, the basics of the signal corps, and aviation mechanics. Interactions between the cadets and the military trainees were fairly smooth, with an occasional problem such as when the Navy recruits rooted for TCU instead of A&M at Kyle Field. The clash of cultures was soon overcome as events around the world placed demands on all in training at A&M and across the nation.[53]

In March, the quartermaster corps and the ordnance branch units were added to the ROTC program at A&M. A total of 200 new ROTC contracts—in reality, new second lieutenants slots—were made available to the college. The Army training staff on campus was increased to 27 officers, 16 noncommissioned officers, and 35 privates. With these additions, Texas A&M became the only university in the nation to offer training and commissions in nine branches of the regular Army: field artillery, infantry, engineers, chemical warfare, coastal artillery, cavalry, signal corps, quartermaster corps, and ordnance.[54]

DISPATCHES FROM THE FRONT

Given the large scale of training and officer production at Texas A&M, the Aggies on active duty had an immediate impact on the war effort. In a featured editorial, the *Dallas Morning News* noted, "Texas A&M has attained an eminent position as a training ground for defense leaders. . . . it has all nine branches of the services represented . . . and the foresight in its preparations is one reason why the names of Texans appear so often in dispatches from the fronts."[55] And the dispatches flowed in. Aggies were center stage in the first great battles in the Pacific Theater, with most of the attention directed toward the defense of the Philippines and its two prime lines of battle—the Peninsula of Bataan and the Island of Corregidor in the mouth of Manila Bay. Following the fall of the Philippine mainland, all eyes were directed to the "Rock," as Corregidor was known to the defenders, under the command of Gen. George Moore '08. The island fortress was the last line of defense. Gen. Douglas MacArthur left the island on March 11, 1942, directing the former A&M Commandant to "Hold Corregidor." Surrounded and outnumbered by the enemy, low on ammunition and medical supplies, the garrison was hopeful that a relief force would soon arrive. By April, the situation was critical as General Moore and many of the handpicked group of Aggies he had assigned to his command observed the anniversary of the Battle of San Jacinto on April 21. Moore had spoken via shortwave radio to the crowd at the November, 1941, Bonfire; however, this time, news of the Aggie Muster electrified concerned Americans not only in Texas but nationwide. The *Houston Press* reported, "Word of the gathering rang 'round the world." U.S. forces could not relieve Moore, and the island was captured by the Japanese invaders in early May. All defenders were killed or taken prisoner.[56] In response to the contribution of the Aggie defenders on Corregidor, General MacArthur issued the following statement:

"Texas A and M is writing its own military history in the blood of its graduates not only in the Philippines Campaign but on the active fronts of the Southwest Pacific. Texans daily emblazon the record with outstanding feats of courage on land, on the sea, and in the air. No name stands out more brilliantly than the heroic defender of Corregidor, General George F. Moore. Whenever I see a Texas man in my command I have a feeling of confidence."[57]

Aggie Muster would be forever linked with the events on Corregidor, as Association of Former Students Executive Secretary E. E. McQuillen established a network of Muster observances worldwide annually on San Jacinto Day. Meanwhile, during the balance of 1942 thousands of Aggies, mostly recent graduates or draftees, were in action in all branches of the armed forces. Maj. John A. Hilger '32, second in command to Col. Jimmy Doolittle, along with former cadets Robert M. Gray '41, William M. Fitzhugh '36, and Glen C. Roloson '40, bombed Tokyo for the first time on April 18 in a secret mission using land-based B-25 bombers launched from the aircraft carrier USS *Hornet*.[58] All were awarded the Distinguished Flying Cross. In early June,

The 1946 Aggie Muster on Corregidor Island in Manila Bay. The Muster was held at the mouth of Malinta Tunnel, headquarters of Gen. George Moore '08 during 1941–42. Courtesy Texas A&M University Archives, Cushing Library

Ensign George H. Gay '40, the sole survivor of Torpedo Squadron Eight during the Battle of Midway, was shot down and spent twenty-four hours floating amidst the battle, the only eyewitness to the U.S. Navy's first major victory over Japanese sea forces. Hospitalized with two wounds, Adm. Chester Nimitz personally decorated Gay for his actions, noting that "there are many Texans who, like Gay, are tough and full of fight."[59] In July, 1942, Colonel Hilger and Ensign Gay returned to Aggieland to receive a hero's welcome and to address the Corps.[60]

Stories of Aggies in action filled the Texas and national press as well as columns of the *Texas Aggie, Time,* and *Life.* On-campus training was further intensified by Colonel Welty and his staff as the War Department demanded more officers. The Class of '42 became the last class for a number of years to have a somewhat traditional, albeit early, graduation, final review, and commissioning on May 16, 1942. All the 605 new Aggie lieutenants, one graduate recalls, "finished on Saturday and reported for duty on Monday morning!"[61] Enrollment remained steady at about 6,500, yet there was an ongoing shortage of teachers and staff. The draft inductions and those who volunteered increasingly took more personnel away from the college.

Cadets were subject to wartime rationing regulations that resulted in a shortage of brass uniform insignia, rubber-soled shoes, and sugar in the mess halls.[62] Corps trips in the fall of '42 were canceled due to fuel rationing and a shortage of trains; however, the Aggie Band was allowed to travel "space available" (i.e., hitchhiking!) to out-of-town games but only on a voluntary basis. Notwithstanding, Colonel Welty reported that by fall, 1942, 1,364 cadets were in the advanced course

and only months from commissioning. In a feature article issued statewide, the Commandant explained the role and duty of Texas A&M and its cadets if called to active duty: "That the Reserve Officers' Training Corps has a distinct place and is a valuable link in our scheme of national defense cannot be denied. . . . A&M has and will continue to surpass all other educational institutions in this respect without lowering its traditional high standards." The rate of mobilization or call to duty concerned Colonel Welty and administrators on campus regarding the wartime role of ROTC. Some feared that the campus could become nothing more than an Army training site, void of any academic purpose. By October, 1942, more than 9,000 Aggies were on active duty in the armed services.[63]

HOLLYWOOD COMES TO AGGIELAND

In the midst of all the campus changes and wartime preparations, a Hollywood production company from Universal Pictures headed by producer Walter Wanger arrived in Aggieland in 1942 to film a movie, *We've Never Been Licked,* billed as depicting the traditions and history of Texas A&M. The notoriety of the college and its alumni in the national press had caught the attention of the producer. The all-American campus was a perfect location to film. A true novelty to both cadets and faculty, the production caused both excitement and annoyances.[64]

Classes were canceled and during filming the Corps staged reviews on the main drill field and formed the Block "T" in Kyle Field, with each shooting session lasting for hours.

Cadets and faculty as well as coeds from Texas State College for Women were used as "extras." With the backdrop of A&M campus life, the movie tells the story of a cadet, Brad Craig (played by Richard Quine), and his experiences as a new cadet prior to the war in the Pacific. Though panned by most East Coast movie critics upon its release in mid-1943—like many "war movies" made during this period—the movie was a tremendous hit on the A&M campus. The story line aside, the movie provided an excellent look at the Corps of Cadets and campus that few outside of Texas had ever seen. Unique film footage of the Bonfire in November, 1942, Yell Practice, cadets in Sbisa Dining Hall, cadets reciting campusology, the Aggie fight songs, and Reveille I and stand-in give an important glimpse of the campus and Corps life in the early 1940s. "It is much better than the Corps expected," recalled Head Yell Leader John M. "Jack" Knox '46. "The story was woven in very well and typifies the real Aggieland. It is a picture to be proud of and . . . to see what lies behind Aggie tradition," he said. While big city reviewers in New York and Hollywood may have criticized the movie, all had unqualified praise for the Texas Aggies' contribution in the war.[65]

The excitement of moviemaking on the campus did not lessen the seriousness of the nationwide mobilization for war. Somewhat like the SATC program established during World War I, the War Department announced the formation of the Army Specialization Training Program (ASTP) in late 1942. Along with the ASTP, an Enlisted Reserve Corps was also established; taken together they bypassed the normal operation of the ROTC system to produce officers.[66] Furthermore, in spite of efforts by the college administration and the Association of Former Students to obtain a clear understanding of just how eligibility and education deferments applied to the cadets at A&M, the primary trigger for induction during the balance of the war hinged on when a cadet turned eighteen years of age. Once again cadets were preoccupied with rumors of when the next large-scale call-up would be announced. Speculation ended in late 1942 when the entire junior and senior class, totaling 1,306 cadets, were called to active duty and assigned to Army units or the ASTP. Each new inductee in the ASTP would receive monthly an enlisted man's pay as well as be issued regular Army uniforms and equipment, en route to either officer candidate school or directly to a replacement center for deployment overseas.[67]

THE CLASS OF '47

President Walton and the Commandant throughout the war endeavored to squelch the numerous rumors and misinformation, but they often found this task difficult because firm information was so sparse from the War Department on the process by which cadets would be called to active duty. They were, however, repeatedly told by the Army to keep training. Walton attempted to inform parents on how the War Manpower Commission was setting priorities and enlistment timing, yet his letters often proved only a short-term comfort given the continually changing demands of the War Department.[68] Efforts to streamline the mobilization only caused more confusion throughout much of 1943. Dr. Walton informed the A&M Board in mid-May that the Army had officially notified him that the advanced (junior and senior) ROTC was "discontinued for the duration of the war." In mid-1943, the number of ROTC branches of instruction at Texas A&M was reduced from nine to seven—eliminating the quartermaster corps and ordnance. Total cadet enrollment (mostly fish and sophomores) in the "Corps" in June, 1943, not counting the 6,000 enlisted personnel in the various auxiliary Army and Navy military programs, was 1,655.[69]

New fish cadets entering the Corps of Cadets in the fall of 1943 were told to be ready for induction when they turned eighteen. Total fall enrollment was only about 2,100. The 1,300 fish in the Class of 1947 would have a unique experience at Texas A&M. If not volunteered or deferred, each new Aggie was called to active duty when he reached draft age. Some were able to finish one, two, or three semesters. The Class of '47 had no "formal" graduation or final review because most members of the class were called to active duty and finished their education and graduation (most as civilian veterans) upon their return from the war. Their A&M graduations would be spread from 1948 to 1952. Notwithstanding, this disruption to the Class of '47—those fish that entered in the fall of 1943—along with all wartime former students, would result in an uncommonly strong bond and identity of comradeship through the postwar years that followed.[70]

With three trimesters replacing the traditional two semesters, there was no real beginning or ending of the school year. The traditional highly structured class system was greatly modified. All freshman entered as "frogs" (so named for those cadets that "jumped" into A&M without completing the traditional two-semester freshman academic year) their first trimester, became "fish" the second, and sophomores the third. Those enrolled

in January, 1943, became the first Class of '47 and were sophomores by the time their fellow '47ers entered as frogs in September. Duke Hobbs '47, a fish during this period recalled the dynamics of campus life in wartime:

By the fall of 1943, the Corps of Cadets consisted of freshman and sophomores in the Class of '47, underage juniors in the Class of '46 and a handful of seniors. The few seniors were either waiting to be called to active duty or were classified 4-F, physically not qualified to serve. The cadet officers were juniors. Army branches had been eliminated and all units drilled as Infantry; however, cadets were permitted to wear the branch brass of their choice. The Corps lived in Walton Hall and Dorms 14, 15, 16 and 17, the "New, new Area," dorms near Walton, Sbisa Dining Hall and Northgate. The sailors and marines occupied the "New Area" (the Quad) and the ASTP units were assigned to the older dorms in the center of campus. The cadets got along fine with the ASTP, most of whom were in the Class of '45 or '46 and had been Aggie cadets a few months earlier. The sailors in the V-12 training program were a different story. Some had been in other colleges while others had gone into the Navy right out of high school. Most professed to dislike the Army, particularly cadets in Army uniforms. Furthermore, they often supported Aggie rivals at Kyle Field!

Navy trainees were seated in the horseshoe of Kyle Field for football games. The Cadet Corps traditionally was assigned seats by class with the freshmen occupying the front rows, sophomores behind them, followed by the juniors and few seniors in the best seats looking down on all the others. Just before the t.u. game began on

Thanksgiving Day, 1943, a group of sailors unfurled a huge banner across the top of the horseshoe. It read, "Beat the hell outa A&M!" An upperclassman yelled, "Freshmen, take it down!" The Corps of Cadets, probably 1,500 strong, poured over the railing and across Kyle Field to the horseshoe. The U.S. Navy rose to meet the challenge. Just before the cadets reached the far railing, the Aggie Band struck up the "Star Spangled Banner." While the cadets and the sailors stood at attention, the "Kampus Kops" removed the offending banner. Thus began the game for the 1943 Southwest Conference Championship.[71]

Cadets, faculty, and staff continued to leave for duty—only those below draft age, with an approved occupational deferment, or classified "4-F" as physically unfit for duty remained on campus.[72] Fernando Zuniga '47 of Laredo recalled, "The military training at that time was really 'military!' We were of course in the middle of the war and the Army tac officers were tough."[73] The constant turnover of cadets took its toll on most traditional functions of the Corps. For example, membership in the Aggie Band was open to *any* cadet in *any* branch as the band dropped from its all-time high of 210 members in 1942 to barely 60 in 1943. However, the biggest surprise in the fall of '43 was the success of the football team. The youngsters left on campus—primarily fish and sophomores—became known as the "Kiddie Korps." The average age of Coach Homer Norton's 1943 cadet football team was seventeen. Nonetheless, their 7-2-1 record was good enough for a post-season trip to the Orange Bowl.[74]

THE WALTON TURMOIL

The enthusiasm over the 1943 gridiron success could not hide the tremendous jolt that war mobilization had on Texas A&M. In the summer of 1943 the A&M Board of Directors faced declining enrollment, War Department demands to train more and more men, and the task of improving the college as a "technological institution" of higher learning. The board expressed full support for the war effort but gradually began to lose confidence in President Walton's ability to manage the change. In the midst of this almost complete disruption to the normal academic schedule, a clash of personalities erupted. The turmoil overshadowed the magnitude of A&M's adjustments that would be necessary to fulfill war demands. Walton's endorsements from board members continued to dwindle. The board in August voted five to three not to retain him as president—in essence firing him. At first no direct reasons were given, and he was even lauded in the press for his contributions to the college. Though replaced by an interim president, Frank C. Bolton, issues concerning the Walton dismissal proved much deeper than mere enrollment problems and adjustments to wartime training. That his departure was orchestrated at this junction of turmoil and wartime pressure belies the underlying seriousness of the actions taken by the board.[75]

The seemingly sudden dismissal of Walton was quickly questioned by many. At issue was the future of the Corps of Cadets and the military tradition within Texas A&M's scheme of higher education in a college environment. Numerous charges began to surface concerning his role in the administration of the college. Walton was vocal in his belief that the board placed too much emphasis on the Corps of Cadets and military training. In response, the board members circulated an unofficial story that he had been unable to curb hazing in the Corps, thus raising the recurring debate over compulsory military training, all-male versus coeducation, and the status of such nonmilitary items as civilian graduate students, new funding for research, and a review of the liberal arts curriculum. At the heart of the turmoil was a call for the assessment of the future mission and role of the college. In this situation the Corps was to some degree both villain and hero. The debate, dating from the founding of the college, once again revolved around the definition and emphasis to be placed on the Morrill mandate to provide "military tactics." As the debate escalated, the Texas State Senate General Investigation Committee, headed by Sen. Houghton Brownlee, voted to inquire into the charges leveled by all parties. An internal review in early March, 1944, ended quietly only to be reopened following new allegations in the form of a formal hearing in Austin in April.[76]

It may seem odd that at the height of the war, Texas A&M, of all institutions—one that provided so many men for active duty— would become embroiled in turmoil. At issue was the postwar role of the Corps of Cadets and concern over A&M's time-honored traditions. Texas A&M *Centennial History* historian Henry Dethloff concluded that

> Walton appears to have gotten caught on the horns of [an] Aggie dilemma—"interests" that involved rather intangible attitudes, rather than literal programs:
> The attitudes had to do mainly with the Corps of Cadets, and coeducation at Texas A&M. The Board of Directors in 1943–1944

wanted to perpetuate the all-male military tradition of the College, while at the same time renovating and modernizing the academic program. The Corps of Cadets and many former students wanted to preserve the tradition, if necessary at the expense of academic modernization. Some former students, faculty, and administrators wanted to modernize, and, if necessary, at the expense of traditions. No interest group had genuinely clear-cut concepts about what it would or could not do. The question became whether substantial academic changes could be effected without altering long-established procedures and traditions.[77]

ABOVE AND BEYOND

The balance of the war during 1944–45 witnessed an easing of confusion over the draft guidelines, if for no other reason than all fit-to-serve were subject to the draft.[78] While the Corps numbered between 1,600 to 2,100 during this period, training by the special units of the U.S. Army, Navy, Marines, and Air Corps continued uninterrupted. The ASTP contingent numbered 2,500. Between February, 1943, and June, 1944, the Army Air Corps trained 4,092 new airmen and the Marine trainees numbered 2,380. While the cadets of the Corps often complained that the visiting trainees rooted for the opposing football team and shunned traditional Aggie hitchhiking etiquette by "upstreaming," all went as well as could be expected. In all, over 15,000 noncadet military trainees received wartime instruction at Texas A&M.[79]

However, many cadets were concerned by what they felt was an erosion of Aggie traditions due to all the turmoil and changes on campus. In May, 1944, cadets protested the reduction of upperclass privileges and significant changes in the demerit system by camping en masse overnight on the lawn in front of the Academic Building. The next morning they calmly submitted their concerns to the administration and went to breakfast and class. These concerns were jointly reviewed and modifications made to the demerit system and the late-night "lights out" policy for upperclassmen. Further signaling both change and the end of an era, the cavalry barns, no longer in use, were ordered taken down. With their social events dramatically curtailed in 1944, the cadets in September made a request that when Germany was defeated they wanted to have a large victory party.[80]

News of such trivial items as the administrative infighting and campus sit-ins must have seemed strange to Aggies stationed all over the world in every theater of operation and branch of service. By 1944, the full magnitude of the Aggie impact on the war effort was becoming known. Like World War I, Texas A&M far outdistanced any other college or university in supplying active duty officers and men to the war. One measure of their worldwide presence was the incredible number of Aggie Musters held by former students on San Jacinto Day in 1943, 1944, and 1945. Muster coordinator E. E. McQuillen noted that the more than six hundred 1944 Muster programs "would follow the sun . . . through the South Pacific, Australia, India, the Mediterranean and England."[81] Holding Roll Call for those who were absent, singing the "Aggie War Hymn," telling old college stories, and leading a Yell Practice for a fleeting moment allowed a brief escape from the war. "Where an empire is to be built," Maj. Gen. Harry H. Johnson '17 remarked, addressing a cheering crowd of Aggies near Naples, Italy, in April, 1944, "and wherever an enemy is to be knocked out, we Aggies will do our part." Reporting from the Naples, Italy, Muster, a staff correspondent for the *Dallas Morning News* commented on the gathering:

Jerry [the Germans] must have heard them when someone put the men through some Aggie cheers. A few minutes later the ack-ack began chunking steel into the low-cast clouds at unseen enemy planes.

But if Jerry had hoped he could break up the celebration he was dead wrong. If anything, the raid, followed by another an hour later, appeared to have been part of the program.

There was more singing, interspersed with more cheering and backslapping. They put a lot of vigor into one of their favorite Thanksgiving Day tunes, Goodbye to Texas University, I believe it is called.[82]

However, all the Muster good cheer could not account for the tremendous wartime contribution of the Aggie former cadets nor the extensive combat losses. McQuillen, executive secretary of the Association of Former Students, estimated that of the nearly 40,000 living Aggies in 1945, more than 20,000 served in the armed services during the war. Of this number 14,000 were officers. The impact and result of decades of training at Texas A&M became very apparent. By war's end, the Aggies had provided 29 men who reached the rank of general officer. As early as 1944 newspapers nationwide began to chronicle the A&M contribution.

A&M men made an indelible mark. Innovation, leadership, tenacity, and an all-out will to win marked those who served. The names of the famous and not so well known fill this roster: Moore and the defenders of Corregidor; the Connally brothers, both aviators—James '32, winner of the DSC, killed in action near Mindanao, and Clem '38, winner of the Navy Cross at the battle of the Coral Sea; Hilger, Hughes, and Gay flying and fighting against great odds; a young Ranger commander, Lt. Col. James Earl Rudder '32, among the first to land and scale the cliffs of Pointe du Hoc at Normandy on D Day; Gen. Andy Bruce '16, bold leader and founder of the tank killers; and an Aggie postal worker turned highly decorated commander and future U.S. congressman—Olin E. "Tiger" Teague '32. Hundreds of young officers and soldiers served like James F. Hollingsworth '40, who fought across North Africa and Italy and spearheaded the attack into Germany— the recipient of three Distinguished Service Crosses, four Distinguished Service Metals, four Silver Stars, three Legions of Merit, three Distinguished Flying Crosses, four Bronze Stars for valor, and six Purple Hearts.

Hollingsworth was destined to be one of the most decorated general officers in the annals of American history. O. P. Weyland '23 covered the Third Army push into Germany, winning the accolade from Gen. George S. Patton as the "best damn general in the Air Corps." And of A&M's ten World War II fighter Aces. Two Aces—Jack Ilfrey '42 in Europe and Jack Landers '42 in the Pacific—were among the earliest flying legends, both scoring their fifth confirmed kill on the same day—December 26, 1942. Entire families were drawn into the war, as witnessed by the Cokinos brothers of Beaumont—Pete '38, Jimmie '40, Mike '43, and Andrew '44. At war's end Aggie Lt. Albert "Buck" Kotzebue '45 was one of the first Americans to cross the Elbe River to link with the Russians on April 25, 1945, during the final assault on Berlin. The list goes on and on. Also remembered were those who did not return— more than 950 Aggies. Former students ranging from the classes of '06 to '48 served, with the Class of '45 alone losing more than 100 members. On every battlefront worldwide

TABLE 6-3
World War II Officer Production

School	Officers	Other Ranks	Total	General Officers	Killed or Died
Texas A&M	14,123	6,106	20,229	29	953
VPI	4,472	2,512	6,984	16	301
Clemson	4,142	2,333	6,475	16	370
Citadel	*	*	6,300	*	279
VMI	3,159	943	4,102	50	182
Norwich	1,270	425	1,711	16	86t
N. Georgia	870	2,081	2,953	0	100
Penn. Mil.	395	518	918	6	37

* Unknown or unavailable
t Total casualties

Sources: *Texas Aggie;* Kraus, *Civilian Military College,* p. 487; VMI Museum; *Charleston Post,* Dec. 7, 1945, p. 12; Land Grant College ROTC Survey, June 3, 1959.

Aggie lives were lost. Special notice was given to the loss of A&M's first two-time All-American football player, Company Commander and Capt. Joe Routt '37, killed in action at the Battle of the Bulge in late 1944.[83]

Particular recognition fell to the seven Aggies who received the nation's highest decoration, the Congressional Medal of Honor. One newspaper referred to these heroes as "These Are the Brave," and brave they were. Four of the recipients received their honor posthumously. Army Air Corps 2nd Lt. Lloyd H. Hughes '43 of Corpus Christi received his award posthumously for bravery in a bombing mission against the refineries at Ploesti, Romania, in August, 1943. Following action at the Anzio beachhead in Italy, Lt. Thomas W. Fowler '43 on May 23, 1944, spearheaded a heroic tank attack near Corano. Surviving this intense action, he was killed ten days later at the head of the drive on Rome. Sgt. George D. Keathley '37 was also decorated for action above and beyond the call of duty in Italy on September 14, 1944, near Mount Altruzzo. When all his unit's officers were killed in action, Keathley assumed command and rallied his company; in the process he was himself mortally wounded. Lts. Turney W. Leonard '42 and Eli L. Whiteley '41 received the Medal of Honor for action in Germany. Both men led attacks in the face of great odds and after being wounded numerous times. Each was able to rally his men to protect their position and prevent defeat. The recommendation for Leonard was written by his company commander, Capt. Marion C. Pugh '41, who called him "the bravest man he ever saw." Leonard survived the battle but was later reported missing in action. Whiteley returned to Texas A&M to become a professor of agronomy. He died

on December 2, 1986, in College Station and was accorded full military honors in a campus ceremony. And two of these highest awards for bravery were given to Aggies in the Pacific. Marine Sgt. William Harrell was personally decorated by President Harry S. Truman at the White House for his actions on Iwo Jima on March 3, 1945, and Army Air Corps Maj. Horace S. Carswell, Jr., was posthumously recognized for his heroic action to save his crew following a bombing raid over Japan.[84]

Jack W. Ilfrey '42 of San Antonio with his P-38, "Happy Jack's Go Buggy," was one of the first Aggie fighter aces of World War II, with eight confirmed enemy kills in North Africa and Europe. Courtesy Jack Ilfrey

TABLE 6-4
Aggie World War II Fighter Aces

	Kills	Theater	Unit	
Jay T. Robbins '40	22	Pacific	8th Gp	Air Corps
Glenn E. Duncan '40	19.5	Europe	353 Gp	Air Corps
David Lee "Tex" Hill '38	17.25	Pacific	AVG*	Air Corps
John W. "Jack" Landers '42	14.5	Pacific	357 Gp	Air Corps
Jack W. Ilfrey '42	8	N. Africa/Europe	1 Gp & 20 Gp	Air Corps
Edwin J. Herman	8	Pacific	VMF-215	Marines
Charles R. Bond, Jr., '49	7.5	Pacific	AVG*	Air Corps
Lewis W. Chick, Jr., '38	6	Europe	324 Gp	Air Corps
William B. Freeman '38	6	Pacific	VMF-121	Marines
Kenneth D. Smith '38	5	Pacific	VF-17/84	Navy

* The AVG (American Volunteer Group or "Flying Tigers") disbanded in early 1942 and reformed as the 23rd Fighter Group.

HOMECOMING

On May 7, 1945, Gen. Dwight D. Eisenhower, Commander SHAFE (Supreme Headquarters Allied Forces in Europe), advised his commanders that as of "zero one four hours Central European Time" notice of the unconditional surrender was given to "cease active operations."[85] The cadets in Aggieland and Aggies and Americans nationwide prepared to celebrate. All attention now turned to Japan, which would surrender in August. A happy nation was eager to return to a less stressful time. At the surrender of the Japanese empire on the deck of the USS *Missouri,* former Aggie Yell Leader, defender of Corregidor, and POW Col. Tom Dooley '35 was a member of the official party to witness the signing of the Japanese surrender documents. And when the first American occupational forces under Aggie command of the First Cavalry Division prepared to move into Tokyo after the surrender, the lead tank was flying the Texas flag and emblazoned with the slogan "We've Never Been Licked."[86] Henry Dethloff in the *Centennial History* notes, "The chronicle of the fighting Texas Aggies will never be fully written: but the sense of achievement and of accomplishment has undergirded the Aggie tradition with a quiet, proud sense of duty and purposefulness."[87]

On campus the tribute to the "last full measure" shown by the Texas Aggies in war took place at the April, 1946, Victory Homecoming Muster. The day-long event, hosted and chaired by Lt. Col. Bill Becker '41, featured a speech by General Eisenhower in the north end of Kyle Field, on the same spot where—nine years earlier in 1937—President Roosevelt had extolled the Aggies to train hard and be prepared.[88] The event was one of Ike's first major presentations since returning to the United States from Europe. The general that Easter Sunday morning was quick to hail the tremendous contribution of the A&M former students and cadets: "No more convincing

Medal of Honor Winners of
World War II

(Drawings by Loraine Blount)

Lieutenant Lloyd H. Hughes '43

Lieutenant Thomas W. Fowler '43

Sergeant George D. Keathley '37

Lieutenant Turney W. Leonard '42

Lieutenant Eli L. Whiteley '41

Sergeant William G. Harrell '43

Major Horace S. Carswell, Jr., '38

Seven Texas Aggies won the Medal of Honor in World War II. They are (left to right, from the top) *Lt. Lloyd H. Hughes '43, Lt. Thomas W. Fowler '43, Sgt. George D. Keathley '37, Lt. Turney W. Leonard '42, Lt. Eli L. Whiteley '41, Sgt. William G. Harrell '43, and Maj. Horace S. Carswell, Jr. '38. Drawings by Loraine Blount, courtesy* Texas Aggie testimony could be given to the manner in which the men of Texas lived up to the ideals and principles inculcated to [in] them during their days on this campus. . . . I can feel only a lasting admiration for Texas A. & M. ROTC. This admiration extends to the individual as well as to the institution that produced you."[89]

The war years brought Texas A&M and its former students worldwide attention. While the response of the college to the national crisis brought national acclaim, it also by 1944 created grave concerns regarding the postwar role of the Corps of Cadets and the enhancement of higher education programs at the institution. No longer would the campus and its programs be a provincial little school. New ideas and peacetime demands to enhance the academic programs were viewed by many on the Board and former students as a threat to the time-honored traditions of Texas A&M and the Corps of Cadets.

Elements of the 1st Cav Division at the city limits of Tokyo as part of the first official entry of U.S. troops. Brig. Gen. Hugh Hoffman is surrounded by a group of Aggies in the lead tank, emblazoned with the Texas flag and motto "We've Never Been Licked." This picture was a gift to Gen. Earl Rudder '32 from Wick Fowler. Courtesy Texas A&M University Archives, Cushing Library

Gen. Dwight D. Eisenhower was the 1946 Homecoming Muster speaker for ceremonies held in the north end of Kyle Field. Pictured with Eisenhower are Association Executive Secretary E. E. McQuillen '20 and Medal of Honor recipient Capt. Eli L. Whiteley '41. Courtesy Sam Houston Sanders Corps Center

In the months and years following the 1946 "Victory Homecoming" celebration, Texas A&M braced for the largest growth in student enrollment in the college's history, primarily due to the return of thousands of veterans. The Aggie Corps of Cadets would once again be the focal point for those who wished to transform the college into a modern, postwar academic institution.

7

JUDGMENT DAY IN AGGIELAND

The Corps is not what it used to be. In fact—it never has been.

Haydon L. Boatner, Commandant of Cadets, September, 1948

The men of Texas A&M can stand up to any men in the world and compare favorably their education and training for leadership—leadership in the pursuits of peace, and if it comes to war, leadership in battle.

Omar Bradley, General of the Army, 1951

In humble reverence this building is dedicated to those men of A. and M. who gave their lives in defense of our country. Here is enshrined, in spirit and in bronze enduring tribute to their valor and to their deep devotion. Here their memory shall remain forever fresh, their sacrifices shall not be forgotten.

Memorial Student Center dedication, April 21, 1951

ALTHOUGH THE WAR was won, few at Texas A&M were prepared for the postwar impact on the campus and the Corps of Cadets. The steady growth of the Corps from 1938 to 1942 to a strength of more than 6,500, followed by the dramatic reduction in enrollment to a "bunch of 17 year olds" numbering fewer than 1,900 in 1944, was not nearly the shock to the campus as the more than 9,000 cadets and veterans who invaded Aggieland by 1946. The new mix of students would mark the first time in the college's history that the Corps of Cadets did not completely dominate the campus. The large majority of returning vets had been former cadets, but their perspective and wartime experiences gained in all four corners of the world forever changed the direction of the college, its academic programs, and the Corps. With their education financed by the Serviceman's Readjustment Act of 1944—better known as the GI Bill—these veterans returned with an intense sense of competition and purpose and were described as "serious players" by Richard Alterman '49, editor of the *Battalion*. They returned to a campus that had no facilities to house them or—for many—their new wives. Furthermore, A&M's outdated academic system used teaching methods that had not kept pace with the technological advances realized during the war. Most importantly, the Corps of Cadets in a peacetime environment searched for leadership and an identity.[1]

Most veterans of World War II did not intend to return to A&M as cadets. The Corps held little attraction to those hardened by war and supported financially by the GI Bill. Although many returning vets were former cadets, the scope of their interest did not include involvement with the on-campus cadets. However, those vets who returned to A&M returned to a campus that had been controlled by and centered around the Corps of Cadets for more than seventy years. Thus tensions surfaced between the cadets and vets, as the Corps, supported by the nostalgic alumni, looked to reclaim its primary role on campus. The vets had a different view.

For the most part, the Corps and the returning veterans were not much concerned with the feud between former president Walton and the Board of Directors that had escalated in charges and countercharges over a thirty-six-month period from 1943 to 46. However, this lingering dispute and debate over the postwar direction of Texas A&M

would have a direct impact on changes in the organization of the Corps, the drafting of new rules and regulations, and the separation of each new freshman class to a remote site, away from the main campus. The Texas Legislature threatened investigations that would, over time, embroil many factions—students, veterans, alumni, administration, Board, parents, politicians, and vocal citizens of Texas—as the college worked to redefine its future as an academic institution as well as determine the role that on-campus all-male military training would play.[2]

POSTWAR PLANNING

Planning for the postwar era at Texas A&M had started in early 1943, during what became the last months of the Walton administration. In addition to surveying and assessing local feedback on the transition of the college, Walton, in his nineteenth year as president, solicited data and ideas from many of the leading educators of the period. These efforts resulted in drafting a formal set of documents entitled the "The A. and M. College of Texas Committee on Postwar Planning and Policy." This document was actually a series of reports and estimates on all facets of the college: enrollment, student life, academic concerns, and facilities. The section on student life would have the greatest bearing on the future of the Corps of Cadets.[3]

In the section "Cadet System of Military Training," it was recommended that the Corps of Cadets be continued, "even if ROTC is abolished." There was concern that in the postwar federal military budget cutbacks, ROTC would be significantly curtailed or canceled altogether. A great deal of attention was placed on efforts to reduce hazing; thus, the administration was advised to consider housing freshman "apart from other students." For the returning veterans, it was recommended that it would be acceptable for them to be in the Corps if they desired, yet they should be housed "apart from other cadets" except for required military formations and reviews. Those veterans who elected not to be members of the Corps would be allowed to wear civilian clothes anywhere on the campus as well as be permitted to eat in the mess halls. Concern over postwar discipline was a key element for management of the student body. It was suggested that the Commandant's Office be placed under the "direction of College authority as distinct from Army authority." Also the day-to-day interaction by administrators with cadets and veterans concerning the disciplinary policy would also need revamping. Discipline had historically been administered by the Office of the Commandant. Thus the veterans who did not elect to return to the Corps were not considered under the control of the Commandant. To address this change the administration created the "Student Environment Control Committee," which evolved into the Office of Student Life to address specifically the large influx of non-Corps students. This new department was administered by a director with a "rank comparable to that of Dean of Men."[4] Furthermore, it was suggested that as a result of the new department the use of regular Army personnel as "tactical officers" in matters of discipline be discontinued.

These recommendations, drafted by the Walton administration, were submitted to the A&M Board for review at about the time he was voted out of office, somewhat supporting—if the plans were greeted as detrimental to the Corps—Walton's claim that the Board was anti-Walton and too pro-Corps. In the minds of many pro-Corps advocates, changes such as the ones outlined could and would alter the scope and importance of the Corps of Cadets as they had known it.[5] The Board and Walton found themselves at a major crossroads in the college's seven-decade history—how to best enhance the overall academic postwar programs of the college while at the same time not hampering the traditional role of the Corps of Cadets. The Corps of 2,000 "yelling" cadets would prove to be recalcitrant. Enthusiastic about wartime restrictions being lifted, the 1946 *Longhorn* offered some indirect insight into the perspective of cadet priorities: "No matter what time of the year it is, weekends mean time to travel. Someone once remarked that if there was a Corps Trip to Hell some Aggies would leave two days early." Their enthusiasm, pride, and vigor would soon be tested.[6]

By the time Col. Guy S. Meloy, Jr., arrived to become Commandant in the fall of 1946, plans were under way to implement a large part of the Postwar Planning Committee's recommendations. In May, 1945, the outgoing Commandant, Colonel Welty, was instructed to turn over all non-Corps discipline to the new Office of Student Life, headed by John W. "Dough" Rollins '17, former major, along with his assistant, Durward B. Varner '40, former Cadet Corps Commander. In November, 1946, the Office of the Dean of Men was formally created to oversee all student life. The professor of military science and tactics (PMS&T) would still remain the prime contact for cadet discipline, including hazing, referring only college-related infractions to the dean of men. These changes reflected the administration's concern over how best to handle a reduction in hazing as well as manage the needs of the large number of returning veteran "civilian students." Additionally, new emphasis was placed on academics. The Postwar Planning Committee concluded, "Pride (driven by a number of factors) in high academic achievements should be motivated by: Insisting that scholarship hold the position of first importance in the selection of persons for positions carrying 'Corps prestige.'" By late 1946, the veterans outnumbered the cadets 3-to-1, out of a student body of 9,200.[7]

In addition to efforts in reconstituting the Corps of Cadets and placing the college back on as normal a footing as possible, steps were taken to bring back as many of the Aggie customs, traditions, and events that had been reduced or canceled during the war. Corps trips, a full regular Southwest Conference schedule, regularly held Yell Practices, Bonfire, and a more normal class routine of five and one-half days—cadets still attended Saturday morning classes and labs—quickly took hold. Cadets worked to train a new pup dubbed "Rusty," a Reveille look-alike with a white-tipped tail, selected by a vote of the cadets to replace the famed mascot that had died of old age and was buried at the north end of Kyle Field in January, 1944. After two decades as director of the Aggie Band, Col. Richard J. Dunn retired, and direction of the band was placed in the hands of Col. E. Vergne Adams '29 on January 7, 1946.[8] At the 1946 Mother's Day ceremonies the first Gen. George F. Moore trophy, named for A&M's first general officer, was awarded to the best cadet company based on military proficiency, grades, athletics, and general excellence—I Com-

pany, commanded by Robert B. MacCallum '47. Edward D. Brandt, Jr., '48 of Houston was selected Cadet Colonel of the Corps for 1946–47 and, for the first time, Corps members used a center pole in order to build a bigger Thanksgiving Bonfire. And upon receiving Board approval, Colonel Meloy reactivated the Ross Volunteers beginning February 1, 1947.[9]

After the departure of Dr. Walton, the Board selected Frank C. Bolton as acting president for nine months. Effective May 27, 1944, engineer and University of Texas alumnus Gibb Gilchrist was named president of the college. No stranger to A&M and fully aware of the importance and role of the Corps of Cadets, Gilchrist recommended that a four-year curriculum leading to a degree in military science be established and that an application be made to bring Naval ROTC to the campus. The military science degree plan and major were approved in November, 1944, and appeared in the official *Bulletin* or catalogue announcing the 1946–47 session. However, no degree in military science was ever conferred. Colonel Meloy in August, 1946, announced the new requirements for the advanced course in ROTC. The reinstatement of the advanced reserve officer program—halted in May, 1943, for the last two years of the war—now included two sets of requirements: first, guidelines for cadets with no prior military experience; and second, special criteria for veteran candidates.[10] The veterans would be given credit for any ROTC training completed prior to the war as well as a waiver for the basic course in ROTC if they had "twelve months honorable service" in any branch of the armed forces. All students enrolled in military science either as an elective or with a contract "must wear the

The Gen. George F. Moore '08 trophy was created in 1945 to recognize the outstanding cadet unit of the year. I Company, commanded by Cadet Capt. Robert B. MacCallum '47 (center under flag), *was the first unit to receive the award. Courtesy Robert MacCallum*

prescribed uniform between Reveille and Retreat."[11] The secretary of the Navy rejected Texas A&M's applications to introduce NROTC in 1945 and again in 1949 and 1968, on the grounds that there were ample programs to produce Naval officers. NROTC finally came to A&M in 1972.[12]

Fully aware of the need to expand the scope of the academic programs at Texas A&M, Gilchrist would prove the postwar catalyst to provide a new vision for the research and academic programs at A&M. His plan of action included the Corps, but it

required changes in a number of entrenched cadet customs and practices that he thought detracted from the image of the institution. Notwithstanding, the administration in its official statement on the "Objectives of the A&M College of Texas," included a positive assertion on its support of military training at the college: "A strong and effective system of military training for male students of the main College or of any of its branches was compulsory for all except those to whom credit may be granted for active military experience or equivalent training."[13] However, creating a new image for postwar Texas A&M and developing a progressive operation struck a blow at the heart of what many felt was the bedrock of Aggie tradition.

THE MIDNIGHT MARCH OF '47

One of the most extraordinary episodes in the history of the Corps of Cadets was the assignment of all new freshman cadets to the Bryan Army Air Field in 1947. The base, renamed the "Annex," was obtained for use by the college from the War Department during the postwar deactivation of military facilities in 1946. At first, the Annex was used to house an overflow of more than 700 returning veterans during the 1946–47 school year. Marion T. Harrington '22, chemistry professor and future A&M president, was named an assistant dean and put in charge of the facility some twelve miles northwest of the main campus. In the fall of 1947, all new fish, about 1,500 in the Class of 1951, were sent to the Annex. John Paul Abbott, an English professor, replaced Harrington to oversee the academic schedule; Colonel Adams organized the fish band; and Army Lt. Col. William Becker '41 was dispatched to be the assistant PMS&T for freshman training and moved into quarters on the old base. "I was in Washington working after the war," recalled Becker, "when President Gibb Gilchrist came up to D.C. and recruited me and a number of Aggies to come back to A&M."[14]

Relocation of the fish to the Annex removed them from daily contact with upperclass cadets on the main campus. Concerned with hazing and the image of the college, the Board of Directors was directly involved in an effort to break old habits and customs that adversely affected retention of the fish class. Postwar change within the Corps first began during the fall of 1946 with upperclassmen being prohibited from having any cadet providing "room service, personal service or run errands." All forms of drills or activities were prohibited unless approved in advance by the Commandant. The new guidelines were quickly protested—by the cadets, veterans who had been former cadets, and numerous former students—as a frontal attack on all Aggie traditions. The veterans attempted to remain out of Corps objections and protest; however, their influence was a factor. Cadets received a tremendous amount of advice from former students and veterans on how "it" used to be done in the Old Army days. The new regulations—along with the revised discipline structure to be administered from the Office of the Dean of Men—sent a signal that more emphasis was to be placed on the civilian, rather than the historical, purely military, or Corps regimen. Cadet reaction was swift.[15]

The document that officially outlined the new Corps regulations, *The Articles of the Corps,* was presented to the cadets on January 21, 1947. In a meeting with Colonel Meloy, this set of rules confirmed the cadets' worst fears. On the next night, the twenty-second, the Corps of Cadets marched en masse in protest to President Gilchrist's campus home across from Sbisa Dining Hall. Colonel Meloy attempted to stop the cadets as they formed-up in the Quad but to no avail. The president greeted the marchers on the front porch in his pajamas and bathrobe.[16] Denouncing the new rules, they told the president that these actions would hamper the effectiveness of cadet officers and reduce their responsibilities for training and developing the fish into Aggies.

It was a fish privilege to shine the statue of Sully each fall. Courtesy Texas A&M University Archives, Cushing Library

JUDGMENT DAY IN AGGIELAND • 173

In a dramatic climax near midnight, more than 200 cadets removed their brass and rank and submitted their resignations as cadet officers by dropping their insignias at the feet of President Gilchrist. The president had very little to say, stating simply, "I accept with regret."[17] After singing the "Spirit of Aggieland," the cadets dispersed.

The uprising then went into its next phase. Returning to the dorm area, an all-night meeting resulted in the drafting of a list of demands to the administration and Board from the seniors to correct the situation. While the Commandant was initially praised, the cadets cast Gilchrist a no-confidence vote. Looking to seek broad public approval for their actions, the seniors released their recommendations to the newspapers statewide:

1. that cadet officers be granted more voice in issuing orders

2. that seniors retain privileges over freshmen

3. that two weeks notice be given for impending changes in regulations

4. that the system of selecting faculty to the panels to "try" students charged with infractions of the rules be changed

5. that the Senior Court be allowed to try minor cases

6. that no regular military officers be housed in the dormitories

7. that the present uniform be retained if desired by a majority vote

8. that seniors have the right to ratify cadet officers nominated by College authorities

9. that threats to deprive officers of advanced ROTC contracts be removed

10. that extra drill be permitted when company commanders desire

11. that freshmen be removed from the Annex to the main campus.[18]

In the wake of the administration's refusal to reinstate all cadet officers, the students sent a second list of demands to the president calling for removal of the discipline officer, Lt. Col. Bennie A. Zinn, and the resignation of Gilchrist. The administration, in turn, hoped to sort out the leaders of the uprising, interview all the cadets involved, and on a case-by-case basis reinstate cadets who met the approval of the Commandant and dean of men. The Commandant advised the cadets that they would only consider each cadet's request on an individual basis. Colonel Meloy began a three-day marathon of interviews and meetings with the cadets, refusing to bargain. Media attention statewide resulted in mail and telegrams supporting Gilchrist and the administration.[19] Citing the historically high dropout rate of 48 percent in each freshman class, the president laid blame on a system of hazing that could not be accepted or perpetuated. The cadets countered with the argument that if the system was so bad, how could the administration account for the splendid war record by those who had been former members of the Cadet Corps system? Furthermore, to take their story to the public, the cadets were able to visit a number of Aggie clubs and Mothers Clubs around the state. Yet, still the situation remained unresolved.[20]

The vast majority of the public and the Board were solidly behind Gilchrist and his new policies—there would be no compromise. Bent on what they felt was a valiant effort to uphold the traditions of the Corps as well as "emulate their predecessors," the cadet protest began to break down in February.[21] At the February Board meeting, the Directors of the college concluded, "hazing is the principal trouble." However, those familiar with the situation knew it was not that simple. The Commandant was instructed by the Board to convene a four-member panel, over which he was to preside to review the enrollment status of those cadets who failed to comply with the directives of the college and Commandant's Office. Appropriate penalties were to be swiftly administered. Although now angered at the Commandant and Cadet Corps Commander Ed Brandt for standing with the administration, the cadets had little recourse. After more than a week of hearings, fifty-seven cadets were reinstated as cadet officers and eighty-seven returned to the ranks as cadet privates.[22]

In March, the cadets were still not completely calmed, casting a no-confidence vote against Corps Commander Brandt and Cadet Lt. Col. Jack Nelson. It was falsely reported that both cadets tendered their resignations to the Commandant. Colonel Meloy would have categorically ignored such a request.[23] For the most part, the veterans and their official campus organization—the Veteran Student Association (VSA)—had stayed out of the early stages of the dispute between the cadets and the president. However, the VSA began to express concern that their interests as students were being harmed and that the solution, due to their low level of confidence in the administration, was to replace the president. In the spring of 1947, the veterans would elevate the dispute to an entirely different level of confrontation as they zeroed in on a completely new set of issues far beyond those raised by the cadets. The vets "got stirred up" and brought into

question the administration's policies, management style, and heavy-handed treatment of the cadets.[24] They felt Gilchrist had proven to be inadequate in fostering the peaceful postwar growth and improvement of the college. What the veterans wanted and what the cadets demanded were largely unrelated. In a mass meeting of the VSA, the veterans posed six questions or concerns dealing with what they felt were substantive issues concerning facilities and the management of the institution: (1) concerns that funding for new classrooms and labs had been mismanaged; (2) questions were raised on the "high price" paid by the college for bottomland along the Brazos River; (3) concern was expressed for what was felt were excessive profits earned by the campus Exchange Store at the expense of the students; (4) why the new $100,000 wind tunnel was not in operation (it was rumored that the new facility was being used to store hay); (5) why President Gilchrist refused to accept the outright gift of Bryan Army Air Field; and (6) why A&M did not have a "definite tenure system" for the faculty.[25]

Unlike the actions and demands of the cadets in January, the veterans' claims against the administration resulted in an immediate response statewide. The "six questions" raised issues that placed the ongoing dispute squarely in the public sector and more importantly prompted action by the Texas House of Representatives to open an investigation into the troubles at A&M. The investigators began in early April, interviewed hundreds of witnesses, and compiled more than 2,000 pages of testimony.[26] Former president T. O. Walton, the State Federation of A&M Mothers Clubs, faculty, cadets, veterans, former students, and legislators were drawn into the process. All factions were able to vent their concerns, questions, and objections to Gilchrist and the Board, including views on hazing and college fiscal policy. For the most part, the administration responded acceptably to the six questions. By Final Review in 1947, the investigation was terminated. The committee concluded that there was no misapplication of funds, that the administration was in compliance with enforcing laws that prohibited hazing, and "having found no evidence of any malfeasance or misfeasance on the part of the administration," the investigation was thus discharged. As for the Corps of Cadets, the report pointedly stated, "While this misconduct was confined to a minority, the entire Corps stands condemned. . . . a great and enthusiastic school spirit is appreciated and is to be encouraged, still it must not be prostituted to the accomplishment of evil or to be nurtured to such extent that it grows progressively worse and assumes a form of fanaticism for its retention."[27]

Robert "Sack" Spoede '48, a veteran returning to the campus who in 1946 chose to pursue a commission as a member of the Corps, was an eyewitness to the march on the president's home as well as the events that followed. In recalling the outcome of the situation, he noted in hindsight, "It gave us no pleasure, but people realized the old times were over, we were going to have civilian students on the campus . . . times change."[28]

Both cadets and veterans tested the will and resolve of the Texas A&M administration. Concern over tradition and time-honored customs in the case of the cadets versus efficiency and accountability on the part of the veterans raised questions on the

future course of the institution. A change of command in the Commandant's Office further indicated modification in the way things used to be.

The strong stand by Colonel Meloy helped calm and resolve the turbulence on the campus. However, in July, 1948, he was transferred to Japan in command of the Nineteenth Infantry Regiment, which became one of the first units to be dispatched to South Korea a few weeks after the North Korean invasion in 1950. Although he had been Commandant at A&M and held a string of staff positions, Meloy had never commanded troops in battle. In command of the delaying action until reinforcements could arrive, Meloy's regiment defended the Taep'yong-ni crossing on the Kum River in mid-July. Heavily outnumbered, the former Commandant was severely wounded trying to rally his men. Left for dead in the confusion, he was finally rescued just as the enemy prepared a counterattack. Awarded the Distinguished Service Cross for heroism, he recovered from his wounds to become commandant of the Army infantry school at Fort Benning in 1953 and was promoted to the rank of major general in 1958 in command of the Fourth U.S. Army, headquartered at Fort Sam Houston. In 1961 he was promoted to four-star rank and appointed commanding general of United Nations forces in Korea. He retired a four-star general in 1961 and settled in Terrell, Texas.[29]

"BULL" BOATNER

One of the most assertive personalities ever to be involved with the Corps of Cadets was Col. Haydon L. Boatner. Prior to arriving at

A&M in the fall of 1948, the former Marine Corps private in World War I, 1924 graduate of West Point (first in the class), and Mandarin Chinese language specialist had compiled a distinguished military career. During World War II he had been a brigadier general and chief of staff under Lt. Gen. Joseph W. "Vinegar Joe" Stilwell in the China-Burma-India (CBI) theater, and for a time was chief of staff of the Chinese army in Ramgarh, India.[30] In the demobilization after the war, he reverted to his permanent rank as colonel. In spite of the fact that many urged him not to take the job of Commandant at Texas A&M, he looked to the opportunity as a challenge. In light of the trouble of '47, the Corps needed and received a new bull-headed style of no-nonsense leadership. While not always liked by the cadets, his brand of organizational skills and demand for excellence proved timely.[31]

Colonel Boatner instantly took a proactive role in dealing with the cadets, the military staff, and the A&M faculty. Having been in the military since he was eighteen years old, he brought a high degree of experience and energy to the Office of the Commandant. In addition to developing the first cadet leadership program and manual to train all cadet officers, he took a direct interest in the fish housed and trained at the Annex. During the first year the "fish regiment" at the Annex was staffed with three assistant PMS&Ts, four tactical officers, twenty-five senior cadet officers, and a few junior cadet noncommissioned officers. By 1949–50, fish rotated leadership positions. Except for a visit to the main campus for an occasional football game, the Fish Regiment was a stand-alone unit. Freshman cadets were placed in leadership positions. To occupy idle time,

cadets organized a fish band, an intramural program, and drill team.

Given the campus overcrowding due to the veterans, Boatner felt it would be only a few years before the Corps would all be back on campus. Boatner, while tough and demanding, was much more at ease with the cadets than his predecessor. And he never missed an opportunity to address the Corps. In September, 1948, the new Commandant began his tenure at A&M with a "talk" (some recalled that when he talked he tended to scream or shout) with the incoming fish dressed in fatigues and in formation on the drill field at the Annex—the Class of '52.[32] "You have heard much of the reputation that our cadets have made," he said. "It has the very finest reputation of any college in the United States. It is not going to be a bed of roses. The Corps is not what it used to be. In fact—it never has been. The Class of '52 is going to be just as good or as bad as you make it. It's your class. You are it. I am starting out with you. And I am willing, unhesitatingly, to risk my reputation on your success."[33]

The Corps of Cadets in 1948–49 numbered 3,768 cadets: 412 seniors, 741 juniors, 915 sophomores, and 1,700 fish. In addition to handling the Fish Regiment of eleven units and a band at the Annex, Boatner worked to streamline the upperclassmen on campus into five Army regiments, one Air Force group, and the band. Five companies in the Sixth Regiment consisted entirely of returning veterans. New units were added, especially in the Air Force group, as the Corps continued to grow during the balance of the decade. In mid-1949, due to the creation of the U.S. Air Force as a separate service, the Pentagon advised Texas A&M that administration of the Corps of Cadets and allocation of "U.S.

TABLE 7-1
Texas A&M Corps of Cadets, 1949–50

Corps Staff: Doyal Avant, Jr. '50 Commanding									
Infantry Regiment		Artillery Regiment			Cav-Eng Regiment		Air Force Group		
1BN	2Bn	1BN	2BN	CA	Cav HQ	Eng HQ	1sqn	2sqn	3sqn
A	D	A	D	A	A trp	A	A flt	D flt	G flt
B	E	B	E	B	B trp	B	B flt	E flt	H flt
C		C			C trp		C flt	F flt	I flt
									K flt

Composite Regiment			Veterans Regiment			Fish Regiment			Aggie Band HQ
Batt	Batt	Batt	1BN	2BN	Jocks	Staff HQ			Staff
ASA	A Qtr	ATrn	A	D	A	1	5	9	Maroon
A	B Qtr	B Trn	B	E Flt	B	2	6	10	White
	A Ord	A Chm	C	A Comp		3	7	11	
				B Comp		4	8	12	
								13	

Fish Regiment housed at Bryan Air Force Base, known as the Annex.

property" will be "conducted in common" to limit duplication and promote economy.[34] By 1950–51, the Corps consisted of eight regimental size units and the band. Furthermore, as the Corps grew, Boatner bore down on the professionalism of his regular Army and Air Force officers assigned to the college. He reminded them that "Military Science is not to be taken perfunctorily . . . at all your drills bear down on the command appearance, posture, and voice . . . require perfection in every little detail."[35] The Commandant encouraged as many senior cadets as possible to qualify for "Distinguished Military Student" or DMS status, a major step toward a regular officer's commission and extended active duty.[36]

The Corps throughout this period not only maintained a "superior" rating "as a pre-eminently outstanding" program from the Department of the Army, but also remained the largest uniformed full-time Corps in the nation. Commissions in eleven branches of the Army and the new Air Force units set A&M apart. Texas A&M, in spite of its on-campus problems during the transition after the war, maintained strong political ties in Washington, D.C., via Congressman Olin E. Teague '32. His ongoing role as a vocal ally of the Corps would prove invaluable. Teague, whose congressional district encompassed the college, worked over a two-decade period to ensure that Texas A&M and particularly the Corps and ROTC programs received the maximum allowable federal support. On many issues, without the congressman's direct involvement funding for the programs at the institution would have diminished. He made sure the new Department of Defense was aware that Texas A&M and its Corps were unique.

One such case, early in his congressional career, was the status, implementation, and impact of the Selective Service Act of 1948 on recruiting at A&M of cadets awaiting ROTC contracts. While not included in the act, Teague, "after repeated inquiries," convinced the Department of the Army to issue orders to the field that provided a special deferment for those enrolled in ROTC. Mindful of the almost complete decimation of the upper classes in 1943–45, this allowed as many cadets as possible to complete training toward a degree and a commission. In advising administrators at A&M on armed forces manpower needs and changes formulated in Washington, Teague warned, "Since *we* were 200 contracts short . . . I urge you to request the largest number you can conceivably use at A&M." The advanced contract quota for Texas A&M in 1948–49 was for 1,029 cadet officer trainees—715 for the Army and 314 for the Air Force. A&M soon received what it needed.[37]

In the wake of Texas A&M's efforts to secure more commissions, the Department of Defense, Congress (due to funding), and the executive branch began a vigorous debate on the immediate postwar role of the armed forces. Sensing that federal funding would diminish, Bryan banker Travis B. Bryan sought the reopening of Bryan Army Air Field by either the active Air Force or the Air National Guard. Citing the benefits of the base to the Air Force, in his lobbying with Teague, to Secretary of the Air Force Stuart Symington, and to the White House was endless. Having just gained use of the Annex for the overflow of new freshmen, President Gilchrist was op-

posed to Bryan's overtures in Washington.[38] While against "planes flying [from the base that would] disturb the students," the A&M administration softened its opposition to support a National Guard fighter squadron at the old base. Gilchrist could remove the fish if needed. He had little defense when President Harry Truman issued an executive order stating, "Every citizen is urged to do his utmost in aiding the development of effective reserve components [National Guard] of our armed forces."[39]

The fish were returned to the main campus in September, 1950, and the Annex reopened as Bryan Air Force Base in early 1951. Elements of the U.S. Air Force training command set up basic pilot training using T-6s, followed by T-28s and T-33s. However, shortly after the Korean War the base was once again closed. In 1953–54, Travis Bryan and the Bryan Chamber of Commerce led a last-ditch effort to have the base considered for the campus of the newly created U.S. Air Force Academy in 1954, yet these efforts faded with the selection of Colorado Springs. The base or annex has since been returned to Texas A&M and designated the Riverside Campus. The disposition of the Annex was not nearly as important to the Corps and Commandant as was the desire of the administration to relocate all freshmen back on the main campus.[40]

Given Boatner's background as a West Pointer and thirty years of active duty experience, he was keen to prepare his military staff to deal with the cadets. Some officers and cadets disliked the short, stocky Commandant, while others remembered him as "firm and fair." Referred to as "the Boat," he was often quick to correct or reprimand; one former student recalled, "His temper and his tactics were feared."[41] Notwithstanding his strong-willed approach, his genuine concern for the cadets ("I am always available") and his insistence on high-quality military training and contact with the cadets were paramount.[42] In January, 1949, the Commandant convened the "Aggie Social Customs and Courtesies" program with cadet membership and the help of the faculty and staff to address the image and perception of the Corps of Cadets to both the campus and outsiders. These efforts were largely responsible for the beginning of the Military Weekend activities each spring to invite supporters of the Corps, distinguished military officers, and representatives from other military institutions to the campus in order to showcase the Aggie Corps of Cadets. In March, 1949, the event spotlighted a dozen visiting general officers and war hero Audie Murphy, who was designated an honorary Cadet Colonel of the Corps.[43]

Boatner's written directives, planning sessions, briefing papers to the Corps leadership, and speeches were frequent and detailed. Hazing and harassment were drastically reduced in the early 1950s.[44] His efforts to define and shape the mission of the Texas A&M Corps of Cadets, following one of the most extensive Corps reorganizations during 1949–50, were a significant challenge given the continued presence of veterans and return of the fish class to the main campus. The Corps in the fall of 1950 numbered 4,320 (65 percent of the total A&M student body). Boatner's commitment to produce quality "citizen-soldier" officers marked all his actions.[45] One of his primary areas of emphasis was leadership:

> Leadership can be developed—it is not an inborn ability. Success in the management of men and in human relations requires the same personal characteristics and trait of character in all responsible individuals. We will be constantly alert for chances to develop latent leadership, initiative and logical reasoning in each student through personal exchange and contacts—constructive and corrective career guidance so our graduates will have had the benefits of the maximum experiences possible in college. Cadet service as private, noncommissioned and commissioned officer will be exploited to provide the experience needed in later life to develop compatibility with and control over other men.[46]

Boatner's emphasis on developing well-trained officers with the basics of leadership served as the formal part of an Aggie cadet's education in the 1950s. A less formal approach to class unity, the buddy system, and "training" on the Quad emerged in the day-to-day dorm life. The ongoing struggle concerning rituals between new fish and upperclassmen became legend in the post–World War II era.

THE GREAT DROWN OUT

Cadets in the Corps often released their tensions in curious forms of controlled combat, such as "drown outs" or "float outs" and the "BAB-O Bomb." For years the cadets had conducted water fights as well as wholesale drown outs of a dorm with water hoses, like the scene featured in the movie *We've Never Been Licked*. Yet, by the 1950s, nothing was more harrowing than the new fish getting even with a mean, over-assertive "pisshead" (sophomore). Direct physical contact between cadets was "officially" prohibited, but those fish who could efficiently execute a late-night raid became legend. One fish account by Joe Fenton '58 highlights this event:

First thing we needed was a screw stick. It was made out of the end of a wooden Coke case. The handle was notched so it could slip over the end of the door knob on the outside and prevent it from opening. Properly placed and jammed it could hold a captive in confinement for days or until released. Possession of one of these tools could get you in a world of hurt so we had certain hiding places like behind the medicine cabinet in our room. The cabinet would slip out of the wall. Next, the stick would be attached to a coat hanger and lowered into the space between the walls and would be removed only when needed.

As the year rolled along, we became more daring, creative, cunning, and clever in our search for the perfect drown out. We practiced with teams. A team was made up of two "Bucketeers," a "BAB-O Bomber," and a "Stickman."

When we thought we were ready, we set the date and time. Our team consisted of fish Maloy—Babo Bomber, fish Purvis—Stickman, fish Coker—Bucketeer #1, and I was Bucketeer #2. It was a Thursday in November, and we were loaded for bear. We slipped down the hall like hostile Indians. With buckets held high, we spilled not a drop. We could hear the soft breathing of the slumbering pissheads. We knew we had to be good. Accuracy was everything. If we slipped, stumbled, or fell, our death would be long and painful. Fish Purvis was shaking like a dog passing a peach seed. Fish Maloy was the cool one. He puffed on the filterless

cigarette that would provide us with a fuse. He flipped an ash, took a puff, and nodded. . . . it was time to bust into the pisshead room.

Fish Coker hit Mr. Silar on the bottom bunk with hydraulic force. He nearly washed him through the wall. While he was gagging and strangling, Carpenter on the top bunk looked down and saw the water coming. He tried to jump it, missed the bed, and came down astraddle Silar's neck and washed all over the room. Old cool Maloy lit the fuse on the Babo Bomb and set it gently on the floor in the middle of the darkened-wet room. The little red fuse sparkled and twinkled brightly. As if in slow motion, fish Coker left, I left, and nervous fish Purvis slammed the door and applied the stick; thus, locking in one room fish Maloy and the two meanest pissheads in the State of Texas. Before they could turn the light on, the room exploded and microscopic granules of scrubbing powder hung in the air. They got the window open and went in search of their tormentors.

Fish Maloy went in the closet and got behind the clothes and stayed very still until the Babo got the best of all of them. When the soap mixed with the water, a slurpy mess was created and you never heard so much slipping, sliding, crashing, and banging in all your born days.

Finally, one of the sophomores made his way to the closet door, which emitted great billowing clods of Babo. When Carpenter opened the closet, fish Maloy hit a brace and shouted as loud as he could, "Howdy Mr. Carpenter, Sir. Howdy, Mr. Salis, Sir!" Later on fish Maloy stated, "I knew we were caught in a drown out, but he couldn't say we didn't speak." The mess we made had to be cleaned up by all the fish in the outfit, but fish Maloy was held captive all night, and he didn't look good the next morning.[47]

KOREA

The campus interclass rivalry between the fish and the upperclass cadets proved petty in the face of the emerging events in the early 1950s. In the spring of 1951, there was a growing concern that Korea would expand into a larger conflict. While the Army once again became concerned over manpower needs, Colonel Boatner launched an all-out effort to ensure that all cadets became more aware of the seriousness of the events in the Far East. Plans for the annual Military Day and Ball were curtailed. Senior cadet P. H. DuVal, Jr., '51 penned the poem "The Last Corps Trip," an instant hit with the cadets. By the 1970s the poem was read as a key part of Bonfire's evening program.

THE LAST CORPS TRIP
P. H. DUVAL '51

It was Judgment Day in Aggieland
And tenseness filled the air;
All knew there was a trip at hand,
But not a soul knew where.

Assembled on the drill field
Was the world-renowned Twelfth Man,
The entire fighting Aggie team
And the famous Aggie Band.

And out in front with Royal Guard
The reviewing party stood;
St. Peter and his angel staff
Were choosing bad from good.

First he surveyed the Aggie team
And in terms of an angel swore,
"By Jove, I do believe I've seen
This gallant group before.

I've seen them play since way back
  when,
And they've always had the grit;
I've seen 'em lose and I've seen 'em win
But I've never seen 'em quit.

No need for us to tarry here
Deciding upon their fates;
'Tis plain as the halo on my head
That they've opened Heaven's gates."

And when the Twelfth Man heard
  this,
They let out a mighty yell
That echoed clear to Heaven
And shook the gates of Hell.

"And what group is this upon the side,"
St. Peter asked his aide,
"That swelled as if to burst with pride
When we our judgment made?"

"Why, sir, that's the Cadet Corps
That's known both far and wide
For backing up their fighting team
Whether they won or lost or tied."

"Well, then," said St. Peter,
"It's very plain to me
That within the realms of Heaven
  They should spend eternity.

And have the Texas Aggie Band
At once commence to play
For their fates too we must decide
Upon this crucial day."

And the drum major so hearing
Slowly raised his hand
And said, "Boys, let's play the Spirit
For the last time in Aggieland."

And the band poured forth the anthem
In notes both bright and clear
And ten thousand Aggies voices
Sang the song they hold so dear.

And when the band had finished,
St. Peter wiped his eyes
And said, "It's not so hard to see
They're meant for Paradise."

And the Colonel of the Cadet Corps
  said
As he stiffly took his stand,
"It's just another Corps Trip, boys,
We'll march in behind the band."

Administrative officials began to discuss the steps needed for an accelerated year-round, three-semester program. And plans were made to reissue rifles to the personal care of each cadet, a policy that had been discontinued in 1947. Boatner noted, "In view of the current international situation . . . a more serious attitude in the Infantry Regiment will have a salutary effect on the entire Corps."[48]

In July, 1951, Boatner was promoted back to the rank of brigadier general and reassigned to immediate troop duty in the Korean conflict. It took four and one-half days for him to go from the relative calm of College Station to his Korean assignment as Assistant Division Commander of the Second Infantry Division. His years of experience in Asia proved invaluable.

Somewhat concerned about the caliber and fitness of new Army troops in combat, he set up a program to give a brief, five-day indoctrination for all newly arriving personnel on how to stay alive. Furthermore, he remained in close contact with officers at Texas A&M, exhorting the Army cadre to train hard, very hard: "Training, training, training must be the guts of any ROTC program . . . this type of warfare now being waged in Korea is a special one."[49] In addition to commanding frontline combat units at the battles at Bloody Ridge and Heart Break Ridge, he was given the unenviable position in mid-1952 of being placed in command of the prisoner-of-war camp at Koje-do. In this role, he had full responsibility for more than 80,000 hard-core Red Chinese and North Korean POWs. Once again the "Bull" demonstrated his ability to "control other men." In response to questions about how he was able to handle the riotous prisoners and such a difficult assignment Boatner would often quip, "After dealing with the Texas A&M Mothers Clubs during my assignment as Commandant of Cadets at A&M, I could handle anything!"[50]

For Americans, Korea has often been termed the "forgotten war." The so-called war remained officially a "police action" or "conflict." From June, 1950, to July, 1953, it is estimated that 1,900 Aggies representing all the A&M classes from 1936 to 1953 served during the harsh and bloody test of wills over the spread of communism. Shortly after the armistice was signed in late 1953, the *Texas Aggie* reported that 58 former students had been killed in action and at least 6 were listed as missing in action. The Korean War and the Vietnam conflict that followed in the 1960s were flash points of communist expansion in Asia during the not so cold Cold War.[51]

SEVENTY-FIFTH ANNIVERSARY—
1951

In 1950–51, Texas A&M would observe a year-long celebration of its seventy-fifth anniversary. The milestone came at a period of relative calm for the campus and the Corps. Administratively, Gibb Gilchrist survived the "March of '47." The Board rewarded him, naming him the first chancellor of the Texas A&M System (September, 1948–August, 1953)—a direct outcome of the legislative investigation after the 1947 unrest. In his place, Frank C. Bolton served as president from 1948 until mid-1950, at which time Marion T. Harrington '22 became president. Events during the observance would reflect on the contributions of the past and focus on the emergence of a new era for the institution. The 1950s postwar boom in jobs, technology, automobile sales, and housing demand due to low interest rates was further felt in the demand to provide educated and well-trained graduates to assume roles in the new American economy.[52] The reduction in armed services manpower needs would result in fewer cadets, with or without a commission, entering active duty on an extended basis. As Cold War–era Washington planners and Congress debated the future role of what was being called "Universal Military Training" or UMT, President Harry Truman envisioned a large trained reserve and National Guard force. The dream of the World War I veteran was to create a permanent "citizen army," in which every male from eighteen to twenty would serve one year in the military and then be placed in a reserve unit—"ready in case of emergency." However, the UMT concept was rejected by Congress, which chose to rely on the ROTC program and the Selective Service (draft) process.[53]

In the wake of the UMT debate, it seemed likely that Texas A&M and the other 200 colleges and universities that offered ROTC would continue, most likely on a reduced scale, to produce officers. In concert with the other seven predominantly "military oriented colleges"—Clemson, North Georgia, Norwich, Pennsylvania Military, The Citadel, VMI, and VPI—the goal of Texas A&M was to broaden its academic curriculum and research programs, as well as obtain special recognition of the Corps of Cadets in order to receive as many officer contract slots as possible. Showcasing the institution and the Corps during the seventy-fifth anniversary observance of the opening of the college was a timely means of "tell[ing] the A&M story."[54]

Aggie Muster in April, 1950, was one of the first functions of the anniversary. The distinguished former Cadet Corps Commander in 1904–1905, Marion S. Church '05 of Dallas, delivered his address in a packed Guion Hall. In June Gen. Omar N. Bradley, the chairman of the Joint Chiefs of Staff, presented the commencement address to the Class of '50, which included more than 700 veterans. The World War II field general thanked them for their sacrifice and contribution, stating, "The men of Texas A&M can stand up to any in the world." Additionally, the general commissioned 306 new second lieutenants, bringing the grand total of A&M officer production in 1950 to 600. On October 4, Gov. Allan Shivers spoke at a Kyle Field event, followed in early November by the formal inauguration of M. T. Harrington as the thirteenth president of the A&M College. The featured speech was given by Gen. Dwight Eisenhower, then president of Columbia University. Judge John W. Goodwin, one of the first cadets to enroll at the college on October 4, 1876, was a special guest. Furthermore, author George Sessions Perry '40 released the first full-length history of the college, *The Story of Texas A and M*. Texas A&M and the Corps of Cadets were once again in the limelight.[55]

The spring of 1951 again highlighted the Corps with the now traditional observance of Military Day. However, the event that most captured attention was the dedication of the Memorial Student Center (MSC) on April 21. For three decades, since President Bizzell first proposed that a suitable memorial building be established as a living tribute to those who gave the supreme sacrifice in World War I, plans were debated and delayed. Not until 1943 was the idea reconsidered by T. O. Walton. After a number of site visits to facilities on other campuses, the hiring of J. Wayne Stark (three years before the center was completed), and a somewhat heated debate on exactly what the facility would be used for—student center or conference center—did President Bolton turn the first spadeful of earth in August, 1948. Budget concerns further delayed the building, which in the end cost $2 million. Completed in September, 1950, the center stands in honor of those former students of Texas A&M who died in World Wars I and II.

With the MSC considered primarily a "student center," tentative plans were begun in the late 1950s to build a continuing education center adjacent to the main structure. J. Earl Rudder '32, then vice president, began working with the architects to study the feasibility of renovating the aging Guion Hall and at the same time extending the east end of the MSC to join the two buildings. Far ahead of its time, the idea was eventually abandoned. And unfortunately—to allow room for Rudder Tower—Guion Hall was razed in 1971 before such plans could be further evaluated.[56]

COLONEL "JOE"

The fall of 1951 proved to be a little-noticed turning point for the Office of the Commandant and the Corps of Cadets. Colonel Boatner would be the last non-Aggie to be Commandant at Texas A&M. For seventy-five years the position of Commandant had been filled predominantly by West Point alumni, seventeen in number, except during the founding years when a series of former VMI officers and Southern Civil War veterans molded the early foundation of a military system and training. A&M graduates Col. Charles C. Todd '97 and Col. George F. Moore '08 were exceptions, with Todd serving on an interim basis in 1899 and then as Commandant from 1923 to 1926 and Moore in the late 1930s. Other A&M former cadets who had served as interim Commandants in addition to Todd and Moore were Harry Martin '95, E. J. Kyle '99 in 1899, and F. E. Giesecke '86 in 1890. Each held the position only until a permanent Army officer was assigned to the college. That it would take nearly eight decades before Texas A&M would be able to provide an unbroken string of qualified former students is not surprising. Prior to World War I, while all cadets received training in military tactics and many joined reserve or guard units, only an average of two graduating cadets per class—given the overall low national manpower requirements of the standing army and the emphasis to retain West Point—

trained officers—were offered regular commissions in the U.S. Army. Nonetheless, when there was an immediate need for more officers, as was demonstrated in both the Spanish-American War and World War I, A&M former students were quick to respond. And for most of these citizen-soldiers who answered the call to arms, once the conflict for which they were called was over, they exited the armed services to return home to their civilian vocations. In this regard, the framers of the Morrill Act, who in the early 1860s envisioned a ready reserve of citizen-soldiers in time of national emergency, accomplished their objective.

Not until the full implementation of ROTC in 1920 did Texas A&M, as well as other institutions nationwide, begin to develop a pool of senior officer candidates that in time could return to their respective alma mater to fill the role of PMS&T and Commandant. And in the case of Texas A&M, even this path was not fully the route Col.

Joe E. Davis took to become Commandant of cadets. Colonel Davis, a native of Foreman, Arkansas, was a member of the Class of '29, a cadet in C Company Infantry, and graduated from Texas A&M in 1930, majoring in rural education. Shortly after graduation, he was employed by the college in the Commandant's Office under Col. Charles Nelson as Corps property custodian and supervisor of day students or "casuals." In 1931–32, Davis was given the title of Assistant Commandant with the reserve rank of first lieutenant. He would remain in this position for a decade until the outbreak of World War II, by which time he was a major and in charge of both Corps administrative functions and discipline. In addition to periodic summer camp duty, Davis served brief reservist temporary duty assignments (TDY) at Fort Hood in Central Texas and at Fort Chaffee, Arkansas. In 1944, he attended the Command and General Staff School at Fort Leavenworth, Kansas. And each time he

TABLE 7-2
Corps of Cadets Fall Enrollment, 1950–59

Year	Fish	Soph	Juniors	Seniors	D&C*	Corps	College
1950	1659	1225	780	670		4320	6675
1951	1861	1964	980	782		4687	6593
1952	1684	1201	812	881		4578	6277
1953	1798	1109	670	668		4245	6198
1954	1507	1027	581	755	50*	3920	6257
1955	1618	886	685	570	108	3867	6837
1956	1804	848	646	655	119	4072	7200
1957	1853	937	642	554	108	4094	7474
1958	1822	755	548	401	128	3654	7077
1959	1751	995	399	444	214	3803	7094

* In the fall of 1954 compulsory training halted and cadets in a non-ROTC status were termed "Drill and Ceremonies" or D&C cadets.

returned to his job in the Commandant's Office at Texas A&M.[57]

Arrangements for Colonel Davis to become Commandant began in February, 1951, when Boatner advised the president that his assignment at A&M would be ending in the summer. Boatner expected no extension of duty at A&M and had already notified the Department of the Army he desired a "more active or combat assignment."[58] President Harrington was in full agreement of taking the necessary action to confirm Davis as PMS&T and Commandant upon Boatner's departure. Unsure of the full impact of the "national emergency" (it was seldom referred to as a "war") in Korea, there was a sense of urgency due to the growing shortage of senior Army officers. Boatner, who was instrumental behind the scenes in sealing the final confirmation of Davis as Commandant, had advised Harrington that in addition to his departure, the PMS&T slots at the University of Oklahoma, Oklahoma A&M (State University), and Louisiana State University would all be vacated in the summer. Thus, timing and confirmation of the next Commandant was deemed important.[59]

Concerned with what appeared to be the possibility of a major call-up of all reserve officers, Harrington in phone conversations with Gen. LeRoy Lutes, commanding general, Fourth Army at Fort Sam Houston, effectively received clearance to relieve Colonel Davis of any active duty obligation in order to be assigned to A&M College as the Commandant of Cadets beginning August 1, 1951. Because General Lutes could not confirm the appointment of Davis as PMS&T, both Harrington and Boatner wanted the assurance they controlled at least a part of the transition and continuity by

naming him the next Commandant. By Final Review, 1951, the plans had been confirmed with the adjutant general of the Army, Harrington noting, "Since the Army and the Air Force are on equal basis, it is best to have the Commandant and coordination for the School of Military Science to be an officer other than the Professor of Military Science and Tactics or Professor of Air Science and Tactics." The aim was to act swiftly to ensure the A&M Commandant's position was filled for the sake of the Corps.[60]

When Davis became Commandant, he had been on the Texas A&M campus for more than 25 years, having known and worked for ten former Commandants of Cadets—representing fully a third of all the senior officers or PMS&Ts assigned to the college since 1876. By the time of his retirement as Commandant in 1963, Colonel Davis had come into contact with more Aggie cadets than any other individual, with the possible exception of J. Malon Southerland, during the first 125 years of the Corps.[61]

The outbreak of the war in Korea proved to be a reality check for those who were eager to cut the military budget and drastically reduce manpower needs. American commitments worldwide during the Cold War era dictated that the United States maintain larger than normal standing armed forces. Once again, the buffer to field a large force rested with manpower planning that reinforced the importance of the reserve–National Guard forces and thus ROTC. By 1952, twenty institutions had curtailed ROTC instruction. The debate over Universal Military Training, which included provisions for a draft, centered on the amount of the federal budget that would be allocated to the armed forces and training. Congressman

Teague noted in House debates, "UMT is the long range program of military insurance for our security." Efforts to secure a larger U.S. military presence at the Bryan Air Force Base and the drive to secure ROTC contracts were but the first of many battles waged by Texas A&M, with the aid of Teague and others, to maintain the viability of the Corps of Cadets.[62]

Texas A&M administrators and alumni were determined to protect the image of the college and adjust to the growth and diversity of the institution as new programs, curriculum, and faculty were added. To prepare senior cadets better for Corps command positions, an annual leadership conference was begun in the fall of 1952 and held at the Lakeview Methodist Assembly campgrounds near Palestine, Texas. Colonel Davis wanted the Corps to move away from the class-dominated system run at the unit (company and squadron) level and instead emphasize rank and academic excellence. Corps Commander Weldon Kruger '53 appointed First Division Commander Joe Wallace '53 to spearhead a committee to review unit commander grievances. Many cadets felt the Commandant's Office was rushing to change the regulations in order to diminish the effectiveness of the units. Davis, in response, wanted to ban the

Cadet uniforms during the 1950s began to be gradually more distinctive. The senior on the left wears a tailored serge shirt and the one on the right wears standard issue. Note that both seniors wear black belts. Courtesy Texas A&M University Archives, Cushing Library

use of the "board," the use of required "physical inspection" to determine if hazing exists in a unit. He sought to reduce freshman harassment significantly. Discipline was to be administered via the demerit system and the "bull ring"—an area where cadets under arms would walk a prescribed tour in order to eliminate demerits. If necessary, Davis informed the cadets, all means within the "constituted authority" would be used to enforce discipline. In response to the hard stance taken by Davis, the *Battalion* printed a poll that stated that the reduction in "hazing and such customs can result only in the downfall of the Corps." Nonetheless, the Commandant continued to tighten the rules. The survival and viability of the Corps of Cadets in the 1950s hinged on a growing debate of how to expand the college's programs in the face of bad publicity concerning the all-male compulsory Corps environment. Enrollment and retention during the decade would plateau, arming proponents for a noncompulsory Corps and admissions for women.[63]

Furthermore, the armed services gradually needed fewer active duty officers. Many active duty assignments in the armed forces, which had traditionally been a minimum of two years, were altered by the Reserve Forces Act of 1955 to six months' active duty for training (ADT) only. Colonel Davis was no stranger to these new demands, taking a proactive role with the cadets, faculty and administration, and the former students to garner support for the military program. In the process, Davis, the A&M Board of Directors, and Presidents M. T. Harrington (1950–53 and 1957–59) and David H. Morgan (1953–56) were faced with an uphill battle. One of their primary concerns was that there not be a rush to judgment that could damage the Corps of Cadets. Morgan was

instrumental in contacting the other civilian military institutions to lobby for the support of a strong on-campus military program. Morgan advised the Office of the Secretary of Defense, "The morale of our Corps of Cadets, your future officers, is being seriously affected by doubt and uncertainty arising from the failure of the armed services to maintain the traditional ties of long standing with the Military Colleges."[64] The perceived image of the Corps of Cadets at Texas A&M, due to such items as low retention rates, repeated flare-ups with hazing, and reports of secret on-campus organizations, as well as increased external pressure to change, placed the Corps and its former students on the defensive.[65]

The conclusion reached by Morgan and the other heads of the eight leading military institutions was to promote actively their contributions in both war and peace—to separate and highlight their uniqueness. In an extensive assessment by the member colleges, it was proposed that the definition for a civilian military college read as follows:

Units established in essentially military colleges or universities which confer baccalaureate or graduate degrees and which are fully accredited by their regional accrediting associations; at which the average age of the students at the time of graduation is not less than 21 years; which require, except in the case of state supported agricultural and mechanical colleges, all students to pursue military training throughout the undergraduate course; which require all members of the ROTC to be habitually in uniform; which organize their student bodies as corps of cadets under constantly maintained military discipline; which have as objectives the development of the student's character by means of military training and regulation of his conduct in accordance with disciplinary principles. Veterans or foreign nationals may be exempted under this definition.[66]

THE IMAGE OF THE CORPS

Many seemingly small events to outsiders and those not involved with Texas A&M and the Corps became pivotal issues over time. One early example was an assault on the cadet uniform. The vintage World War II–style O.D. (olive drab) uniform including the "pinks and greens," the Sam Browne belt, khaki day wear, and the famous senior boots and boot breeches were questioned by non-Aggie ROTC military headquarters personnel as being out of step with the 1950s. A hint of such a problem over maintaining the distinctive uniform had surfaced as early as 1948–49, when the Department of the Army notified Texas A&M that they could no longer provide uniforms for "foreign" (non-U.S. citizen) students due to a "shortage of instructor personnel available for ROTC duty."[67] Incensed by shallow Pentagon rationale and swift policy changes, both Bolton and Gilchrist went directly to the then Army chief of staff, Omar Bradley, with what amounted to a "vigorous protest" from the A&M Board and demanded that this be corrected. Although the directive involved only seventy-five foreign cadets primarily from Latin American countries, it was felt that such a limitation or prequalification could next result in a directive against noncontract cadets—known as "drill and ceremonies" or D&C—making them the next to be excluded.[68]

The initial subtle attack on the Aggie uniform did not fully reach a boiling point until the U.S. Air Force sent what amounted to an ultimatum that its ROTC cadets at A&M would wear only the regulation Air Force blue issue uniform. The newly created Air Force, formerly the U.S. Army Air Corps, wanted to demonstrate and claim its uniqueness. Similar attempts had been made at other colleges with some success, yet the eight predominantly military colleges—each with a distinctive uniform—did not wish to alter their uniforms. Texas A&M fought back, concerned that there could be further federal erosion for support for on-campus military programs. President Morgan in testimony before the House Armed Services Committee, identified his concern impacting all eight remaining military colleges: "The rapid expansion of ROTC units during recent years has weakened the traditional ROTC programs in land grant colleges in general and in military colleges in particular."[69] To avoid further hassle, Clemson dropped its full-time military program and began to admit women.[70] Stressing efficiency and budgeting requirements, the Air Force forced the uniform issue.[71] However, a coordinated response by President Morgan, Congressman Teague, Sen. Lyndon B. Johnson, the A&M Board, (and behind-the-scenes maneuvering by Major Generals Meloy and Boatner) prevailed on the Defense Department in January, 1953, to approve the Aggie uniform as "distinctive and appropriate."[72]

The U.S. Air Force tried again in February, 1954, to require all AFROTC cadets at Texas A&M to wear the blues by the 1954–55 academic year. This time a letter from Assistant Secretary of the Air Force H. Lee White fully disregarded all the previous agreements and lumped Texas A&M cadets in with "all the other 188 institutions in accordance with Air Force policy." Officials at A&M were getting tired of the constant confusion and mindless Pentagon letters invoking "policy."[73] In a very pointed, detailed letter of reply, President Morgan responded that the Corps of Cadets and Texas A&M were not like the other "187 civilian colleges . . . which students wear the uniform only on drill days and reside in fraternity houses and private boarding houses." The Corps at A&M, "*is a way of life . . . our whole program is built on the basis of one corps.*" Going on the offensive, both Colonel Davis and President Morgan testified, at the invitation of Rep. Olin Teague, on the state of the Corps of Cadets and ROTC to the U.S. House Committee on Armed Services.[74] Furthermore, in response to veiled comments from the Pentagon on the appearance of the Aggie uniform with senior boots, Morgan clearly stated the position of Texas A&M: "The seniors *will* wear officer-type boots and boot breeches, items which are not distinctive of the Army, but do present an attractive appearance to the general public and a strong motivating factor since it is the ambition of many elementary and high school students in the state to achieve the *distinction* and *privilege* of wearing these boots."[75] Morgan's message was simple: Don't mess with Texas A&M!

While the initial battles left the uniform and distinctive identity of the Corps substantially intact, it became apparent to both Morgan and Davis that the Department of Defense no longer had the same level of interest in not just Texas A&M, but also the other military colleges. Morgan, Davis, and the cadet officers of 1953–54 began to review the options. The Corps was informed by the

administration that it was "responsible for its reorganization and revitalization." The president advised the cadet leadership, "The job will never be complete. Every year the new officers hold the destiny of the Corps in their hands."[76] Morgan stressed that image was everything. He further noted that poor retention could not be tolerated if the college were to grow and prosper. He reminded the cadets, "Until two years ago [1951] the 'fish' were kept in a separate area under their own officers, a practice that further lent itself to the practice of abusing underclassmen [sophomores]."[77] Morgan felt that the Corps of Cadets was at a major crossroads. Pressure to adjust to new federal guidelines as well as concern over the direction of the college resulted in an extensive self-assessment during the 1953–54 academic year. Cadet Corps Commander Fred H. Mitchell '54—citing difficulties in the transition from all-fish units at Bryan Field back to the campus—noted that his class was concerned with getting the fish back into the mainstream of the Corps without undue harassment. Additionally, studies were conducted on the status of the "military college," student attitudes toward change, reorganization of the School of Military Science, and a special academic council committee evaluation on the role of the Corps. Collectively, these studies and reports set the tone for the course and the future of the Corps over the upcoming years.[78]

During 1953–54, the *Articles of the Corps of Cadets* were rewritten to update the Corps organization and policies. Morgan, an eager visionary, informed Cadet Colonel of the Corps Fred Mitchell '54 that the new articles would lay the groundwork for the "ideal Corps for the second half of this century."[79] Events were moving rapidly at A&M. A faculty committee appointed by Morgan recommended that the best course of action was to discontinue the compulsory military training program in the fall of 1954. This action was viewed in the press "as a last resort" and drew severe criticism from the former students. Morgan strongly felt that making the change would enhance, not detract, from the growth and viability of the Corps of Cadets. In an address to the entire Corps in March, 1956, President Morgan answered those who questioned the noncompulsory decision as well as the recent attitude of Washington toward military colleges and particularly Texas A&M: "The Corps at A. and M. will go on *if the students want a Corps of Cadets.*"[80] With emphasis on a cadet-governed Corps of Cadets, the future of the organization would be literally in the hands of the students. Based on their actions and activities over the next few years the president's admonition was both challenged and put to the test a number of times.

Weekly Yell Practices in The Grove during the 1950s were often highlighted by a rousing pep talk by P. L. "Pinkie" Downs '06. Courtesy Texas A&M University Archives, Cushing Library

CADET REACTION: THE MOLE MEN AND THE TTS

Aggies would face responsibility or a lack thereof in various fashions during the 1950s. Strange events, tragedy, and secret groups, mixed with Aggie enthusiasm, marked a unique time in the institution's past. The students, mainly the cadets, never failed to keep A&M in the news.

The emphasis on change was greeted in different fashions by the Corps. Surely a large segment, especially the fish each year, were removed from the debate—until they became upperclassmen and developed a sense of responsibility to "protect Old Army" and

the traditions of the Corps. "Mr. A&M," P. L. "Pinkie" Downs '06, the official greeter for the college, was a pivotal character in constantly being a source of encouragement for the cadets—he was considered "the heart and soul of the Aggie Spirit." His enthusiasm for the cadets and the college was unbending, and most have credited Pinkie as the originator of the "thumbs-up" and the slogan "Gig'em Aggies." In the fall of 1953, cartoon character "Cadet Slouch"—created by James Earle '54—first appeared in the *Battalion*, poking fun at daily campus life. Cadet enthusiasm for the college, the Corps, and the sports program never waned—in fact, with

"Can you beat that? My chili unstopped our sink!"

The "Cadet Slouch" cartoon about campus life, created by Dr. James H. Earle '54, was a major feature in the Battalion *for nearly three decades. Courtesy* Battalion

a larger alumni and student body, the spirit intensified.[81]

The fortunes of the Aggie postwar gridiron program floundered until the arrival of Coach Paul "Bear" Bryant in 1954. Texas A&M wanted a winning football program, and the Bear was seen as the man who could make the difference. Yell Leader Joe West '54 remembered how, during the first Yell Practice in the Grove, he electrified the student body. "Bear escorted the whole team on the stage at the Grove and the head yell leader intro-

duced Bear," West recalled. "He slid off his tie and threw it out to the Corps. Then he slammed his jacket down on the stage and the Corps went wild."[82] A new era in Aggie football had begun! Most of the athletes were members of the Corps throughout the 1950s, but demands of recruiting and the growing intercollegiate competitiveness eventually led to the athletic program's inclusion of non-Corps students. One defining point was the infamous 1954 Junction, Texas, preseason football camp, with which Coach Bryant intended to send a message that A&M was serious about football. One writer noted, "Bryant separated the men from all other life forms. Or, as Gene Stallings '57 put it: 'We went out there in two buses and came back in one.'"[83] Survivors of Junction gelled into a great team, resulting in a 9-0-1 season in 1956. Three All-Americans—Jack Pardee, Charley Krueger, and Dennis Goehring—helped the Aggies win the Southwest Conference Championship, the first in more than a decade. Aggie superstar running back John David Crow earned the Heisman Trophy in 1957.[84]

Corps trips, Bonfire, the annual Military Ball, widespread recognition for the Fish Drill Team, the annual Twelfth Man Bowl Game—pitting the Army cadets against the Air Force cadets, and the increased size of the Aggie Band continued to stand out as unique features of the college.[85] The 265-member Aggie Band had gained national attention during their appearance in the Los Angeles Memorial Coliseum when the Aggies were the gridiron guests of the UCLA Bruins. Under the stern direction of Col. E. V. Adams '29, the Fightin' Texas Aggie Band, contrary to conventional wisdom that it was "impossible to have a good large marching band," gained national renown.

Adams stressed that the goal was to have a band that displayed "military dignity with audience appeal." Responding to media questions on how cadets responded to his management style, he replied, "I could walk into the dormitory at two o'clock any morning, blow my whistle and the band would fall out in marching formation!"[86]

There was no lack of excitement. Bonfire became a major project of the Corps each November. Built on the main drill field in front of the MSC, it became the major event of each fall semester. The Corps approached Bonfire as one of the main occasions to have cadets train in leadership and responsible positions, yet for the most part the yell leader–directed project was built by the fish of each new entering class. Outfits were divided into units for cutting and hauling in the trees, stacking logs, and twenty-four-hour guard duty to prevent any premature lighting.[87]

Twin tragedies struck in 1954-55. In November, 1954, a Naval T-28 trainer made a diving low pass at the bonfire the week before completion. Unable to pull out of the steep dive, the plane came apart and large sections of the aircraft fell on the drill field, covered with fish and crews stacking wood. Miraculously, no one on the ground was hit by the flying debris. Both the student pilot, Aviation Cadet Anthony Verduzco '54 (who attended A&M only two years) of Laredo and the instructor were killed when the main body of the plane crashed west of the railroad tracks.[88] The next year, a sophomore cadet, James E. Sarran '58, was standing guard duty on Wellborn Road west of the campus when he was accidentally hit by a truck. At the time Sarran was the only known on campus student death associated with the annual Thanksgiving event. A plaque in his memory was placed on the east edge of the main drill field.

THE ROLE OF "OLD ARMY"

A backlash surfaced among hard-core cadets concerned that the changes would fundamentally undercut the Corps' traditional base of privileges and activities—primarily the autonomy of the upperclassmen to both train the fish and choose (with the approval of the

Commandant) their own leaders. They abhorred any and all change from what they *thought* was the proper course of "Old Army" for the Corps. One concern of many cadets was that the military ROTC departments would try to turn the Corps into a narrowly defined military academy of sorts, thus diminishing the fraternal-traditional "Old Army" base of the Corps. Reaction to an overmilitarization of the Corps surfaced in the mid- and late 1950s.[89]

The primary underground movement was a resurfacing of the mysterious TTs or True Texans (Tonkawa Tribe)—first briefly in 1952 and then again in 1954–55. Rumors had circulated for years that such secret organizations had somehow "influenced student elections and selection of cadet rank." The administration moved quickly to identify those who were members of this unapproved, seemingly secret organization. During 1954, leaders of the TTs addressed a meeting of the senior class with a mixed message that they "did not exist" and if "they did exist," they were social in nature and posed no threat to the Corps or the college. Administrators felt otherwise and thoroughly investigated the TTs' activities and disciplined more than twenty-five members.[90] The extent of their impact on the Corps in the 1950s will probably never be fully known. A second semi-secret organization, the "Mole Men," were widely rumored to circulate through the steam tunnels under the Corps dorms in the Quad during the late 1950s. While most were never identified, they printed a periodic newsletter, *The Mole,* which would appear scattered throughout the dorms and campus, protesting the "creeping militarism in the Corps of Cadets." One member recalled, "We were just kicking up our heels, to poke

fun at the Trigon—mainly directed at some of the tac officers such as the 'Goose' and 'Porky.'" By 1959, the Mole Men had faded away.[91]

Morgan, who had become president in September of 1953, began addressing the reform of the Corps, complementing the administration's efforts statewide and in Washington, D.C., to protect the Corps. Reforms and a noncompulsory Corps were difficult to convey to hard-core Aggies. Morgan had not graduated from A&M. Thus, his actions and sometimes heavy-handed boldness ruffled relationships with Chancellor Harrington, who had recommended him, the Board, and the former students. In one of his last efforts to improve the Corps, he had the Commandant develop an oath of office for all senior cadet officers and then had the entire 4,000-member Corps witness it being administered in a public formal ceremony in G. Rollie White Coliseum.[92] Cadet officers were given wallet-size cards of the oath to be carried at all times. President Morgan addressed the gathering, outlining the items that he considered detrimental to "preserving the Corps' strength." Reflecting on the changes and events since the early 1950s, the Corps—in the president's estimation—must not "permit irresponsible persons to usurp authority, permit groups or cliques to form within the Corps, make raids on other schools, spread rumors, or fail to follow the will of the majority."[93]

The half-hour program was taped and rebroadcast by the college radio station WTAW. Copies of the tape were distributed to the Mothers Clubs and Aggie clubs statewide. The oath of office was reported in the press as an "anti-hazing oath."[94] However, hazing did not come to an abrupt halt. There were

TABLE 7-3
Texas A&M Corps of Cadets, 1959–60

Corps Staff: William B. Heye, Jr., '60 Commanding				
First Brigade			Second Brigade	
1st Battle Gp	2nd Battle Gp	3rd Battle Group	4th Battle Gp	5th Battle Gp
A-1	E-1	A-2	E-2	I-2
B-1	F-1	B-2	F-2	K-2
C-1	G-1	C-2	G-2	L-2
D-1	H-1	D-2	H-2	M-2

First Wing		Second Wing	
1st Group	2nd Group	3rd Group	4th Group
1	5	9	13
2	6	10	14
3	7	11	15
4	8	12	16
			17

Consolidated Aggie Band	
Maroon	White

those on campus who did not agree with the new changes. The shadow of the Corps's heritage and perceived past appeared in a *Battalion* editorial titled, "Leaders Trained by Responsibility." The editorial writer expressed guarded concern:

All students and faculty members at A&M should be worried about the future of the Corps of Cadets if it continues to fall more and more into the hands of the military departments. From conferences with former students and others who lived in "old army" days, it has been learned that the Corps of Cadets formerly was more of a brotherhood of students living under a semi-military organization. . . . Remember that leadership training is fostered only through placing responsibility in the hands of those who would learn to lead."[95]

In December, 1956, Morgan resigned after three whirlwind years. He had moved too fast to begin reforms—changes that were in fact under way at many colleges nationwide. In view of what had occurred years earlier during the Bizzell, Walton, and Gilchrist eras when administrators opposed hazing and attempted reforms of the Corps structure, Dethloff in the *Centennial History* concluded that, in spite of the fact that he was encouraged to leave, "Morgan's more radical reforms were implemented with remarkable success."[96] And the 1957 *Aggieland* commented that Morgan "took the reins of the presidency in troubled times." Change was difficult at A&M. David W. Williams, appointed acting president by the Board in early 1957, once again polled the faculty members on their view of military training. They voted overwhelmingly for an optional, noncompulsory Corps. The more liberal attitude of the faculty did not fit well with the wishes of the alumni, cadets, or the A&M Board of Directors, which voted 5-4 to reverse its 1954 policy and in November, 1957, restored compulsory military training for all freshmen and sophomores, beginning in September, 1958. And though faced with "vigorous" alumni opposition, the question of coeducation loomed as the next issue on the horizon.[97]

In the fall of 1958, Col. Frank L. Elder became PMS&T. While Colonel Elder and his staff concentrated on the ROTC program, Colonel Davis and newly appointed Vice President J. Earl Rudder focused on the image of the Corps, freshman retention, and a plan to encourage "higher academic achievement."[98] By 1958–59, the Corps of Cadets discussed phasing out Army branch identification among the cadet units. Implementation began in September, 1959. Cadets labeled the changes to the program "New Army"; no longer would infantry, field artillery, or cavalry be distinct units. The process would not be completed until 1959–60. One reason for the change is that it had become very difficult to match accurately ROTC graduates who had spent two years in a particular Army branch at A&M with active duty manpower needs and assignments in the same branch. To reduce the confusion, the Army determined that it would be more manpower effective to switch to general military science training and have a branch selection made after a generalized summer camp and just prior to commissioning. New freshman cadet orientation endeavored to blend the Corps into a more homogeneous organization. While Saturday morning classes were temporarily halted, most week-

day afternoon military drills were canceled and scheduled for Saturday—except when there was an out-of-town Corps trip. Cadet response to the changes was mixed. "Old Army" traditionalists noted this was yet another attempt to undercut the Corps. The fifty-nine company and squadron size units in 1958–59 were consolidated in 1959–60 into thirty-seven units. The new organization dropped the designations of "regiments and battalion" in favor of "brigades and battle groups." The Air Force units maintained the "wing and group" scheme. And few noticed or objected that 1959–60 Corps Commander William B. Heye, Jr., '60 (selected the outstanding fish in 1957 and sophomore in the Corps in 1958) was married during his senior year.[99]

Throughout the late 1950s, concern persisted over the attrition of each new entering fish class. A *Battalion* headline in October, 1958, stated, "'Fish' Drop Outs Plunge to Third of '57 Number," yet only 50 percent of the freshmen returned for their sophomore academic year between 1956 and 1959. Despite the addition of a faculty advisor system in the late 1950s, only about a third of all entering cadets would graduate. Overall student enrollment—including non-Corps veterans, graduate students, and fifth-year seniors—averaged 7,136 from 1955 to 1959, with the total college enrollment peaking at 7,474 in the fall of 1957. While the Corps returned to compulsory status for all entering freshmen, civilian students would gradually outnumber cadets. By 1959–60, the Corps of Cadets was at its decade low of 3,600 cadets. While other universities in Texas were also experiencing meager enrollment growth, the irregular enrollment statistics at A&M were cited as a primary reason why the college should revert to a noncompulsory Corps and open its doors to women. However, the strict, traditional, hard-core alumni, cadets, and Board of Directors would be reluctant to move too fast.[100]

AGGIELAND'S ORDEAL

Up until 1959, Texas A&M and the Corps of Cadets had remained vestiges of their own making. Isolated in terms of its distance from the growing urban centers of Texas, the college and its alumni had fought to preserve what they felt was the essence of the spirit of Aggieland. In an insightful series of articles in the *Houston Post* in early 1958, staff writer Leon Hale delved deeper than any previous outside observer to capture the gist of the dynamics swirling around the unique nature and role of Texas A&M. An indication of the problems of the college was the difficulty the institution had in defining the path of growth and image for the next decade. The A&M Board of Directors—all members but one former cadets—was a key element in this passage into the future. For the first time, the Board began to split over a number of pivotal issues. In the case of the reinstallation of compulsory military training, the 5-4 vote was indicative of a division of opinion at the college. Those voting against compulsory military were Board President Wilfred T. Doherty '22 of Houston, Vice President J. Harold Dunn '25 of Amarillo, Herman F. Heep '20 of Austin, and A. E. Cudlipp of Lufkin (not an Aggie). Following the motion made by Reginald H. "Jack" Finney, Jr., '38 of Greenville, he and the following vocal group of the Old Guard were in support of the mandatory Corps: L. Howard Ridout, Jr.,

'27 of Dallas, Eugene B. Darby '25 of Pharr, Henry B. Zachry '22 of San Antonio, and Price Campbell '13 of Abilene.[101]

The 5-4 vote would change the course of the college. The faction led by Finney, Zachry, Ridout, Campbell, and Darby, joined by new Board member Sterling C. Evans '22 of Houston in 1959, were not voting against progress, but for the maintenance of the fundamental, traditional roots and military heritage of the institution. All knew change was needed, but not at the explicit expense of the Corps or the "military image" of the college. After six years on the Board, Doherty and Dunn were not reappointed. Beginning in 1959 a progression of former cadets took the helm of the Board of Directors. Zachry served as chairman of the Board for two terms (1959–61), followed by Darby (1961–63), and Evans (1963–65). Old Army had prevailed.

At issue was how A&M would address the "formidable task in preparing men for the space and 'jetomic' age," while at the same time preserving the essence of the traditional value at A&M. Leon Hale, in his *Houston Post* series on "Aggieland's Ordeal," cut to the heart of the concern with an article titled "'Spirit' Feared Imperiled by Any Change." The academic community at A&M strongly questioned whether "this strict regimen of military life is compatible with high academic achievement." Many faculty assumed that the stagnant enrollment figures were caused by the image and role of military training at Texas A&M, when in fact the low wartime birth rates from 1939 to 1944 had reduced the number of potential college students in the late 1950s, not only at A&M but at universities across the state and nation. This demographic reality ran headlong into the A&M traditionalists who saw no need to expand the college and those who felt that growth, and only growth, was the hallmark of a good institution of higher learning. Hale concluded, "Aggies fall into two highly indistinct categories—those who are 'Old Army' and those who are not."[102]

The 1950s was a decade of transition for the Corps and the concerns of those who wanted to expand both the enrollment and the scope of the programs offered at the college. The growth of the postwar Corps of Cadets was tested in the mid-1950s when membership was made optional. The Board's reversal marked the last significant effort to limit growth (by restricting non-Corps enrollment growth) in favor of a traditional all-male Corps. In spite of the trend toward change, traditionalists liked the vintage Texas A&M. Thus, the overriding issue for the college by the early 1960s was how to handle the evolution of both the Corps and the college from their traditional roots as the nation entered the space age. A most unlikely leaer in he person of James Earl Rudder was expected to facilitate the metamorphosis of the old A&M College of Texas to a uiversity of the future—while maintaining alumni and cadet support and avoiding doing irreparable damage to the image and traditions of Texas A&M.

8

THE RUDDER ERA

Rudder restructured, revitalized, and revolutionized the institution.
He built a university where a college had been.
  Henry C. Dethloff, *A Centennial History of Texas A&M University, 1876–1976*

Military training and Cadet Corps life instill into the individual
student . . .Solid and desirable character traits, and will stand the
A. and M. graduate in good stead in whatever walk of life he may
find himself . . . the unique concept of the development of the "citizen
soldier" is tested and proven sound.
  Report of the Century Council, 1962

I may have been naive in my love for and feeling of obligation to my
country. . . . I had a duty to fight. That was enough for me.
  Michael Lee Lanning '68, *The Only War We Had,* 1987

DURING THE DECADE of the 1950s, Texas A&M succeeded in remaining insulated from many of the events of the time. Some talked about change at the all-male southern military college, but they accomplished only a few alterations in the basic fabric of the institution. Steeped in its tradition and proud of its past, A&M and the Corps of Cadets would implement during the 1960s many of the far-reaching ideas and innovations that stalled during the Gilchrist, Morgan, and Harrington administrations. Texas A&M, like many college campuses nationwide, would be a barometer for the dramatic events of the decade. For the first time, children of the post–World War II baby boomer generation began to arrive on campus. Their world in the 1960s was a much faster paced reality. Elvis and rock-and-roll music, color TV, the race for outer space, the Cold War, Woodstock, hippies, JFK and LBJ, and Vietnam intertwined to start a transformation in the basic culture and fabric of America. Furthermore, attitudes toward campus military training would also be challenged. The Corps of Cadets proved to be at the crossroads of the changes sweeping the country. Unlike the service academies at West Point or Annapolis, the civilian military colleges, of which Texas A&M was the largest, recognized that their all-male military program would gradually be overcome by forces that ushered in coeducation, a more open attitude toward scholarship and research, and the de-emphasis of on-campus military training. After the reversal of the brief noncompulsory period in the mid-1950s, students at Texas A&M in 1965 would once again be given an option merely to walk across campus and away from the military style of living. Many hard-core Texas A&M traditionalists from that period rued the day.[1]

The assessment that the Texas A&M Board of Directors began in the late 1950s to improve the overall quality of education at the college proved a slower process than many expected. Active alumni, A&M Mothers Clubs, conservative Board members, and, most importantly, the cadets were reluctant to accept the change most saw looming on the horizon. The issue that became even more volatile than noncompulsory military training was the full admission of women to the college. A majority of former students of the college were overwhelmingly against coeducation throughout the decade of the 1950s.[2] Texas A&M would be the last of the fifty-six land-grant universities across the nation to admit coeds on a full-time basis.

The vision of Presidents Morgan and Harrington for Texas A&M would in time set the tone of the relationship between the Corps of Cadets and the progressive movement to broaden the scope and makeup of the campus. The war in Vietnam and campus response to the anti-war movement would be a key factor in allowing the Corps during the mid- and late 1960s to focus on its military-fraternal lifestyle and mission. The Corps had been torn between a fraternal "class system" and a quasi-military body, but once change was under way both the alumni and administration endeavored to cast the Corps of Cadets as a unique and "elite" organization. Amid this mixture of crosscurrents, leadership was needed to balance all factions. The governor of Texas handpicked the man to lead Texas A&M into the late twentieth century—James Earl Rudder.[3]

The former Army major general became president of Texas A&M on June 1, 1959, and for a decade would shape the blueprint of progress for the institution that would be implemented over the balance of the century. Bridging attitudes held by the generation from the 1930s, early 1940s, the depression, and World War II with the new ideas of the 1960s space-age generation placed Earl Rudder at the moment of truth for both A&M's short-term and long-term future.

General Rudder had attended John Tarleton Agriculture College at Stephenville for two years prior to graduating from Texas A&M in 1932 with a degree in industrial education. Nicknamed "Spike," he played center on the A&M football team, was a "Sbisa Volunteer" (waiter), campuswide intramural wrestling champion in the 175-pound class, and a cadet captain on the First Regiment staff.

At graduation he was commissioned a second lieutenant in the Army.[4]

Returning to ranching after graduation, he also coached football at Brady High School, winning two district championships. In 1938 he returned to Tarleton State as head football coach, compiling a three-season record of 23-7. Throughout this period he remained active in the Army Reserve and was called to active duty in 1941. He organized the Second Ranger Battalion in 1943 and commanded a pre-invasion D-Day mission at Omaha Beach to silence the German guns at the top of the hundred-foot cliffs at Normandy's Pointe du Hoc. One of the most heralded Aggies in the history of the institution, General Rudder was the recipient of the Distinguished Service Cross, Legion of Merit, Silver Star, two Bronze Stars, and the Purple Heart with Oak Leaf Clusters as well as numerous foreign decorations. Gen. Omar Bradley, commander of the U.S. Ground Forces in Europe at the time of the Normandy invasion, said, "No soldier in my command has ever wished a more difficult task than that which befell the 34-year-old commander of this provisional Ranger Force."[5] After the war, Rudder was elected mayor of Brady, Texas, and remained active in the Army Reserve, commanding the Ninetieth Infantry Division (USAR). In January, 1955, Gov. Allan Shivers appointed him commissioner of the General Land Office of Texas, to clean up the fraud- and scandal-ridden operations that plagued the land office. By late 1957, Rudder revamped the veterans' land program, placing the agency back on a solid footing. Thus, by the time of his arrival at A&M as a vice president in early 1958, in essence as the chief administrator of the college reporting to Dr. Harrington, who functioned as both president and chancellor of the institution, Rudder was battle tested.[6] The summons to his alma mater would be Rudder's greatest reward and one of his greatest challenges. Ray Bowen '58, the 1957–58 deputy Corps Commander, recalls that when he and Corps Commander Jon Hagler '58 met with Rudder upon his arrival on campus, "It was clear to us he was in charge."[7]

Except for a number of brief visits, such as delivering the 1956 campus Muster address on the steps of the MSC, Rudder had had only limited contact with the campus and the Corps. Upon arrival he was surprised to find an air of unrest among the cadets, as well as discontent among a scant number of nonregs (civilian students), faculty, and staff. His concern quickly turned to the Corps. At the center of the unrest, Corps freshman retention had dropped to an all-time low. In May, 1958, in one of his first addresses to the Corps he was blunt about the dissension in the ranks. A feeling of "Last of the Old Army" prevailed among the cadets. Many believed that the hint of a noncompulsory Corps and the possibility of full-time coeds jeopardized the Corps. Rudder confronted the cadets with the facts: "The average company enrollment throughout the Corps [1957–58] is 70 and the average freshmen per company 35—one company lost 7, three companies lost 9, seven companies lost above 20, and one company lost 24." He lay much of the blame on the class system, inquiring of the cadets, "Do you want a Corps or a class system? We have many units instead of a Corps in a great many respects!"[8]

In the fall of 1958, he met with all Corps upperclassmen to review the upcoming year and assess retention. The so-called "New Army" image was deemed appropriate. During the summer, compulsory military training was reinstated, the Army abandoned the branch system for the general military science curriculum, all cadets were issued black shoes (to replace the old brown style) and "pinks," pearl gray slacks.[9] Outfit drill was moved from Thursday evenings to Saturday morning. A clear charge was given to the "active duty" cadre in both Army and Air Force detachments to be more involved in the cadet dorm area or Quad.[10] And for a time the famous freshman brace disappeared along with "cush" questions at mealtime—a quiz by upperclassmen to determine if the fish would be permitted to have dessert. Lamenting the

change at A&M, the 1959 *Aggieland* bewailed the deactivation of "branch outfits . . . the most colorful tradition in A&M's history. . . . Branch pride has become one of the more memorable A&M experiences. Now, the Jocks, Groundpounders, Buzzard Busters, Wig-wags, Blanket-stackers, TC's, Flaming Onions, Big Guns, and Test-tube Cleaners will all be gone. In their place will be a new and progressive organization of the Corps. But what can take the place of the cry: 'Beat the hell out of the Jocks!'"[11]

In the fall of 1960, the distinctive Army

Officers (in dress uniform) and colors on Simpson Drill Field. Photo from author's collection

The Corps brass was designed to replace the branch designation insignia: Per Unitatem Vis *(Through Unity, Strength). Courtesy* Texas Aggie

branch insignia was replaced by the Office of the Commandant with a design developed by the cadets. The new Corps-wide "Corps Brass" carried in Latin scroll the motto: *Per Unitatem Vis*—Through Unity Strength.[12] The new president urged the Corps leadership to take the new motto to heart in order to meet the challenges facing them on campus: "I want you to guide the freshmen with common sense—not hazing."

In his assessment of the state of the Corps and the future course of the organization, Rudder was blunt:

The primary objective of your lives on this campus is an education. Or, are you looking forward to privileges, authority, to prolonging for awhile indulgence in immaturity and irresponsibility, to indoctrination of young men with a philosophy of "take it now so you can dish it out later." We today are facing a serious, troubled world with a timetable of uncertainty—and not a world of carefree irresponsibility.

One thing that is important to watch—the attrition of freshmen—be sure that their resignations from your ranks are not due to pressures from upperclassmen. Out of a total of 1,000 freshmen in the Air Force last year, we registered 375 sophomores as of 4 P.M. on Saturday, September 13th. Out of a total of 840 freshmen in the Army ROTC last year, we registered 360 sophomores. From October 1, 1957 to February 10, 1958, we had a loss of 393 freshmen in Military Science (44.3%), and a loss of 419 freshmen in Air Science (43.4%). In 1956–57 we lost 34.5%, in 1955–56 29%, and in 1954–55 19.9%.

Pausing to stress the seriousness of retention, Rudder concluded, "At this rate the Corps will eliminate itself!"[13]

The Corps in 1959–60 numbered 3,803 cadets, representing more than half of the total 7,094 student enrollment. Rudder's goal over the next decade was to double the size and diversity of the student body. Opinions about the best way to achieve this goal were varied. From across the state vocal alumni weighed in with their opinions, and political pressure from all quarters was brought to bear on the planning for A&M's future.

The issue of growth at A&M would be debated for more than a decade until it was eventually accepted as a positive indicator for the institution. In the meantime, Rudder—the the governor's hand-picked man—had been told in 1958 "to go over there and straighten that place out." The reaction by the faculty, alumni, and cadets would dictate his success or failure, yet none should ever fail to discern that for the most part the "general" was in total control at A&M. At no time in the past 125 years—with the lone exception of the Ross administration—had the fortunes, image, and viability of the institution rested so squarely on the shoulders of one man. Burdened by no academic baggage, his mission was to make the necessary changes at A&M to expand the institution, while at the same time preserving the conservative traditional values of the college. Practicing the old military adage of "lead, follow, or get out of the way," Rudder initiated an incredible period of adjustment for the college and the Corps of Cadets. The 1958–59 Cadet Corps Commander, Don Cloud '59, summed up President Rudder's tremendous influence in facing the challenge regarding coeducation and a noncompulsory Corps: "Being an A&M graduate, Rudder was one of the few people on the face of the earth at the time that could pull this off without a general blood letting."[14]

"COEDS" AND THE "UNIVERSITY"

The path to the "full" admission of women at Texas A&M started some four decades before it became a central issue in the 1950s and early 1960s—and would not end until the early 1970s. Women had attended Texas A&M classes from the earliest years of the college, but not as full-time students, and none had been eligible for a degree.[15] For example, Professor Charles W. Hutson's daughters attended A&M—Ethel in 1893–95 as well as twin sisters Mary and Sophie in 1900–1903. In 1921 Mary Evelyn Crawford, sister of engineering professor Charles W. Crawford '19, enrolled with the special permission of President William Bizzell. She finished the requirements for a degree in English in 1925; however, she was not permitted to receive a diploma at the annual commencement. And during the Great Depression era, a handful of women were admitted as hardship cases on a one-time, limited basis only—none received degrees.[16] Texas A&M, like other military colleges and service academies, had established a mandate to offer a college-level education and military training to male cadets who may or may not become full-time soldiers. The emphasis on a southern military style of training and educational environment at VPI, VMI, The Citadel—and Texas A&M—did not, due to its mission and mandate, include coeducational students. Traditionalists were quick to point out that women-only colleges flourished nationwide and that their exclusive enrollment was not any different from the all-male military colleges.[17]

In spite of the fact that the "all-male military program" concept flourished at Texas A&M, periodically women did attend Texas A&M but only as special students during the regular term or during summer school sessions. Daughters of the college professors or staff and coeds from near by Bryan were generally the most likely to attend on a limited basis. The issue of unlimited attendance was not fully tested until the early 1930s, when a group of women from Bryan filed a lawsuit demanding that they be admitted full time, just like male cadets. The girls from Bryan were represented by Col. Charles Todd '97, former Texas A&M Commandant, who had entered the practice of law after retiring from the Army in 1925. His brief to the court—the foundation document on the issue of coeds at A&M, while overlooked for decades by those who felt they were raising the issue for the first time in the 1950s—was simple and direct. Todd argued that, notwithstanding the role and emphasis of the Corps of Cadets, a state-funded college was obligated to allow equal educational training for coeds—in spite of the fact that there were ample all-female colleges and other institutions in which they could enroll. On January 9, 1934, Todd's efforts were rejected on appeal. The all-male Corps of Cadets survived.[18]

After a brief expression of interest by coeds to enroll in the post–World War II period, the issue of full coeducational admissions did not resurface in earnest. During the early 1950s, the "Bull of the Brazos"—Sen. Bill Moore '40—and others were determined to make Texas A&M coeducational. President Rudder (at first against coeds) found himself between factions on both sides of this issue, but he was fully aware of the tremendous impact on the institution. Moore, the state senator from the district that included Texas A&M, introduced a coeducational resolution in the Texas Senate on September 3, 1953. It was first approved by a hasty voice vote, but Sen. Searcy Bracewell '38 of Houston prevailed upon the Senate to reconsider the measure, thus rescinding the resolution by a 27-1 vote. After the final defeat Moore predicted to all that A&M would be coed within ten years. His prediction was accurate.[19]

In the late 1950s the courts once again heard the case, only to be over turned on appeal to the Texas Supreme Court. Twice the case went to the U.S. Supreme Court, and twice they refused to consider the appeal, each time by the margin of one vote. The issue was addressed once again in the Texas Legislature to no avail. The *Battalion* in April, 1959, hailed the turn of events: "No Coeds for Aggieland; This Time It's for Real—So A&M Remains An All-Male Citadel." However, the legislature strongly suggested that change was needed and that steps should be taken in the future to admit women. Due in large part to powerful opposition of the alumni, the issue would linger, with no binding action taken.[20]

BLUEPRINT FOR PROGRESS

Shortly after his formal inauguration in March, 1960, Rudder, fully aware of the controversial issues facing the college and his administration, felt it was time for the institution to go beyond the standard self-study so often hastily convened by organizations looking for a quick fix. Instead, he convened a process to conduct a broad analysis that could be used to plan the future direction and mission of the institution. On the eve of the hundredth anniversary of the passage of the

Morrill Land Grant Act, the A&M Board of Directors on April 20, 1961, concurred with Rudder and launched what would be the most extensive series of studies and long-range planning conducted to that date on all aspects of the college. This multifaceted initiative began with a twenty-four-man committee of faculty and staff known as the Committee on Aspirations; second, one hundred outstanding citizens conducted the Century Study; and third, the Board of Directors concluded the series of studies with a report that resulted in *The Blueprint for Progress*. Rudder assigned four primary questions to the Committee on Aspirations that struck at the heart of the recent trials of the past, while focusing on developing a plan for the future: (1) What kind of student does Texas A&M seek to produce? (2) What is the mission of the College? (3) To what degree of academic excellence should the faculty and staff aspire? (4) What should be the scope and size of the A&M College by the centennial anniversary in 1976?[21]

The Corps of Cadets was at the epicenter of this evaluation project. Rudder was aware that the study would provide those against the Corps a semi-official soapbox to question its worth and mission. Given the flat enrollment figures, poor campus facilities, and the overall dominance by the Corps and their alumni—the conclusions of the 1962 Committee on Aspirations resulted in a frontal attack on the Corps. After stating the obvious, "The Corps determines the students' habits, attitudes and ambition," the committee pointed out that the Corps lifestyle had become more important than academic life. Students came to Texas A&M to be Aggie cadets first! Those who had not been members of the Corps pounced on the negative conclusions. Furthermore, the committee argued that

the compulsory Corps policy had resulted in a low retention rate, attracted very few transfer students, did not help the graduate studies program or research, and "was the primary factor" for the relatively low enrollment growth during the past two decades. Antagonists felt the Corps had limited the "true pursuit of scholarship" at the college. The faculty-staff report concluded with the following recommendations: military training should be voluntary for all students, the Corps should no longer exist as a residential organization, and adult supervisors should reside in each residential unit.[22]

The lagging enrollment figures and poor retention were factual. Even so, Rudder, the Board, many cadets, and former students were not pleased with the attack on the Corps. Although the report about the Corps was only one part of the final study, it clearly intended to discredit either the Corps or Rudder or both. General Rudder moved cautiously. The *Report of the Century Council* was less specific and returned a guarded view regarding admission of women, while strongly endorsing the military training program of the Corps, its role, and the importance of producing "citizen soldiers skilled in the arts of both peace and war." President Rudder found himself in the middle of a whirlwind. The general knew that to implement unpopular changes he would need support from not only the Board, the press, and a select few legislators but also the backing of the "old guard"—the former cadets of the A&M alumni. Rudder increased efforts to brief the Association of Former Students on the new plans, even while the group was reluctant to endorse them. Furthermore, in the fall of 1962, Rudder recognized former students by specifically honoring four outstanding former cadets with the first Texas

A&M Distinguished Alumni Awards at the Century Study Convocation to commemorate the contributions of past graduates as well as highlight the new direction and vision for the college: Gen. Bernard A. Schriever, '31, commander of the U.S. Air Force Systems Command; John W. Newton '12, general manager of Magnolia Petroleum Company; W. W. Lynch '22, president of Texas Power and Light Company; and Dr. Edward F. Knipling '30, agricultural research scientist.[23]

Total enrollment in the fall of 1962 exceeded 8,100 students, of which 4,060 were in the Corps—the largest overall campus enrollment since the fall of 1948. The growth seemed to refute claims that the college could not attract new students. In addition to extensive state-wide recruiting and a new emphasis on academics (often lampooned by the cadets as the "Grade Point Army"), Rudder initiated what would become over two decades of new construction and facilities expansion campuswide. To enhance the campus facilities for students, the Board approved $3.5 million for five new dorms as well as a project to "air-condition the campus" beginning with Dorms 14, 15, 16, and 17 in the North area.[24] The Wofford Cain Olympic Swimming Pool was completed in 1962 along with plans to construct a new residence for the president. Cadets, alumni, and the administration seemed pleased with the progress and all settled in for the fall, only to be jolted by a highly derogatory *Time* magazine story, which referred to Texas A&M as "Sing-Sing on the Brazos . . . the school is a cluster of penal-looking buildings flying the flag of Texas . . . and Aggies are strictly 'onion packers.'"[25] The author, in his endeavor to discredit Texas A&M and the Corps of Cadets, only solidified the already fierce loyalty of its alumni and students—a result that few

non-Aggies expected or could identify with. In an era of rampant Aggie jokes, this exposé did not prove humorous. The portrayal of the campus "100 miles to anywhere" and its cadets as "proud look-a-likes" would not soon be forgotten. The editor of the *Battalion,* H. Alan Payne '63, gave the best assessment: "[The author] probably summed up his beliefs when he said, 'the prime requirement may not be scholarship, but the prime blessing [at A&M] is belonging.'"[26]

"Belonging" to the Corps of Cadets and Texas A&M was something those who had been cadets understood and something those who had not, often scorned. In the early 1960s, Texas A&M was very different from any other college or university in Texas. This uniqueness, as much as any aspect of the pending changes, is what alumni and students feared would be cashiered in the name of growth. Incoming A&M students could attest to the difference. The daily routine as a fish was a shock to most who came into contact with the Corps for the first time. One fish described his experience and impressions upon arrival at A&M in the fall of 1962:

Texas A&M was still the Agricultural and Mechanical College of Texas in September 1962, when I hugged Mother, then shook hands with my father in front of those long, barrack-style dormitories of stark red brick. We "fish" or freshmen, were lining up for our first formation as Daddy put on his Stetson and pointed his 1961 Oldsmobile back to the rolling mesquite plains of West Texas.

We wore Levi's, T-shirts, and pointed cowboy boots or tennis shoes that Sunday afternoon, but the next morning we'd draw uniforms and rifles.

A&M was steeped in tradition, mostly military and macho. The A&M Corps of Cadets was far more army than the *real* army. Assignment to companies in one of the three brigades, or to squadrons in the two air force wings, was compulsory for [only] the first two years.

We wore khaki uniforms in the fall and spring, and fatigues and jump boots when it rained. In the winter, we switched to World War II–style "pinks-and-greens"—salmon-colored woolen trousers and, for dress, olive blouses.

Seniors strutted in knee-high boots of shiny brown leather, like Gen. George Patton in old photos. Every fish dreamed of the day he'd wear those boots. Many think the boots meant cavalry, but they were the mark of an officer, of any branch, in the pre–World War II Army. And A&M prided itself in being *Old Army.*

Juniors were called "sergebutts" for the permanently creased rayon uniforms they were privileged to wear. The main job of the "pissheads," or sophomores, was to whip the new class of fish in shape. They'd been fish a year before, and accepted the responsibility with a vengeance.

Hazing was officially forbidden, but it was part of daily fish life for a year. A complaint bought you a one-way ticket onto Highway 6. The pissheads [cadet sophomores] had ways of doing that.

The entire Corps stood formation and inspection, then marched to chow, three times a day. Those days began at 6:00 A.M. with the bugle call *Reveille,* blown by a real cadet bugler, not a recording. That's when the "whistle jock," a rotating duty for the fish, threw himself at attention against the wall in each Corps dorm, blew shrilly, and shouted the first of three calls before formation:

> *Spider-D. First call to chow!*
> *Fall out in forty-five minutes.*
> *Uniform is: piss pots and jump*
> *   boots.*
> *Menu is: bullneck: bacon and*
> *   cackle.*[27]

From this brief look at life as a fish, it is easy to understand how those unfamiliar with the Corps would be both skeptical and critical. However, in retrospect, the extensive Century Study of 1962—totaling more than 600 pages of data, commentary, and recommendation—helped to defuse the twin issues of compulsory Corps and the admission of women. With the administrative and Board reports of the early 1960s calling for change, the Corps of Cadets posed an easy target, both in the state legislature and in the media. Some considered the Corps responsible for many previous shortcomings of the college, including its slow development of academic programs and stagnate enrollment. To change drastically the basic fabric of the relationship between the Corps and the college would in reality undermine the military program in spite of its long history—in favor of a more liberal, generic institution that could potentially attract more students. Rudder, the Board, the alumni, and cadets were not ready for such drastic action. However, many pro-Corps proponents realized that the image of the Corps, enrollment growth, admission of women, and the enhancement of the academic program would indeed have to be addressed.[28]

The issue of admission of women was allowed to cool off briefly until the spring of

1963, at which time the Board of directors (with the reluctant support of the alumni) approved at their April meeting a plan to admit women on a "limited" basis, effective for the fall of 1963. News of the shift in policy was unwelcome. Only weeks away from Final Review, many in the Corps felt betrayed. At a called meeting of all cadets in G. Rollie White Coliseum, shortly after the Board's decision, General Rudder was openly booed by many of the angry cadets. Remaining calm, he reviewed the impact of the new ruling on the college and the Corps and encouraged all to work together. His presentation and appeal to the Aggies to keep in mind the larger ramifications for the image of the college only slightly consoled the cadets' worst fears. To many, "old Army had gone to hell . . . again."[29]

A small group of cadets countered with the organization of the "Committee for an All Male Military Texas A&M," with the hopes of carrying their message against coed admission to the alumni and "the people of the State of Texas." Yet for the most part the transition was smooth, with more than 150 coeds enrolled in the fall of 1963. Paul Dresser '64, former Cadet Corps Commander, recalled, "The Corps generally was stoic about things they did not like." Nonetheless, limited on-campus housing and only a modest effort to attract women "discreetly had held their numbers to a trickle" until the early 1970s, at which time women were admitted on an unlimited basis.[30] Women would not be allowed to join the Corps until later in the decade. The inevitable had at last happened and alumni statewide voiced mixed reactions, with most against the change.[31]

Why the reluctance to admit women on a full-time basis? In part it was a fear of the unknown and a concern that somehow the traditions and spirit on the institution would be altered or damaged. Rudder, too, was reluctant about women at A&M, yet after a number of conversations with Lyndon Johnson, he realized that A&M needed to admit coeds in order to prosper. One former student captured the essence of alumni concern: "We'd always been equal at A&M, no matter if Daddy was a big oil man or a sharecropper. It didn't matter. Aggies all wore the same clothes, ate the same food, lived in the same quarters. I was afraid all these changes would upset the fellowship."[32]

Thus, the "official" admission of women was a milestone for Texas A&M. And in retrospect, Rudder was the only one who could have presided over the transition. While many did not agree with him, most realized he had the best interest of the institution at heart. More change was forthcoming. In August, 1963, the Texas Legislature approved a Century Council recommendation to change the name of the college from the Agricultural and Mechanical College of Texas to "Texas A&M University." The school retained "A&M" (unlike many land-grant university that substituted "State" for "A&M" in their official name) to tie the university historically to its land-grant roots. No longer would the primary educational roles of the institution—the agricultural and mechanical (engineering) arts—be spelled out in the official name of the institution. A *Houston Post* writer commented that when people started tampering with the name, they were "over-stepping their bounds." Although many agreed, the concerns of the majority of cadets were more basic—What about our old Corps collar brass and shoulder patch with "AMC"? Would there be any changes in the uniform? What name will we have on our Aggie class ring: college or university?[33] The "AMC" was phased out with the entering fish class, and the administration decided that graduates from 1963 through 1966 could choose college or university. Beginning in 1967, only "university" appeared on the Aggie ring.

CORPS REORGANIZATION

Cadets returned to class in the fall of 1963 to an enrollment of 3,502 in the Corps. The statewide publicity about the changes had only increased interest in the university. During the mid-1960s, Texas A&M was deeply involved in enhancing the university's operations and outreach statewide. The media reported on new A&M programs and facilities as the university endeavored to position itself favorably. On campus, measures were commenced to add three new colleges—liberal arts, science, and geosciences (eliminating the School of Arts and Science)—as well as to expand the data processing center as Aggies explored the opportunities of computing. Research became a prime intent of the university with the construction of a Cyclotron at the cost of $6 million and the completion of the $2.5 million Space Research Center, named in 1965 for Congressman Olin Teague—chairman of the NASA Oversight Subcommittee of the House Committee on Science and Astronautics. Off campus, the possibilities of adding James Connally Air Force Base in Waco as a research and development center and technical institute was balanced with the somewhat contentious transfer of Arlington State College—located between Dallas and Fort Worth—from the A&M System to the University of Texas

System. In hindsight, the transfer was a grave mistake that limited A&M's future presence in a major Texas metropolitan area.[34] The cadets and former students, however, were more focused on what was happening with the Corps in College Station. The editorial staff at the *Battalion* proclaimed, "Change Expresses University Growth," and concluded, "We feel sure that A&M is here to stay, and to grow. As for the Corps, we have said before that only the Corps will decide its fate."[35]

Knowing that the changes on campus would impact the Corps, Rudder also had plans for the enhancement of the Commandant's Office in the fall of 1963. In 1951, the college had hastily secured Col. Joe Davis as a "civilian" Commandant on the college payroll, out of fear that the Corps would be without leadership if there were a massive call-up of military personnel during the Korean War. The plan worked well during much of the 1950s, but Davis and his staff gradually became further removed from the changes in the active military. Furthermore, the civilian Commandant scheme developed in the early 1950s isolated the Commandant and his staff of "college employees" or "tac officers" (primarily reserve and retired military officers) from the mainstream of Army and Air Force operations within the two ROTC detachments. Because the PMS&T and the PAS&T had only to be concerned with their individual programs, they did not always grasp overall operational issues concerning the Corps outside of routine ROTC classes and activities.

To reverse this scheme and breathe life into the Commandant's Office, Rudder submitted a new organizational plan at the June, 1963, meeting of the Board. The new structure would reinstate the incoming active duty regular Army officer in the dual (pre-1951) role as both the senior Army officer (PMS) and Commandant of Cadets. Colonel Davis was transferred to a job at Tarleton State, and members of the Commandant's staff were replaced by active duty military personnel. In lieu of the "tac" officer concept, three "civilian Corps advisors" were employed by the dean of students to assist the Commandant with day-to-day affairs concerning the cadets. Rudder stressed that the changes in the Trigon—a cadet nickname for the Military Science Building that housed the Office of the Commandant and active duty officers in each ROTC detachment—"will strengthen the Corps by providing more realistic training and will save A&M $60,000 annually." Moreover, the general felt that one key element to preserving the Corps of Cadets was to make it more "elite": "Cadets will receive more leadership training in a military Corps run by [active duty] military officers."[36]

Following the departure of Colonel Davis and the retirement of Col. Frank Elder (PMS) in mid-1963, Rudder contacted Rep. Olin Teague and the secretary of the Army in order to make arrangements to select the next PMS and Commandant. In the process, Rudder talked with Col. Denzil L. Baker '33, who was at the time the chief of Special Review Division of the Office of Army Personnel. With access to hundreds of potential candidates to be Commandant via Baker's department, the process moved quickly. Extended conversations with Colonel Baker confirmed that the Mathis, Texas, native fulfilled Rudder's requirements for the position. The necessary orders were "cut" to have him arrive for duty in College Station by August 1, 1963. In announcing the selection of the fifty-one-year-old Baker, Rudder noted that the colonel's record was "among the most distinguished of our graduates." A veteran of both World War II and Korea, he held the CIB (Combat Infantry Badge), two Silver Stars, and three Bronze Stars. Baker, whose son Darrel had also been in the Corps, was eager to return to his alma mater.[37]

Having been away from the campus and Corps for most of the three decades since he graduated, Colonel Baker was both surprised and pleased with the progress and growth of the campus—a far cry from the remote campus of the nearly 2,000-member Corps of Cadets in 1932–33. Both the Corps and university had changed and, furthermore the new Commandant arrived on the scene as both the institution and Corps adjusted to even more scrutiny. Former cadets recall that Colonel Baker was "all military and ramrod straight." While not liked at first, the "by-the-book" Commandant soon gained the full respect of the Corps. Thus, it was Rudder's intent to allow Baker to be the catalyst to reshape and motivate what was to soon be an all-volunteer "elite" Corps. Rudder, Baker, and the A&M Board walked a narrow line under the watchful eyes of the vocal pro-Corps Aggie alumni. Gradually Aggies statewide agreed with Rudder that retention, academic excellence, and discipline should be stressed—yet still not at the expense of the traditions and esprit de corps that had identified the institution over the past eighty-five years. The cadet leadership in the Corps slowly welcomed the change and fresh approach. Many cadets and former students had other concerns and priorities.[38]

The changes at A&M in the early 1960s were made even more irritating due to the fact that the 1960–63 Aggie football teams

TABLE 8-1
Corps of Cadets Fall Enrollment, 1960–69

Year	Fish	Soph	Jrs.	Srs.	DNC*	Corps	Univ.**
1960	1839	1067	488	311	376	4081	7221
1961	1901	1185	407	344	377	4214	7734
1962	1856	1076	439	305	384	4060	8142
1963	1450	952	366	391	353	3502	8174
1964	1456	889	282	279	250	3156	8339
1965***	1317	611	408	408	252	2750	9521
1966	1380	572	485	418	259	2838	10677
1967	1134	665	483	465	271	2739	11841
1968	1207	653	512	499	209	2871	12867
1969	1171	717	442	423	165	2918	14034

* The DNC cadets include both juniors and seniors who do not have a contract to be commissioned in the armed services.

** The name was changed in August, 1963, from the Agricultural and Mechanical College of Texas to Texas A&M University.

*** The Corps of Cadets was made noncompulsory for the second time in April, 1965, to be in effect for the fall semester of 1965.

won only ten games in four seasons. The 1963 season under Coach Hank Foldberg was a disaster. The season opened with an 0 and 3 start, with losses to LSU, Ohio State, and Texas Tech. The Aggies beat Houston 23-13, then dropped all but one of the balance of the games for a 2-7-1 finish. In spite of the bleak record, the Aggie Twelfth Man by tradition wholeheartedly supported the team in victory or defeat—the adage at College Station was that Aggies never lose, they only get outscored. And the "salty" public address announcer C. K. Esten, an assistant professor of English, had by the mid-1960s become the much recognized "Voice of Kyle Field."[39]

In like fashion, numerous rituals accented the fall gridiron season. Corps trips, weekly Yell Practice, and Bonfire were the big public events.[40] Back on the Quad a whole set of time-honored customs were ingrained in Corps lore. The fish were required to learn the names and positions of the entire Aggie

team roster, the schedule, and key gridiron stats. Furthermore, fish were expected to know the name, hometown, and academic major of all the upperclassmen in the dorm by Thanksgiving. In addition to answering endless "campusology" questions, fish made "spurs" out of coat hangers and bottle caps to be worn the week prior to the game with the SMU Mustangs. Upperclassmen were not exempt. Juniors were routinely required to stand in their chairs during evening chow and "call the hogs"—Soooeee, Pig!—prior to the University of Arkansas Razorback game. Aggies even poked fun at themselves via Jim Earle's "Cadet Slouch" cartoons in the Batt. After one dismal scoreless game, Cadet Slouch wondered if the Aggie tradition of kissing your date after a touchdown could be adjusted to "kissing after first downs!"[41]

These routine annual traditions were eclipsed midway through the 1963 season when a group of cadets developed an elaborate

Ol' Sarge, the 105 Howitzer first used in the north end zone of Kyle Field at the 1963 Thanksgiving Texas–Texas A&M game. Loading are cadets Ken Koch '66 (left) *and Malon Southerland '65. Courtesy J. Malon Southerland*

plan to "mascotnap" all the rival Southwest Conference icons—alive or inanimate—the (stuffed) Owl from Rice, the Pony from SMU, the Horned Frog from TCU, and the 1,700-pound Longhorn steer from Texas. As the informal and fully unofficial scheme unfolded, mascots would appear on the A&M campus and then be returned. The *Battalion* ran a headline, "Owl Home—Steer Gone," as the cadets prepared after some delay and negotiation to return the Texas mascot, Bevo, to its Silver Spur handlers. Colonel Baker and Dean of Students James P. Hannigan worked to downplay the seriousness of the incidents, indicating to the media that the involved cadets would "probably" be placed only on conduct probation. The antics of the

fall of '63 proved a tremendous morale boost. A&M officials and the Texas Ranger in charge, O. L. Luther, were assured by the cadets that the current "Bevo-napping was brought to an end" and that they would put a halt to all future episodes during the football season.[42] Only time would tell!

RANGER

In the mid-1960s, A&M mascot Reveille II had become a permanent part of all Aggie events. Raised and cared for by the cadets of Company E-2, the fourteen-year-old collie mascot was nearing retirement when cadets launched a veiled attempt to select Ranger II,

Boot line in Kyle Field. Photo from author's collection

President Rudder's second pet bulldog, as a viable successor. Ranger I, named for the Ranger Battalion Rudder commanded during World War II, died shortly before the Rudder family arrived in College Station in 1958. Ranger II was a favorite of the cadets as well as the "campus dog and unofficial mascot." The feisty canine, nicknamed "Earl" by the cadets, often sat in the middle of campus streets to stop cars, securing rides around the block or across campus. In the fall of 1965, Ranger, in a surprise appearance, had a brief moment of glory at Kyle Field: "When Reveille became sick this fall, Ranger filled in at the first home game. After the Aggie Band took the field at halftime, Ranger exploded from the chute at the north end zone and romped hell-for-leather past thousands of cheering fans. Before the evening was over, the adventuresome animal tripped half a dozen band members, attacked the bass drummer, and assaulted the University of Houston cougar."[43]

Ranger's bid for A&M mascot never materialized. The popular campus favorite died on December 9, 1965, and was buried in front of the president's home. Reveille II, A&M mascot from 1952, was retired at Final Review in May, 1966, and was succeeded by Reveille III in the fall of 1966.[44]

BONFIRE

Yearly in November, the Corps of Cadets look forward to the ritual of building the annual Aggie Bonfire, symbolizing the burning desire Aggies have to "beat the living hell out of the University of Texas" or t.u., their arch rival, on or around Thanksgiving Day. The early 1960s was no exception. The cadet-organized project, which by tradition is built by each new fish class, would require the involvement of the entire Corps. After a period of preparation in the cutting area near campus, safety training, and site preparation of the field behind Duncan Dining Hall, the construction would begin under the direction of the senior Yell Leaders. As the trees were cut, the logs were transported to the campus stack area. A majority of the work was done the week prior to Thanksgiving. In 1962, the game with Texas was to be played in Austin, thus the bonfire would be lit the preceding Tuesday. The use of a "center pole"—in the case of 1962, two sections of seventy-five and thirty feet lashed together—allowed for higher construction. The center pole of '62 Bonfire was ninety-five feet tall. While some equipment and cranes were used, the majority of the work was done by manpower. Built like a layer cake, each log was lifted by block and tackle (pulley blocks with rope for hoisting), then tied vertically in place with bailing wire. *Sports Illustrated* described the Aggie Bonfire as "a pyramid of logs." Once construction was complete, a symbolic outhouse painted burnt orange was placed on top of the stack. An evening of yells, fight songs by the Aggie Band, and speeches preceded the torching.[45]

The 1963 Bonfire followed the same routine. However, for the first time, the traditionally all-Corps project incorporated the use of "non-regs," or civilian students, in both the cutting area and at the stack. Civilian Yell Leader Royce Knox '64 had hoped to reduce friction between the civilians and the Corps and was instrumental in organizing the non-regs. An elaborate network of guards and checkpoints prevented the premature lighting of the stack. With the excitement of stealing the mascots, spirits were high on campus.[46] The *Battalion* editorialized the significance of the annual Aggie Bonfire tradition:

> We will readily admit that the construction of the Aggie Bonfire doesn't do much for academic quality of our University. But one thing is for sure, the construction of the world's largest bonfire—which the Aggie Bonfire most certainly is—demonstrates one of the characteristics that have won Aggies respect around the world. That one characteristic is the Aggies' ability to take on the biggest of jobs and then do it well.

> We want to promise the fish, who have never worked on an Aggie Bonfire before, that they are about to undergo an experience they will never forget and one that will seldom be matched—regardless of what they might accomplish during their lives. For many of the Class of '67 the Aggie Bonfire will be their first experience at accomplishing the apparently impossible. Learn the lesson well and it will go with you through life.[47]

On Friday, November 22, as workers prepared to put the finishing touches on the bonfire, tragedy struck in Dallas when President John F. Kennedy was assassinated as his motorcade rolled through the city. Shock gripped the nation, and scheduled events on the A&M campus came to a halt. Work on the nearly completed bonfire was also halted.

Out of respect, the cadets held a memorial service in Guion Hall and decided to cancel the pregame annual Bonfire to honor the slain president. The '63 Bonfire was dismantled and hauled off.[48]

THE CORPS STANDARD

Colonel Baker and his new staff worked closely with the cadet leadership to enhance all aspects of the day-to-day life and training in the Corps. One of the first steps was an extensive review of the Corps's policies and procedures. Over a two-decade period the "rules" had taken on a number of forms: "The Cadence," the "Aggie Code of Honor," "Cadet Code of Conduct," and the *Articles*

of the Cadet Corps.[49] In the fall of 1964, these many components were revised and updated into one document, *The Standard.* In the transition from the 1950s style of Army branch designations for the cadet units to the revised organizational structure without regard to Army specialty, cadet units had created an inconsistent mix of "outfit policies" and procedures that had been passed down through the years—some under the heading of tradition, others only what the cadets termed "good bull" or fun. Many of these practices did not agree with the *Standard* or, for that matter, the old *Articles of the Cadet Corps.* While mindful of military procedure and protocol, the Corps had long been a fraternity of sorts, with many smaller fraternities at the outfit level within the organization. The

TABLE 8-2
Texas A&M Corps of Cadets, 1964–69

1964–65 Corps Staff					
1st Brig	2ndBrig	3rd Brig	1st Wing	2nd Wing	
1BN 2BN	3BN 4BN 5BN 6BN		1GP 2GP	3BN 4BN	
	Plus: Combined Band		Total co/squdns: 40		

1965–66 and 1966–67 Corps Staff				
1st Brig	2ndBrig	Air Division		
1BN 2BN	3BN 4BN	1Wing 2Wing 3Wing		
	Plus: Combined Band	Total co/squdns: 30		

1967–68 and 1968–69 Corps Staff				
1st Brig	2ndBrig	1st Wing	2nd Wing	
1BN 2BN	3BN 4BN	1GP 2GP	3GP 4GP	
	Plus: Combined Band	Total co/squdns: 30		

Trigon or "tool shed" or "squirrel cage," as it was nicknamed in the 1960s, hoped to reel in those not in compliance. However, the cadet company and squadron commanders wielded a great deal of power and influence within their units and the Corps. This autonomy, while it created uniqueness, was hard to monitor in the Commandant's Office. Outfit policies were not the only issue addressed in the new guidelines.[50]

Concerned with the retention of freshmen, a close review was given to the role and authority of sophomores to train and discipline freshman cadets. The third classmen were advised they would not "give orders, instruct or discipline privates of the class junior to them (fish), except to cadets of their own respective units . . . and then only as specified by their unit commander when acting as a squad leader or assistant squad leader."[51] Juniors were to ensure compliance, and the senior class would be held responsible for the actions of those in their units. At the heart of these new policies was a need to improve the sagging retention, generate esprit de corps, and reduce any unfavorable publicity about the Corps due to hazing. Both Colonel Baker and President Rudder had good reason to be concerned. Of the 1,450 fish who entered the Corps in the fall of 1963, only 889 returned for their second year—a loss of 39 percent of the Class of '67 in the first year. Because retention was viewed by most cadets as secondary to preservation of the Aggie traditions, such numbers were merely statistics to most. Often during cadet bull sessions on the merits of changes in the Corps, retention, and hazing, someone would conclude that "Highway 6 runs both ways"—a veiled reference to the feeling that those who did not or would not fit into Corps lifestyle and norms were free to depart.[52] Responding to reports and petitions from the cadets that the new rules would damage the Corps and that the organization was "headed for extinction," Rudder emphatically refuted the short-sighted allegations: "I will do all in my power to see it [the Corps] strengthened and preserved . . . I want to see the Corps generate so much esprit de corps that incoming fish are struggling to get in, instead of to get out!"[53] And the change would continue.

NONCOMPULSORY CORPS

On April 24, 1965, the A&M Board of Directors eliminated for the second and final time

In the early 1960s each cadet unit designed distinctive outfit signs and logos that were displayed outside each dorm. This one is for Squadron One. Courtesy Texas A&M Photographic Services

the requirement for compulsory enrollment in the Corps of Cadets for entering freshmen.[54] This action followed a debate over the status of on-campus ROTC and a trend of "elective" attendance nationwide. Between 1961 and 1964, twenty-two major universities and colleges had eliminated the requirement for mandatory enrollment in basic ROTC.[55] The Commandant knew the new voluntary Corps would not be greeted as a positive change. Colonel Baker was frank in assessing the reaction of the cadets in a statement to the Board and the media:

> The change will be difficult for the Corps, but it could be done. You are not going to change to an elite corps overnight or in one semester. It takes time. I believe we have the student leaders to make it work.
>
> I believe the decision to change to a voluntary Corps will result in no major change in the attitude of the cadets toward discipline, good bull, and hazing.
>
> I also doubt that the problems in making the Corps into an elite outfit will be much less than they are today. The key to making the Corps into an elite outfit is to convince the cadets that cadet rank has precedence over class privilege and attainment of this goal will not come quickly.
>
> I believe the Corps will be a happier Corps, provided the Corps does not initially decrease more than 25–30 percent.[56]

President Rudder warned against any repetition of Corps-civilian "animosity that raged during the 1956–58 effort at voluntary military training." Corps Commander Neil Keltner '65 and Student Body President Frank Muller '65, a member of the Corps, both appeared before the Board of Directors and pledged their support to the new changes. However, to some cadets the new rules and changes making Corps membership voluntary fell on deaf ears. The talk of "extinction" and the death of what was perceived to be "Old Army" or the "traditional ways of doing things" caused many cadet units to harden their suspicions and reaction both toward the Commandant and the Rudder administration. The new set of rules was not going to change the Corps overnight. How long could the Corps continue practices that were deemed detrimental to both its viability and image?[57]

By the mid-1960s, funding of the expansion projects envisioned by Rudder at Texas A&M became reality. The 1965–66 operating budget of the university exceeded $30 million for the first time. A large portion of the additional funding went into increased faculty salaries, funds to enlarge the Cushing Library, and new science research facilities. In 1965, Rudder assumed the joint role as both president of Texas A&M University and chief administrative officer of the Texas A&M University System. Parking for student automobiles increasingly became a problem as enrollment was touted as the biggest since the late 1940s. Despite the A&M campus being one of the "largest in the free world," students complained about remote parking facilities—nicknaming them the "Hempstead" and "Navasota" lots! The number of coeds on campus rose from 254 in 1964 to 373 in the fall of 1965. The issue of full admission of coeds was again raised in a ruling by Texas State Attorney General Waggoner Carr, yet the Board chose to maintain its policy of limited enrollment. And no women were to be members of the Corps or "allowed to audit [ROTC] military science courses."[58]

In the fall of 1964, a full decade prior to the admission of women in the Corps, the first African Americans became cadets. The first five to enroll—Cecil E. Banks, Roland E. Corley, Joe Lee Micheaux, Jessie H. Smith, and Samuel B. Williams—only completed the obligatory freshman and sophomore years. While there was some discussion of separate housing, this was abandoned in favor of immediate integration into the Corps units. The Rudder administration was forthright in the admission of black students and expected no significant problems. Some of the early cadets reported encounters with subtle racism but most noted that it was sometimes difficult to separate Corps hazing from racial comments. In the years that followed hundreds of blacks have excelled in all facets of the Corps and campus life. Many have been commissioned and excelled in the armed forces.[59]

By late 1965, the strength of the Corps of Cadets dropped by more than 400 cadets or about 11 percent, while the overall campus enrollment jumped nearly 1,200 to 9,521 students. To address the projected reduction in the number of cadets, the administration and Commandant's Office had developed a plan prior to the June, 1965, Final Review to reorganize and downsize the number of staffs and company/squadron units for the 1965–66 term. During the summer of 1965, over the protest of incoming Cadet Corps Commander Ralph "Fil" Filburn III '66, the number of major unit staffs were reduced from fifteen to ten, and the number of company/squadrons from forty to thirty. The Maroon and White units of the Combined Aggie Band were not changed. The Army ROTC reduced its major units from three brigades with six battalions to two brigades with four battalions. However, the greatest change came in the Air

Force units. Col. Raymond Lee, professor of air science, advocated sweeping changes in hopes of giving the AFROTC more input in the day-to-day Corps operations. He was able to obtain approval for a new air division organizational scheme comprised of three wings (no groups) and fourteen squadrons.[60]

Units were consolidated or eliminated without any advance notice to cadet commanders. Colonel Lee also failed in an attempt to have the cadets in the Air Force units wear the blue Air Force uniform. The Corps returning in the fall did not welcome this rush to reduce units. The 1965–66 organizational chart lasted for only two years. In mid-1967, the old brigade (battalions) and wing (groups) were reinstated. For the balance of the decade, total Corps strength would average slightly more than 2,800 cadets, down 20 percent from 1965. However, the number of U.S. Army ROTC juniors under contract jumped from 141 in 1965–66 to 259 in 1966–67. The increase in advanced Army ROTC was attributed to the prospects of an expanded nationwide draft. By 1969, President Rudder's 1960 goal of doubling the size of the student body was realized as the campuswide enrollment reached 14,034 students. Hazing was never fully eliminated, and retention remained a problem throughout the balance of the 1960s. The Corps of Cadets represented 27 percent of the student body in 1969–70.[61]

The rapid increase in civilian enrollment was not without friction between the cadets and civilian students. Rising demand for campus housing, concern by the cadets about the state of A&M traditions, and the general restlessness of the late 1960s did at times manifest in campus unrest—however, the unrest was among Aggies over campus issues and not due to any reaction or protest to the Vietnam

War or protests on other campuses.[62] One particular incident was vividly recalled by newly appointed Corps Commander Eddie Joe Davis '67. In early May, 1966, Davis had just been notified he would head the 1966–67 Corps and was in his dorm room when he received a call from President Rudder to "get over to the North side of campus and stop the water fight." By the time he arrived on the scene, dressed in full uniform, a confrontation between an estimated 500 students—cadets and civilians—had escalated, with each side pelting the other with rotten eggs, fruit, and buckets of water. Wading into the melee, which started over a late-night "drowning out" and damage in two civilian dorms, Davis was abruptly "doused with water as he appealed for order." The Commandant's duty officer, Maj. Robert B. Moore, just missed being hit and withdrew. The incident soon calmed down and Davis reported to the president that all was in order. "I had just been selected the Corps Commander," recalled Davis, "and I guess the General was just checking me out to see how I would react under fire!"[63]

After four years as Commandant, Colonel Baker decided to retire from the Army in June, 1967, shortly after suffering a heart attack. President Rudder was especially pleased with the contribution by Baker and nominated him for the Legion of Merit, the nation's second-highest decoration for noncombat recognition. Cadet Colonel of the Corps Davis presented Baker with a saber on behalf of the cadets. Eager to continue the use of a regular Army officers in the Office of the Commandant, Col. Jim H. McCoy '40 was assigned to A&M from his job as deputy chief of staff for logistics at the Pentagon. McCoy, like Rudder, attended Tarleton State College prior to graduating from A&M in 1940. The Eddy, Texas, native com-

manded an infantry battalion during World War II, receiving the Silver Star, Legion of Merit, Bronze Star, and CIB. An easy-going traditionalist, he reinstated the old style staff organizational structure and allowed the cadet leadership tremendous latitude in running the Corps. Departing from the strict policy to have active duty military personnel in the Commandant's Office, Rudder approved the hiring of Sgt. Maj. Ken Nicolas to work as the assistant to the Commandant following the retirement of Elizabeth Cook, who had been in the Commandant's Office for forty-four years. This move proved timely for continuity. Nicolas held the job, which included an active role as advisor to the Ross Volunteer Company, from 1966 until 1976. The staff between 1968 to 1972 consisted of the Commandant; Maj. Hal Wandry, discipline and awards; Maj. Ed Solymosy, operations; Malon Southerland, civilian Corps advisor; Nicolas, staff assistant; and two secretaries.[64]

VIETNAM

Nationally, by 1968–69, organized hostile demonstrations and protests against the war in Vietnam and opposition to the presence of ROTC on campus as well as military recruiting programs such as the Marine Platoon Leaders Class (PLC) program, were rampant. In 1968, shortly after the early January Tet offensive in Vietnam, on-campus antiwar protests escalated nationwide. The antiwar protest group, Students for a Democratic Society (SDS) established chapters at more than 300 campuses. In November, antiwar protesters nearly brought the Democratic National Convention in Chicago to a halt. At A&M, the SDS maintained a small office at

Northgate across from the campus and occasionally met in the YMCA; membership, comprised primarily of civilian students, was limited.[65]

The antiwar movement gradually turned more violent. The "Weathermen," a small violent splinter group of the SDS, conducted a number of sensational bombings of draft boards, campus ROTC buildings, and federal installations. In fall, 1969, a black powder bomb tied to the base of a tree exploded in the Quad between Dorms 1, 2, 3, and 4. There were no cadet injuries, but the blast did substantial window damage to Dorm Two, which housed Corps staff HQ and the Guard Room on the first floor.[66] Assuming that the ultra-conservative Texas A&M was the victim of a militant war protest, the event drew national media coverage. Someone at the *Battalion* office placed a "flash" story on the Associated Press news wire. After a brief period of confusion, the incident soon died down when it was learned that four freshman cadets in Squadron Three in Dorm 12 had set off the explosive charge as a prank. A&M officials were relieved; those involved in the incident were expelled from the university.[67]

General Rudder, who had traveled in Vietnam for three weeks in mid-1966 with a special fact-finding team assembled by President Lyndon Johnson, declared there would be no campus protest at Texas A&M and cautioned students to watch out for "kooks and anti-militarists . . . meeting their attack is primarily up to you but I guarantee you won't walk alone! They would have a hell of a fight and this pot-bellied president will be in the front ranks leading." Rudder quietly ordered that the *Battalion* be monitored for any provocative articles and editorials that could cause unrest on campus.[68]

Nationwide, the Selective Service Classification system and ROTC were under major attack. President Richard Nixon attempted to convince the American people that the war would be won with a new emphasis on "Vietnamization," a strategy to turn over much of the combat role to the South Vietnamese. Reports of renewed U.S. bombing of North Vietnam and the use of American troops in the incursion into "neutral" Cambodia in early 1970 was followed by nationwide protest and the shooting by the Ohio National Guard of four Kent State University students during a protest rally. Major violence was reported at 73 colleges. The National Guard was called out at least twenty-four times, and ultimately more than 450 colleges were closed for some period of time. In April, 1970, antimilitary students and faculty at VPI "purposefully disrupted a regular scheduled drill" of the corps. But not at Texas A&M. Although A&M had a number of intense Corps-civilian rivalries, it would remain peaceful through the 1960s and early 1970s, prompting Student Body President Gerald Geistweidt '70 to call A&M a "Bastion of Sanity." General Rudder would tolerate nothing less.[69]

VIETNAM: AGGIES IN ACTION

The cadets were well aware of the change in attitude toward the war, as well as the nationwide antiwar protest. Televisions in the student lounges and the Memorial Student Center were accessible to all cadets. The civilian student body remained a minor campus influence, compared to the authority of the Corps of Cadets. The campus civilians were basically conservative. Students both in and out of the Corps, if they failed to remain academically qualified, were subject to the draft. The upperclassmen in the Corps without Army or Air Force contracts to pursue a commission were vulnerable. Reports and rumors of Aggie officers in action and the mounting losses in the war were routine news on campus. The reputation of A&M's military record, as well as the image and example of Earl Rudder, influenced the attitude of both the alumni and cadets tremendously. Many cadets during the 1960s were from Aggie families, and many had had one or more relatives who had served on active duty either in World War II or Korea. Thus, the Texas A&M reputation as a conservative and fiercely patriotic school permeated the cadet value system through the 1960s.[70]

Nonetheless, there was a concern about the image of the institution across the state. In a confidential study, known as the Reilly Report, commissioned by the Association of Former Students in 1967, a New York marketing research firm confirmed the heavy identification statewide of Texas A&M with military training, ROTC, and the Corps of Cadets. High school students, their counselors, and the general public all had a clear understanding of the image and reputation of A&M. In the late 1960s approximately 45 percent of the nation's adult males were veterans, so the study reflected the views of an older generation that appreciated military training. Thus, the conclusion of the Reilly Report was clear and of no real surprise. The image, as well as the advantages, of membership in the Corps was affirmative: the organization "teaches discipline, respect, responsibility, leadership, and decision-making skills."[71]

Aggie officers excelled at their military duties and were clearly products of their training

and environment in the Corps of Cadets. At the yearly Army and Air Force summer camps, the Aggies annually captured the top awards. The 1968 *Aggieland* boasted that Army and the Air Force cadets "proved their worth by double-timing away with most of the honors."[72] For example, Joe Bush '66, Head Yell Leader, in the summer of 1965 was named the outstanding cadet among more than 1,800 who attended camps in both the Fourth and Fifth Army ROTC training regions. Bush would be killed in action in Laos in 1967. Neal L. Keltner '65, former 1964–65 Cadet Corps Commander and Distinguished Service Cross recipient, returned after his tour of duty in Vietnam to brief U.S. Army ROTC educators on the status of training at their annual meeting at Fort Monroe, Virginia. Furthermore, Keltner was the 1965 recipient of the Hughes Trophy, which recognized the outstanding ROTC graduate in the nation.[73]

The sense of duty and obligation was great. Michael Lee Lanning '68, author of *The Only War We Had* and an infantry company commander in Vietnam, recalled that there was an "obligation to duty." Lanning stressed the idea of duty during the 1960s by noting, "That was enough for me. There was even a fear among some that the war would be over before we could get there." Eddie Joe Davis, 1966–67 Corps Commander and Vietnam veteran, noted that there was something of a shared bond during the tumultuous period: "We had strong agrarian values and roots."[74] During the Vietnam War officers and troops from Texas A&M became high-profile news.

The January 1, 1968, cover of *Newsweek* magazine displayed a vivid picture of wounded American troops in Vietnam scurrying for protection at the Battle of Dak To in the highlands

TABLE 8-3

Vietnam-Era Officer Production at Texas A&M, 1962–73

Year	Army	Air Force	USMC*	Total	Generals**
1962	180	109	3	292	11
1963	190	74	3	267	5
1964	205	104	4	313	2
1965	151	94	11	256	7
1966	193	62	11	266	5
1967	151	96	14	261	5
1968	229	128	18	375	5
1969	259	93	17	369	3
1970	295	118	18	431	3
1971	277	112	15	404	1
1972	247	136	16	399	2
1973	191	146	15	352	1
TOTALS	2,568	1,272	145	3,985	50

* Commissions in the USMC were through the PLC program.
** Texas A&M graduates who attained General or flag officer rank.

Source: Office of the Commandant and Texas Aggie

north of Pleiku. In the center of the picture, taken during one the Army's major engagements with the North Vietnamese regulars or NVA, was Aggie Lt. Larry C. Kennemer '66. Kennemer would be one of an estimated 3,000 Aggies to serve in Vietnam during the decade-long war. The Corps in the 1960s prepared and motivated cadets for leadership in time of war. Three decades after graduation, Kennemer reflected on the impact of the Corps and Texas A&M:

I wish there had been a film crew on the Quad to capture all that was going on. I was 17 years old and was totally unprepared for A&M. My Dad was the Class of '41 and had not said a lot. I was in the Aggie Band. Colonel Adams ran it and the Trigon left us alone. Bonfire was a real turning point. We got to meet a lot of guys from other units. Many I would later see on active duty. On the weekends we had a number of practical warfare exercises to learn tactics and how to deal with certain situations in advance of summer camp. I went to Ft. Sill, we really got a taste for the Army, that's when I knew I wanted to go Infantry. We took all the awards at camp. The reason we took all the awards is the Aggies always wound up being the leaders!

We were prepared, because we had been in the Corps. You are going to be a leader . . . if you can't cut it here, you can't cut it anywhere. I spent 40 months "in country" [in Vietnam] and never had a staff job. I commanded everything from a rifle platoon to a Special Forces "A" detachment. Aggies were where the action was.[75]

The Vietnam War build-up steadily gained momentum by the mid-1960s, but American advisors and combatants had been in Indochina since the early 1960s. Following a seemingly "unofficial" American presence after the French defeat at Dien Bien Phu in May, 1954, during the Eisenhower years, the Kennedy administration created the U.S. Military Assistance Command, Vietnam (MACV), in February, 1962, and quietly committed more than 16,000 "military advisors" to Vietnam.

The *Battalion* began covering the expanding war as early as 1962. Usually articles were buried deep in the paper and nearly all had a common theme: this conflict is a "limited war"; it is about to be won; and U.S. technology is going to make winning a reality.[76] One item of technology touted in late 1962 was a new "turbojet" helicopter, the UH1A—nicknamed the Huey—soon to be one of the most identifiable features of the war. Military public relations sources, wanting to overstate

Hundreds of Aggies from the classes of the 1940s through the early 1970s served in all branches of the armed forces during the Vietnam conflict. On April 21, 1970, these six Aggies (left to right), 1st Lt. Ernie Petrash '67, Capt. Bill Klutz '65, Lt. Col. Bob Bell '53, 1st Lt. Lou Obdyke '67, Capt. Al Rutyna '65, and 1st Lt. Tom McKnight '68, all members of the 8th Tactical Fighter Wing, flew an all-Aggie combat mission in their F-4D Phantom II jet fighters to destroy a bridge in southern Laos. Courtesy Texas Aggie

the role of technology, proclaimed, "It [the Huey] can pack more fire power than any fighter plane that flew in World War II." Such dramatic statements were intertwined with the politics of the era.[77]

Ironically, Lyndon B. Johnson—vice president in mid-December, 1962—delivered the keynote address at the eighth annual Texas A&M Student Conference on National Affairs (SCONA) on the "sources of world tensions." Johnson noted that "economic problems, nationalistic aspirations, and the arms race are causes of the real source of world tension! If we are seriously to understand the relieving of world tensions, a greater degree of political courage, political imagination and political innovation will be required, in both the developed and under-developed worlds."[78] Within twenty months, the limits of such intentions and "political innovation" would be breached, when then President Johnson submitted to Congress the Gulf of Tonkin Resolution to approve and support all necessary measures to repel "any attack" against the forces of the United States. Within weeks of the August, 1964, congressional resolution, full-scale U.S. military operations began in Vietnam, and Congress passed the ROTC Vitalization Act of 1964 to encourage the production of more officers. In announcing his approval of the act, Johnson acknowledged the role of the nation's citizen-soldiers: "The roots of the ROTC program reach back more than a century to 1862 when the Morrill Act required the land grant colleges to offer courses in military training. This vital program [ROTC] constitutes the largest single source of trained officers not just in the Reserves, but for the Regular forces as well."[79]

Col. Ted Lowe '58, 1957–58 Yell Leader, recalled that Aggies were among the first officers

During the decade of the 1960s A&M commissioned more than 1,700 officers in the armed forces. By the turn of the century 50 of these graduates had reached the rank of general or flag officer. Pictured in mid-2000 at Pacific Air Forces headquarters at Hickam Air Force Base are (left to right) Rear Adm. William "Bear" Pickavance '68, Air Force Gen. Patrick Gamble '67, and Army Lt. Gen. Randolph House '67. Photo by Navy Photographers Mate 1st Class Gregory Messier, courtesy U.S. Command Photos

deployed as "advisors" to Southeast Asia in small numbers as early as 1960. Air Force pilot Condon H. Terry, Jr., '56 of Dallas was the first Aggie to die in Vietnam when his T-28 was shot down in June, 1963.[80] In early 1967, Dorsey E. McCrory, the director of development, released details of former 1953–54 Cadet Corps Commander Maj. Fred Mitchell's "two remarkable and successive acts of bravery" late in 1966, resulting in the Silver Star and Distinguished Flying Cross. Such reports had a tremendous impact on morale. Corps members asked for assignments to a combat arms branch—infantry, artillery, or armor. Among Air Force cadets, the number one slot was to be a fighter pilot and, if not available, the next closest thing to "real" action would be a FAC or forward air controller, flying a slow-moving observation plane oftentimes at treetop level. The Corps produced loyal, spirited, aggressive leaders. Their contributions were substantial. Nearly 4,000 Aggies were commissioned between 1962 and 1973; of these 50 former cadets became general or flag officers.[81]

THE TOLL WAS GREAT

Of the hundreds of Aggies who served in Southeast Asia, 160 died or were listed as missing in action. The missing-in-action soldiers have, since 1980, been declared killed in action. The Class of 1966 sustained the highest losses with 15 casualties—13 Army and 2 Air Force. The classes of 1965 and 1969 each lost 11 members. In 1967, Maj. Gen. Bruno A. Hochmuth '35 became one of the few generals killed in action, when his helicopter was shot down. During the decade-long Vietnam era, 1962–73, Texas A&M commissioned 3,985 officers via ROTC and the Marine PLC program (see Table 8–3), while many other A&M graduates enlisted and became officers by way of Officer Candidate School (OCS). In the years since the Vietnam War there has been an ongoing debate about the makeup of the troops that fought the war.[82] The draft deferments and the purposeful avoidance of wartime duty by many college graduates only magnified the proportional contribution of the Aggies. Those who could figure out how to work the system were exempt. Not the Aggies. James Webb, former secretary of the Navy and Vietnam veteran, in the 1980s compiled data that dramatically demonstrate the profile of those who served. Webb contacted Harvard,

MIT, and Princeton and asked for the number of male undergraduates who graduated and the number who died in uniform from 1962 to 1972. Of a total of 29,701 undergraduates from these three colleges, only 20 died in Vietnam.[83]

For those soldiers and airmen assigned to Southeast Asia, it was possible to avoid a combat assignment, because 88 percent of all service personnel were assigned to noncombat "occupational specialties" or rear-echelon support duty.[84] In contrast, more than 70 percent of all Aggies served in direct combat arms and leadership roles either on the land, sea, or in the air. For example, Capt. Robert L. Acklen, Jr., '63 became one of the most decorated veterans of the war with sixty-three decorations, fourteen for valor.[85] Moreover, of those commissioned between 1962 and 1966 from Texas A&M, 30 former cadets rose to the rank of general or flag officer by the 1990s. Carrying on the tradition of the citizen-soldier in war and peace, hundreds of the 1960s Aggie veterans would honorably finish their four- to six-year tour of active duty and return to leadership positions in all walks of civilian life.

On November 8, 1969, even before hostilities had ended in Vietnam, cadets from the classes of 1969 through 1973 helped fund and dedicate the Memorial Meditation Garden located at the south end of the Quad in front of Duncan Dining Hall in tribute to the more than 400 A&M men who had given their lives in defense of the nation since World War II. Gen. Bernard Schriever '31 delivered the dedication address. This memorial was moved from the front of Duncan to the plaza on the north end of the Quadrangle in 1976. The names of those killed in Operation Desert Storm were added in 1991.[86]

THE ESSENCE OF THE CORPS

No assessment of the 1960s in terms of the sacrifice and commitment of the Aggies called to active duty will ever be complete. An entire generation came of age in both the national and world spotlight. The veterans of Vietnam did not experience the fanfare that followed World War II. Instead, they returned home to a nation that questioned the role of the citizen-soldier—a fundamental tenet of the United States. The 1962 *Aggieland* records one observation of A&M cadets' attitudes during the decade of the 1960s:

OBJECTIVES OF THE CORPS OF CADETS
1. To produce officers who have the qualities and attributes essential to their progressive and continued development as officers in a component of the Armed Forces of the United States of America.
2. To lay the foundation of intelligent citizenship for the cadet through a training program designed to be of benefit to him, to the State of Texas, and to the Nation whether in military or civilian life.
3. To provide strong training in basic principles of leadership including promptness, fairness, intelligence, and common sense.
4. To develop a deep sense of responsibility, honor, and integrity, and to cultivate habits of cheerful obedience and precision in the maintenance of high standards of performance, whatever the task.[87]

By the end of the decade a process was initiated to determine the future course of the Corps of Cadets. The Rudder era had tackled a long list of challenges—coeds, voluntary Corps, change in the name from college to university, doubling the size of the student body, support of an unpopular war by men who felt it was their duty to serve, and a growing realization that the Corps of Cadets in years to come could not do business as usual. The process set in motion by the 1962 Century Study and the conclusions reached in the 1967 Reilly Report confirmed without a doubt that the Corps and the university for the balance of the century would be the product of changes made in the Rudder era. Although Texas A&M avoided the antimilitary sentiments rampant on other campuses, the Corps of Cadets faced the issue of how to enhance its image and mission given its traditional all-male makeup and the changing requirements of the armed services.[88] At the heart of the concern was the long-term future of the Corps and the modifications needed to bring it in line with the changes in the overall growth of the university.[89] A 1969–70 internal Corps assessment led to new procedures that molded the Vietnam-era Aggie cadet.

The "system" of priorities and loyalties in the Corps was based on an allegiance and commitment to your "buddies" and your "class." Most cadets felt strongly that General Rudder personified the essence of service to country. Furthermore, a strong unwritten code of loyalty and comradeship had been developed from the first days of the fledgling college, and the post–World War II attempts to deemphasize what had become a rigid structure of loyalties within the Corps had proved difficult. For many, for better or worse, it was the essence of the loyal "Texas Aggie" and the "Aggie system" that produced citizen-soldiers in the Corps of Cadets. The 1969–70 First

Brigade Commander, David Reed '70, captured the quintessence of life and allegiance in the Corps:

The Buddy System and Class System at Texas A&M cultivates a unique soldier for the United States, a unique citizen for the United States. These systems create a unique individual. The Corps never has, and I sincerely hope never will, be noted for producing Little Tin Soldiers. Actually the concept of militarism that has developed at Texas A&M was very accurately defined by *Time* magazine several months ago when they referred to our John Wayne militarism. Perhaps they thought they were being critical of us but actually they rather inflated our egos because we like to think of ourselves as, and I feel our Military Advisors like to think of themselves as creating a different breed of soldier.

What is this breed? Well, it's almost a militia type concept, a rough, hard living, well trained militia that is far differentiated from the standard modern stigma of the mechanical military man. There are very few "Yes Men" that come out of the Aggie Corps but rather people that think for themselves. People that are allowed to express themselves and will not settle for less. People that won't be forced into a slot where they are to be quiet, not express their opinions, and not ask questions. This is not the Texas Aggie.

He's a man that says what he thinks. A man that forces other people to think by his questioning attitude, by his very presence. He's an independent sort of fellow who will walk up and look you in the eye and call a spade a spade. He's a take-charge guy because he's been taught to do this, he's been allowed to do this, by the Class System. He's also been taught loyalty by the Class System. He's been taught loyalty from the foundation up, something the Texas Aggies are famous (or perhaps infamous) for. He's been taught loyalty to his immediate peers, the people he lives with, his buddies. And, having developed this loyalty to them, he's able to realize the beauty and value of transferring this loyalty to school and nation. He's a fellow who's been taught to passionately love things that are bigger than any one-to-one relationship he might have and he has learned to make very great personal sacrifices for these deeper, broader considerations.

He's been taught these things and he's been made this type of a man by the Aggie System. And this system is undeniably intertwined with the Buddy System. Its foundation is the Buddy System. This is what gives us the Texas Aggie. It's not the Rank System and it's not the military system as such. It is this "Aggieized" Military System that produces the type of individual just described. And so before we say that the obvious answer is to do away with the Buddy Class System, we must evaluate the price this will cost us. The price is the Texas Aggie.[90]

For the balance of the century the Corps of Cadets would struggle to balance the "system" and the vestige of what many referred to as "Old Army" with changes in society—largely a byproduct of the societal changes in 1960s—that gradually developed a divergent view of military-oriented training and campus ROTC programs. Neither the Corps at A&M nor its leaders were fully prepared to face the impending changes. Nonetheless, unlike many colleges that discarded and disbanded their uniformed cadets in the late 1960s and early 1970s, Texas A&M resolved to enhance and continue an aggressive (yet diminished) full-time military training program.[91]

END OF AN ERA

Tragedy struck in early 1970 with President Rudder's death at age fifty-nine. As had his predecessor of the 1890s—L. S. Ross—his decade of service at Texas A&M had made a profound difference. The death of General Rudder accented the end of an dynamic era at Texas A&M. Ross and Rudder—both from humble rural backgrounds, heroic soldiers, politicians, and leaders—made their boldest contributions in the realm of education—always with a close eye and obligation to the Corps of Cadets and the traditions it represents. General Rudder helped turn the small agrarian college into the foundation for one of the nation's leading universities. In the process he was able to keep peace among a diverse group of interested parties. Rudder presided over the evolutionary process of reconciling the traditional pro-Corps focus of the alumni with those who desired to expand the academic programs as well as admit women on a full-time basis. He knew the overall growth of the university would, in time, dwarf the Corps of Cadets. Yet he also clearly realized that the bedrock of Aggie traditions and lore would best be maintained in the able hands of a focused and better-organized Corps of Cadets. Thus, the Corps had been preserved.

General Rudder had indeed had a dramatic impact on A&M. In 1973 the east wing of the Memorial Student Center was named Rudder Tower, and a life-size bronze statue was sculpted from the same clay that was used to create the likeness of L. S. Ross in 1919. The Rudder statue was dedicated on October 15, 1994. Annually, the most outstanding A&M graduate is honored with the Brown Foundation–Earl Rudder Memorial Outstanding Student Award.

In campus ceremonies lowering the flag in front of the Academic Building, Gerald Geistweidt '70, student body president, spoke for all Aggies: "President Rudder was many things to many people . . . a war hero . . . a holder of public office . . . a general . . . a president. He wore the same ring I wear and earned the same diploma. But to most of us, he was first and foremost an Aggie.

9

THE
CENTENNIAL
DECADE

*When outsiders think of Aggieland, they think of bands and football,
but they also think of the Corps. At a time of overt anti-militarism and
bombed out ROTC armories, the Aggie Cadet Corps is not only alive
and well, but is expanding . . . the traditional cavalry boots are still
much in evidence here at A&M.*

Lynn Ashby, *Houston Post,* 1972

*We plan to see that the Corps not only remains but prospers. It is abso-
lutely essential to the existence of Texas A&M, not because of what it
has been in the past, but what it can be to the future.*

Gen. Ormond R. Simpson '36, Aug. 30, 1974

*W-1 was just a vision and a written plan. It was a part of the histori-
cal transformation of A&M and the Corps of Cadets. It had been a
male institution, and a lot of animosity was there. We knew we had
our hands full.*

Don Roper '75, First CO of W-1, Twentieth Reunion, November, 1994

THE DESTINY OF Texas A&M Uni-
versity in the early 1970s was largely preor-
dained by the actions and strategic planning
undertaken during the Rudder administration. President James Earl Rudder, as early as
1961, had already begun to shape the image

and focus of the university toward the centennial anniversary in 1976. The Century Study
and the resulting blueprint set in motion the
expansion of the student body, faculty, facilities, and programs. Rudder's untimely death
left to others the task of completing what he

felt was the course for a truly great university. At the onset of the new decade of the 1970s, the Corps of Cadets faced the need to make adjustments to the waning war in Vietnam, public antimilitary sentiment toward the war, and the reduction in the demand for commissioned officers, as well as the need to address one of the last barriers at Texas A&M: the admission of women in the Corps.

Shortly after the death of President Rudder in 1970, Alvin Luedecke '32, Rudder's classmate and a former U.S. Air Force general, was selected as acting president. Upon graduation from Texas A&M in 1932, he began his military career as an Army field artillery officer and then transferred to the U.S. Army Air Corps. During World War II, Luedecke served in the China-Burma-India theater and was promoted to brigadier general at the age of thirty-four, thus becoming Texas A&M's youngest general officer. In addition to his distinguished military record, he served as the general manager of the U.S. Atomic Energy Commission at the Redstone Arsenal and as deputy director of the Jet Propulsion Laboratory in its formative years during the early stages of the Cold War. A quiet and reserved man, in his seven months as chief executive he kept the university on the path set by Rudder.[1] After an extensive search, the Board of Directors selected Virginian Jack Kenny Williams as the eighteenth president of the university. The World War II Marine Corps officer had served with distinction in the Pacific theater, yet he seldom referred to his wartime record or experiences. Dr. Williams was not a surprise to the Board. In 1969 when asked with whom the Board should replace him if anything happened, Rudder had emphatically stated, "Go to Tennessee and get Jack Williams!"[2]

At the time of his arrival at Texas A&M, in fall, 1970, Williams was truly one of a group of progressive, highly experienced, modern-era leaders in higher education. Educated in the post–World War II era, he envisioned a broad role and diversification of academic and research programs at public universities. Assignments at the University of Tennessee and Clemson University (an institution with a strong southern military heritage) and his experience from 1966 to 1968 as the first commissioner of higher education in Texas prepared him well for the job at A&M. A leader in his own right, Williams nonetheless lived and administered in the shadow of Earl Rudder.[3] The two men had known each other as early as 1960, when serving on a national task force on the enhancement of ROTC programs in land-grant colleges. The growth and adjustments experienced by A&M during the Rudder years would be further magnified during the 1970s with the start of a multimillion dollar campus building program.

By early 1973, six major construction projects were near completion or in progress— Zachry Engineering Center; the 1,000-room Krueger-Dunn residence hall complex adjacent to the Corps Quad; the fifteen-story Oceanography-Meteorology Building; expansion of the Memorial Student Center and the addition of the University Center (Rudder Tower and Auditorium Complex); the Liberal Arts and Education classroom complex (Harrington); and the new A. P. Beutel Student Health Center next to the YMCA. The conversion of Military Walk, running north-south between Sbisa Dining Hall and the former site of Guion Hall (razed in 1971 and replaced with Rudder Tower) to a mall area sealed off the heart of the campus into a series of pedestrian walkways. Bolstered by a larger and more diversified student body, Williams determined that Texas A&M should enrich the quality of academic programs as well as enhance outreach efforts of the university's agencies. The vision of Dr. Williams for enhanced academic and research programs at Texas A&M was a hallmark of the centennial celebration and observance held in 1976.[4]

Though many Corps members at the time were concerned about changing policies at A&M, President Williams was very supportive of and comfortable with the Corps. For years he was fond of telling the story that he and his family had barely moved into their new campus residence when the Corps "marched on my home"—not in protest, but instead to welcome them to Texas A&M. This "march" was in stark contrast to the tumultuous antiwar years and campus marches at other universities and colleges nationwide during the late 1960s and early 1970s. Williams had strong words for "activists and protesters" as "the Trojan horse army of militant fanatics in our midst."[5]

THE CORPS, 1970S

By 1970, the Corps of Cadets numbered 2,200 cadets in a university enrollment of 15,000. The strength of the Corps would only once exceed this benchmark level for the balance of the century. In the early 1970s the cadets comprised twenty percent of the overall A&M student body. This percentage declined over the decade as the enrollment of the university experienced explosive growth. For the first time fees and tuition for the academic nine months exceeded $1,000, yet

there is no indication that the increased fees affected recruiting. Assessing the strengths and weaknesses inherent in its historical practices, the Corps conducted an extensive self-study in early 1970. As the Corps gradually diminished in size relative to the growing civilian student body, more emphasis was placed on the image of the cadets both on and off campus. There were many, for example, who questioned the cadet short haircuts—known as "white side wall"—for the fish class, the requirement to be in uniform seven days of the week when on campus, and the low level of cadet interaction with civilian students. Retention of freshman cadets as well as improved academic performance was considered vital to the growth of the Corps. New guidelines were drafted to address these concerns. Responding to cadet concerns, Cadet Corps Commander Matthew R. Carroll '70, in a detailed memo to all cadet commanders concerning changes to the daily cadet routine, concluded with the warning, "Time is running out and the requirement for change is immediate."[6]

Recommendations called for a fundamental adjustment in how freshman cadets would be treated. In an attempt to move away from the boot camp style haircut, cadets were advised to have hair "long enough to comb and part on top." Outside the Quad-Duncan area fish were to "assume modified sophomore privileges"—cadets were not to "wildcat" [cadet yell of approval], "pop to" [stand at attention], or use their traditional answers to such questions as "How are you [fish jones]?" ["Sir, hustlin' as usual, sir!"] or "How's it hanging?" ["Sir, tall and firm, sir!"]. Fish were to address upperclassmen in a normal voice, thus no yelling or "humping" outside the Quad. Meal service—the practice of fish waiting on upperclassmen and not being allowed to eat until after most of the seniors had left the mess hall—was to be discontinued until further notice. Wildcats and outfit yells were limited to three per meal. CQ or call-to-quarters, for the purpose of studying or visiting the library, was to be strictly enforced. And due to the increased number of midday classes and labs, the mandatory noon unit formation on the Quad and march to the mess hall was canceled. At the time, such changes were viewed by many cadets as once again an attack on "how the Corps had always operated." The 1970 *Aggieland* concluded that attempts "to 'professionalize' the Corps . . . had many startling results."[7]

In July, 1971, Col. Thomas Reed Parsons '49 was selected PMS and Commandant of Cadets. Parsons, a native of Arkansas, entered A&M in May, 1945, a week after his graduation from high school, and was a member of "A" Battery Field Artillery. The three-semester World War II academic program placed his fish class on a fast track until they were drafted. At the end of the war, Parsons and his class returned to their regular prewar footing at the school. A participant in the midnight cadet march on President Gilchrist in 1947, the new Commandant lost his cadet sergeant stripes in the aftermath. During his military career, he landed at Inchon and marched to the Yalu River during the Korean War. He was a forward observer for an artillery battery (1951–52), and he then had assignments in Germany, Hawaii, and Vietnam. Returning stateside, he held a number of command positions, including commander of the Fourth Infantry Division Artillery at Fort Carson, Colorado. Upon learning of the search for a new Commandant following the retirement of Col. Jim McCoy '40, Parsons met with Dr.

Williams both on campus and at Fort Riley, Kansas, where the president had gone to observe A&M cadets in summer camp. Parsons noted, "I really wanted the job at A&M and I lobbied for it."[8]

Parsons arrived at A&M to find what he called a "gung-ho organization in transition." The Corps of Cadets, although outnumbered by the civilian students, was the focus of most attention given to A&M. Most observers around the state, not familiar with changes on the campus, still viewed the university as essentially a conservative, military-oriented institution with excellent agricultural and engineering extension outreach programs. And with good cause. The public and potential new students had ample exposure to A&M as a result of the Corps pregame march-ins at Kyle Field as well as Corps trip appearances in either Dallas, Austin, Fort Worth, or Houston each fall. The Fightin' Texas Aggie Band, the Fish Drill Team, and the Ross Volunteers maintained a high profile. For example, public appearances at Fiesta Flambeau in San Antonio, the governor's inauguration in Austin, and the King Rex Parade at Mardi Gras each year in New Orleans received broad media coverage. A local campus tabloid featured the cadet spirit as part of "The Camp—Home of the Fighting Texas Aggies."[9] Emphasis on recruiting, in the wake of increased academic pressure and scrutiny to limit hazing and enhance the attention to military regimen, proved to be a major thrust throughout the decade. To encourage growth in the Corps, a special "replacement badge" was designed to be awarded to those cadets who brought in new recruits.[10]

To further define the period, the armed forces, anticipating the end of the war in Southeast Asia, began in the early 1970s to

The Ross Volunteers are the official honor guard for the governor of Texas. Pictured on the steps of the state capitol in Austin in January, 1975, are Governor and Mrs. Dolph Briscoe. Photo from author's collection

reduce the demand for new active duty second lieutenants. Total ROTC enrollment nationwide experienced a dramatic drop from 218,000 in 1968 to 68,000 in 1974. ROTC would begin to assume a different yet crucially important role in the shift away from the draft to an "all-volunteer" force in the post-Vietnam era. Notwithstanding, ROTC would continue to be the largest, most economical source of second lieutenants. In 1975, the Army commissioned only 9,224 new officers, the lowest number since the end of World War II. With the reduction in ROTC contract cadets at A&M pursuing commissions,

there began a piecemeal rise in the number of noncontract drill and ceremonies (D&C) juniors and seniors. These uniformed cadets had no military obligation, only a "burning desire" to be in the Corps of Cadets. Their desire to be a member of the Corps because of its traditions and fraternal aspects would gradually bring changes in Corps priorities, discipline policies, and recruiting philosophies. With the addition on the A&M campus of modern civilian residence halls, it became increasingly tempting after 1975, unlike at the service academies or the all-male Citadel, for a cadet who no longer wanted to stay in the Corps lifestyle to resign, remain enrolled in the university, and move into civilian housing either on or off campus. Those who chose the Corps wanted to be in the organization, but those who desired to depart had fewer barriers.[11]

In pursuit of the vision President Rudder and Colonels Baker and McCoy had for an "elite" Corps, emphasis was placed on attracting new cadets to a unique experience as a member of the bedrock of Aggie heritage in what was touted as the largest collegiate "leadership laboratory" in the nation. This emphasis on a more elite Corps was not intended to remove the cadets from mainstream campus activities but was instead to be a tool for recruiting new cadets. It was felt that the Aggie traditions, as well as such items as the senior boots, were a strong attraction. The alumni, nostalgic for bygone days, were generally quick to support any measure that enhanced the Corps. However, the Corps, as it annually became a smaller percentage of the overall student body, confronted a dilemma. This double-edged sword in time afforded the Corps an ongoing role as an elite organization but also placed the organization and all its members under greater scrutiny by detractors—both within and outside the university—who questioned the need for such a high-profile, *military*-oriented presence on a college campus.

Within the Corps, the final years of the Vietnam War from 1970 to 1973 witnessed a growing discontent among D&C cadets and fifth-year seniors who, given a low draft

Senior Ross Volunteer officer Bill Weber '76 handles details with Parade Marshall W. W. Young in February, 1976, for the Mardi Gras King Rex Parade in New Orleans. The RVs have appeared in the annual event since 1947. Photo from author's collection

number, could still be involuntarily inducted into the armed forces. Many of these cadets wanted the fraternal experience of the Corps and little or no further military service. Prior to the mid-1970s, D&C cadets were not allowed to hold command positions of cadet units. In 1974, a mild stir was created when a D&C cadet, Charles Scott, was selected as the Fifth Battalion Commander in the Second Brigade. Colonel Parsons ended the unwritten prohibition permanently with the selection of Robert Harvey '77 as the 1976–77 Cadet Corps Commander. Many D&C cadets attended Texas A&M fully aware they would not enter the armed forces. Their enthusiasm and esprit de corps was at the foundation of the institution. In the 1970s, fish, long before making a determination to take a contract, were mindful of one of the key campusology questions of the era: "Why did you come to A&M fish Jones?" . . . "Sir—To get that fightin' Texas Aggie Spirit."[12]

Thus, the post-Vietnam period was yet another era of growth and metamorphosis. The Corps each year fell further into the minority, yet it continued to champion and represent the foundation of the traditions and lore that persist even today. As the keepers of the spirit at Aggieland, the Corps would be both touted as a stalwart of traditional A&M values while at the same time regarded as too insular in its mission and ability to adjust to be a part of the emerging modern multifaceted university. What the conservative Corps had historically represented ran counter to the desires of more liberal academicians who thought the Corps hindered the image of the university. Most detractors of the Corps could neither understand nor appreciate regimented dorm-style living, a military mess hall and family-style dining, uniformed cadets in the classroom,

periodic military reviews and ceremonies, the Corps's high profile in the media, and the intense loyalty to the university and the Corps. Notwithstanding, the public impression of Texas A&M among most older Texans was extremely favorable in the early 1970s, helped along by feature writers like Lynn Ashby of the *Houston Post,* who regularly penned articles touting both the Corps of Cadets as well as the success, growth, and spirit of the university. The limelight on the Corps's traditional focus further attracted attention to A&M. With the ever-growing breadth of A&M programs in research, outreach via the extension and experiment agencies, and enhanced scholarships, it became evident that the Corps would increasingly be held to higher academic standards and expectations.[13]

CAMPUS LIFE

The growth and construction projects campuswide would have their greatest impact late in the decade. The campus of 14,000 students found itself in transition. The alumni, growing faculty, and more diverse student body prospered. In spite of the good times, one observer unaware of the traditions and esprit de corps of the A&M cadets noted in the fall of 1972: "For quite a number of years, A&M, with its unique eccentricity, separated itself from the world. It's been a nice place to come see and to send your son to, but you might think awhile before you sign up for another 48 months of it."[14] Notwithstanding, the Corps reveled in its uniqueness. The gradual transition of the cadets to a smaller overall percentage given the increasing number of civilian students was orderly and without

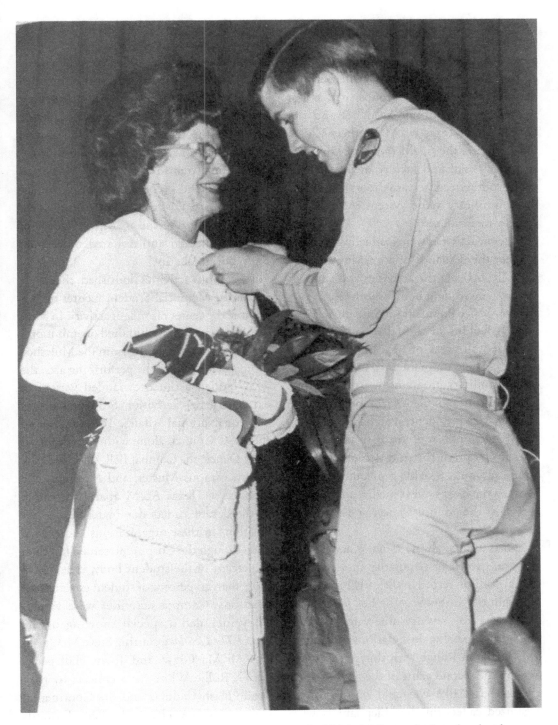

Aggie Mother of the Year for 1973, Ruth Hunt. Known as the "fish lady," for nearly three decades, she supported cadet activities and mentored the "fish." Photo from Aggieland 1972

major incident. One reason for the peaceful shift was the strong base of Aggie traditions, generally conservative atmosphere among all the students—Corps and civilian—and, the proactive nature of groups such as the Association of Former Students and the Federation of Texas A&M Mothers Clubs. Furthermore, the A&M Board of Directors and many staff and faculty members held strong southern conservative roots and shared the promilitary value system. Also, many were former cadets.[15]

Student activities were gradually expanded to meet the demands of a growing student body. To manage the activities and support the growing student body—both Corps and civilian—Vice President for Student Services John J. Koldus in the fall of 1973 greatly expanded all areas of on-campus student services. The Commandant's Office and all campus military programs by 1974 reported to Koldus through Gen. Ormond Simpson, assistant vice president, who had joined the A&M staff upon retirement from the Marines in 1974. For the next decade, General Simpson would provide a pivotal interface between the Corps, news media, and alumni as well as the administration and Board of Regents.[16]

Sports events, always a major activity on campus, gained in popularity in spite of a dismal gridiron record in the early 1970s. The highlight on the era was the Lex James–to–Hugh McElroy seventy-nine-yard winning touchdown pass against nationally ranked LSU in Baton Rouge with thirteen seconds to go in the second game of the 1970 season. On campus, Mike Weynand '72 of Hondo organized cadets for the "Elephant Bowl," pitting the seniors of the Army units against seniors in the Air Force at Kyle Field. Billed

as the football championship of the Corps, the event raised money for Bryan–College Station charities. The annual event supported by the Athletic Department, which provided student athlete coaches and equipment, continued into the early 1980s and was phased out after the reinstatement of the annual spring ritual of the "March to the Brazos." The annual trek to the Brazos has evolved into a weekend event that involves the entire Corps in raising funds for the March of Dimes. During the 1980s, the Corps of Cadets collected and donated more than $400,000 to charity.[17]

As campus activities flourished, the newly expanded Memorial Student Center or "C" became the center of student activity. In addition to more than five hundred organizations, clubs, and events ranging from the Muleshoe Hometown Club to the performing arts, the MSC added a much-expanded variety of extracurricular activities. Student Conference on National Affairs, Town Hall, and the MSC Council along with such events as Ring Dance, the Combat Ball, All-University Night, campus Muster, and Midnight Yell Practice set Texas A&M apart. For much of the decade cadets dominated leadership positions in these organizations and events. Even though the Corps represented less than 18 percent of the student body, cadets held more than 40 percent of student government positions. Campus activities were varied. They included a special showing of the movie *The Getaway*, starring Steve McQueen and Ali MacGraw, and Town Hall packing G. Rollie White for a concert by rock band "Flash Cadillac and the Continental Kids." SCONA, chaired by Chet Edwards '74, future U.S. congressman, examined the ramifications of "The Controlled Society."

Great Issues, a speaker's forum, featured political columnist William F. Buckley, Jr. MSC Director J. Wayne Stark '39 was instrumental in broadening the scope of student awareness for new ideas, trends in the arts, as well as the leading issues of the day. During the 1960s and early 1970s, "Mr. Stark" was instrumental in "placing" Aggies in the leading graduate programs and law schools in the nation. Stark was also a force behind the creation of the modern theater complex in Rudder Center and the formation of the "Opera and Performing Arts Society" or OPAS in the fall of 1972.[18]

Fall would not be complete without Bonfire. In the fall of 1969, the fish of the Class of '73 claimed to have built the largest Bonfire in the history of the annual Thanksgiving activity—109 feet, 10 inches—and maintained their claim with a listing in the *Guinness Book of World Records*. After 1970, the administration limited the vertical height of the Bonfire stack to 55 feet. Held each November on the field behind Duncan Dining Hall, the popular event attracted a great deal of attention. So did freshman cadets of the centennial Class of 1976 for an unauthorized, yet much heralded, event: the kidnapping of the University of Texas Longhorn mascot, Bevo. In early November, 1972, cadets from Squadron One brought the steer to campus. This escapade was eventually resolved by Texas A&M University Police Chief and retired Texas Ranger O. L. Luther (who a decade earlier had confronted a similar situation) and A&M administrators. The mascot was returned healthy and unharmed. The campus was caught unprepared for a major snowstorm in mid-January, 1973. However, larger issues overshadowed campus life and national affairs.[19]

On January 22, 1973, former president Lyndon B. Johnson died of a heart attack at his ranch in Johnson City. On the same day, it was reported that Henry A. Kissinger, national security affairs advisor to President Nixon, had concluded the final negotiations in Paris for a cease-fire of American involvement in Vietnam—"effective 2400 hrs Greenwich Mean Time, 27 January." The decade-long war was once again reported in the back pages of the *Battalion*, informing students that peace was "at hand." Within weeks American prisoners of war, including Aggies Maj. Robert N. Daughtrey '55, Capt. James E. Ray '63, Maj. John C. Blevens '62, and Maj. Alton B. Meyer '60 were released after years of imprisonment in Hanoi. Captain Ray's smiling picture was circulated worldwide on the cover of the Baptist publication *Guide Post*. Returning to Texas, each former POW was presented with a new Aggie class ring from the Association of Former Students to replace the one confiscated while in captivity. Captain Ray, the 1973 campus Muster speaker, addressed a packed house in G. Rollie White Coliseum. In macabre dry humor and detail, Ray, a member of Squadron Six and the 1961–62 president of the MSC Council, recalled how the tough physical and mental training he received as a fish in the Corps played a pivotal role in his survival as a prisoner for more than six years.[20] The returning POWs were honored by the Corps of Cadets as the reviewing officers at the spring, 1973, Parents' Day Review. A&M alumni were quick to recognize the contributions of all Vietnam veterans, at a time when much of the country turned their backs on these citizen-soldiers.[21]

In August, 1973, a glimpse of a bygone era reappeared with the organization of Parsons Mounted Cavalry. Named in honor of the

Commandant, the unit started with fifty mounts and riders. Wearing distinctive cavalry garb, the unit made regular appearances at all campus reviews and home football games. By the fall of 1975, they traveled statewide representing the university and the Corps. Active riders in the new unit were limited to senior cadets who furnished their own horses, tack, and feed; juniors served as support staff. Funded by a grant of $10,000 from the Association of Former Students, the "Cav" was first housed at the Research Annex, formerly Bryan Air Force Base. Initial training and horsemanship were provided by Dr. John Fritz, dean emeritus at Fairleigh Dickinson University. A non-Aggie and member of the U.S. Olympic Equestrian Committee, Fritz provided the first saddle

blankets and regulation McClellan saddles. In 1975, the unit acquired a location west of the main campus on FM 2818, later named "Fiddlers' Green," at the direction of President Jarvis Miller '50, in time to prepare for eight appearances across the state in spring, 1976. Former Army Cav Trooper and avid PMC supporter Hughes "Buddy" Seewald '42 of Amarillo provided the funding to complete the facilities at the Green. The Cav is the only university-level horse-mounted student military unit in the nation.[22]

The fall football season and the back-to-school fanfare were marked by the late summer "fish camp" for incoming freshmen at the Lakeview Methodist Camp at Palestine, Texas. After arriving on campus, the All-University

Aggie Yell Leaders and mascot Reveille at Kyle Field in the fall of 1970. Left to right: Tommy Orr '72, Barrett Smith '71, Head Yell Leader Keith Chapman '71, Tommy Butler '71, and Rick Perry '72. In December, 2000, Perry became the first former cadet to serve as governor of the State of Texas. Photo from Aggieland 1971

Night, generally held early during the first week of classes, set the tone for the fall. Packing G. Rollie, new fish marveled at the echo of the Fightin' Texas Aggie Band in the cavernous coliseum, Reveille's constant barking, and the excitement of the first Yell Practice. With the Corps numbering 2,350 out of a student body of 21,000, the cadets projected a larger campus presence, as noted by the *Battalion* in September, 1974: "Many maroon and white Texas Aggie T-shirts were sprinkled among the khaki-colored uniforms although one didn't need glasses to see that the Corps far outnumbered the non-regs."[23] Following remarks by the student body president, Steve Eberhard '75, and Head Yell Leader Steve Taylor '75, President Jack Williams "made a smash hit" with students by predicting imminent victory and a postseason trip to the Cotton Bowl, telling the enthusiastic crowd, "All those that couldn't get in here have gone to t.u. We've got some [Rice] Owls to pluck, some [TCU] Frogs to gig and some [Texas Tech] Red Raiders to hang by the neck. All I know is that when they come up here, we try to be nice to them and all they do is ring those damn bells. And I know one other thing. At the end of the season, we've got some Longhorns to butcher."[24]

By the fall of 1975, Coach Emory Bellard in his fourth year as head coach beat arch rival Texas 20-10 to be ranked number two in the nation, holding the Longhorns to only 113 rushing yards, and he took the Aggies with a 10-0 record (the first time since 1939) to the hills of Arkansas for a heartbreaking loss against the Razorbacks. While the loss ended the opportunity for a bid for the national championship, it helped place Texas A&M in the spotlight across the nation.[25]

The fall schedule throughout the 1970s was a mix of Corps trips, Bonfire, drill, and campus concerts. But for the Corps, the Quad area was the scene of most day-to-day activity. The Fish Drill Team by 1972 had accumulated five national drill championships at the Cherry Blossom Festival in Washington, D.C. In November each year it had become a tradition for the fish to "haul off" their unit's commanding officer to an off-campus location as well as hide all his uniforms and possessions. A period of negotiations followed between the fish and upperclassmen, which would usually end in the fish being presented their "Corps brass" and, thus, the return to somewhat normal operations by Thanksgiving. Even though the practice of hauling off was officially prohibited by the Commandant's Office, the fish became ingenious in circumventing the rules. Egged on and coached by upperclassmen, the annual fall event was partly the outgrowth of a common daily ritual of "quadding." Cadet Corps Commander Joseph M. Chandler, Jr., in the fall of 1975 attempted to regulate this activity by issuing a "Policy on Quadding." The prime motive of this practice was to allow upperclassmen (in a seemingly harmless manner) to use fish to "get even" with one another. The generally accepted protocol for quadding was that only a junior or senior could order such "treatment," and only on another junior or senior, respectively. Tony Pelletier '75, 1974–75 Wing Commander, recalled that there were a number of "infractions" that would result in an "almost automatic order" for quadding. "These included insulting someone's hometown, mother, girlfriend or date; attempting a move on someone else's girlfriend or date; a refusal to help a fellow Aggie buddy with a legitimate request; and, of course, everybody got quadded on their birthday . . . generally

The Fish Drill Team held the national title for five consecutive years. President Earl Rudder (center) *meets with team members fish David R. Calvert '72, fish George P. Barrientos '72, Cadet Senior Advisor Michael E. Casey '69, and Trigon Advisor J. Malon Southerland '65. Courtesy Texas A&M University Archives, Cushing Library*

speaking, the whole exercise was considered good-natured fun."[26] In spite of the policy, the practice of quadding lasted until the early 1980s when it was officially discontinued due to the fact that the guidelines outlined by Corps staff were seldom followed and difficult to enforce:

1. Quadding will consist of pouring water from the second floor of a dorm onto an individual held on the sidewalk or steps leading into the dormitory.

2. Quadding will only take place during the hours of 1600–1800.

3. No substance other than water will be used.

4. No articles will be used to tie an individual being quadded.

5. An outer garment must be worn by the quaddee at all times.

6. Any quadding activities not mentioned in this policy are prohibited, such as swirlies, mud-holes, etc.

7. Outfit commanders will be held responsible and punishable for a

violation of this policy as well as any staff member passively observing the offense.[27]

The challenge for the Corps of Cadets by the mid-1970s was how to maintain what it viewed as its central campus role, while at the same time adjusting to the transformation to the post-Vietnam de-emphasis on the military. During the early part of the decade two significant changes occurred. First, Texas A&M reapplied for and was granted a Naval ROTC detachment—making the university one of only a few in the nation capable of granting commissions in all four branches of the armed services. The second event encompassed the challenge of how best to bring women cadets into the century-old all-male Corps. The process would demand tremendous planning and attention, because a growing number of cadets and alumni felt that the integration of coeds into the traditional lifestyle of the Corps would signal the end of the Corps of Cadets. Both events altered the composition and traditional character of the Corps.

NAVAL ROTC AT A&M

Shortly after Dr. Jack Williams became president, the former Marine Corps major began, "on a quiet basis," to consult with Lt. Gen. Ormond R. Simpson '36, then deputy chief of staff for manpower at Marine Headquarters in Washington, D.C., about the possibility of approval for an ROTC detachment that would prepare Texas A&M cadets for commissions in both the U.S. Navy and the U.S. Marine Corps.[28] Dr. Williams, aware that an application for NROTC had been turned

down in 1968, worked behind the scenes to ensure that the new effort would encounter no problems. Once the NROTC detachment was complete, A&M would be the only land-grant university in the nation offering officer training in all branches of the armed forces. Williams was advised by the Pentagon that very few new NROTC detachments were slated to be added; however, a detailed application would be given careful consideration at the highest levels in the Pentagon. Adm. Elmo Zumwalt, chief of Naval operations, and Texas Congressman Olin Teague '32 were both notified and supportive. In order to meet the deadline for submission of a detailed application, Williams contacted Colonel Parsons late on the Friday prior to Thanksgiving, 1971, and indicated his plans to bring NROTC to A&M and the urgent need to complete the required documents by the following Monday. During the weekend, the staff of the Army ROTC detachment at A&M prepared the demographic data, facilities assessment, and documents that made it possible to meet the president's deadline. Colonel Parsons recalls, "It is rather ironic that the Army did the Navy's work, yet in fact we all figured into the process."[29]

Following a favorable response by an inspection team from Washington, D.C., the Naval detachment at A&M was approved and scheduled to open in fall, 1973. General Simpson, on campus to serve as the June, 1972, commissioning officer, and Col. E. D. Foxworth, professor of Naval science at The Citadel, were instrumental in briefing both President Williams and Colonel Parsons on the final NROTC details. However, due to the overall preparedness of the university and Corps, as well the immediate availability of facilities in the Trigon for the new detachment, the timetable was advanced to the fall of 1972. Colonel Parsons officially advised both the Army and Air Force detachments plus the cadet commanders on April 20, 1972, of the impending addition of the Navy-Marine unit to the Corps. In the months that followed, General Simpson handpicked Col. Clarence E. Hogan, on assignment in Vietnam, to fill the role as the first professor of Naval science at A&M. He was to report for duty in College Station by July 1, 1972, along with Maj. Dorsie D. Page '56, executive officer, and Gunnery Sgt. Gaudencio Viloria, Jr., '83, better known to the cadets as "Gunny V." The first NROTC students totaled 119 cadets (19 of which were National Merit Scholarship recipients) and began classes in September, 1972. Though the bulk of the new A&M Navy-Marine cadets would not graduate until 1976, Ensign Thomas E. Collins '74 of Waco, a junior transfer from Army ROTC, became the first NROTC officer commissioned in December, 1974.[30]

Cadets in the Naval ROTC program were placed into two newly created cadet units, N-1 Neanderthals and S-2 Marauders. Within three years the Naval staff had grown to 6 officers and more than 500 cadets. In 1974–75, the Navy-Marine cadets, comprising 23 percent of the Corps, were placed into six units (C-2, N-2, H-2, E-2, K-2, and S-2) in the Second Brigade. Interservice rivalry occasionally surfaced, primarily with Army ROTC. With the reduction in Army commissioning slots and the delayed entrance on active duty for reserve officers, the Marines were quick to offer "regular" officer commissions. The NROTC program became so successful in the early years that they repeatedly requested more slots from the U.S. Navy. A&M was given a 500-cadet ceiling that was generally ignored. Dr. Williams and Lieutenant General Simpson lobbied at the highest levels in the Pentagon to lift any restrictions on the number of Aggies commissioned in all branches of service. In the meantime, the Army offset the vigorous Navy-Marine recruiting by offering cadets summer training at either the Airborne course or Ranger School. By 1976–77, the Air Force program had dramatically reduced pilot and navigator slots and tended to compete with the Navy for cadets majoring in technical disciplines such as engineering and science. In hopes of reducing the interservice competition, A&M lobbied for unlimited slots for active duty in each of the armed services.[31]

The Pentagon expressed concern but did little to address the situation. Shortly after being notified that "no longer can active duty be guaranteed" (to graduates of Texas A&M), Dr. Williams in a letter to the Department of the Army was very vocal in his justification for nearly unlimited support of the Corps and its cadets:

> Texas A&M University has remained the one land-grant college in the nation which has maintained a uniform Corps of Cadets throughout its history. We kept our Corps intact through the dark days of the 1950's and 60's when military service and ROTC were synonymous with dirty words. We have paraded from 2,000 to 5,000 uniformed cadets through the streets of cities of the Southwest with great pride; we have offered hundreds of thousands of people the chance to see what clean-cut uniformed young men marching behind a great military band look like and sound like. We have provided the Army with the largest number of second lieutenants from any single university other than West Point.

TABLE 9-1
Texas A&M Officer Production, 1970–79

Year	Army	Air Force	Marine[1]	Navy[2]	Total
1970	295	118	18	0	431
1971	277	112	15	0	404
1972	247	135	16	0	398
1973	191	148	15	0	354
1974	114	138	14	1	267
1975	114	105	11	15	245
1976	94	63	19	23	199
1977	74	56	37	34	201
1978	70	51	30	38	189
1979	88	37	31	49	205

[1]After 1975 Marine officer production includes those in the PLC program.
[2]The NROTC detachment opened at Texas A&M in September, 1972.

SOURCE: Office of the Commandant, Commissioning and Commencement Pamphlets

As long as we can possibly do so, we intend to continue along that course. We will do so despite efforts from a variety of remarkable sources to place roadblocks in our path.[32]

MINERVA'S FINEST

During the late 1960s and early 1970s there was increased pressure on the armed forces to allow women into a broader number of job specialties and make more career opportunities available to them. Pressure had been substantial to make a number of changes, primarily in the Army; thus access to ROTC by coeds soon became a high priority. The traditional all-male programs at the federal service academies at West Point and Annapolis as well as VMI, The Citadel, and VPI had been initially insulated from such change. However, in the case of the service academies and Texas

A&M, public institutions that received large amounts of federal funding, the mandate for admission of women became a growing issue. Furthermore, the peacetime post-Vietnam military had changed its emphasis to an all-volunteer military program in which trained men and women officers and soldiers would fill noncombat roles. Texas A&M, during the interim presidential term of General Luedecke, allowed unrestricted "full open admission" to coeds. A dean of women was hired in August, 1971, and by 1972, 10 percent of Texas A&M's total enrollment of 16,000 students were women. And with the decision to build campus housing for coeds these numbers were expected to more than triple by the centennial observance in 1976.[33]

Col. Tom Parsons in early 1973, in his dual role as both Commandant and professor of military science (PMS), received repeated inquiries from the Department of the Army on

Les Harvey '75, Commander of the Ross Volunteers (RVs), inspects junior class inductees. The RVs are one of the oldest active student organizations in the State of Texas. Photo from author's collection

the status of women in the Corps of Cadets and approached Dr. Williams to advise him of the situation. Parsons recalls, "Dr. Williams was very concerned and stated, 'We have resisted as long as we can—we can do it the way we want to or the government will tell us what to do.'"[34] The president's reference to the government interference was not unfounded. The Civil Rights Act of 1964 had been amended in 1972, adding Title IX, which barred discrimination on sexual grounds in admissions, programs, athletics, and activities offered by colleges and universities.[35] Most college administrators were sympathetic with Title IX in principle, but they resented the increasing government intrusion on higher education policy. The primary Title IX battles in the early 1970s concerned campus support of women's athletic programs. Williams, as well as the Commandant, many alumni, and many of the cadets in command positions knew that it was only a matter of time before the new federal guidelines were broadened and enforced for all college activities. However, the rank-and-file cadets as well as many of the former students were less than enthusiastic. Parsons and the Williams administration knew that the smoothest way to handle this volatile situation was to involve the cadets from the beginning in all phases of the planning, yet, even then, trouble or protest was still very possible.[36]

Thus, the Commandant in mid-1973 advised Corps staff that an extensive plan was needed to address the admission of women into the Corps. After lengthy meetings, it was decided that the lead element in the project would be the newly appointed 1973–74 juniors on Corps staff. Facing an incredible task, the six junior members on Corps staff in late 1973 began to draft one of the most unusual documents in the history of Texas A&M and the Corps of Cadets. Dubbed the "Minerva Plan"—named for the ancient Roman goddess of wisdom—this blueprint for phasing women into the Corps set in motion a process that has taken more than three decades to unfold. Texas A&M would be the first university with a large all-male military student body to address this challenge—a challenge that the all-male service academies did not fully address until the late 1970s, and The Citadel and VMI, well into the mid-1990s.[37] The team of A&M juniors from the Class of '75—John D. Chappell, Corps

sergeant major; Steven J. Eberhard, scholastic sergeant; Daniel P. Gibbs, personnel sergeant; Rickey A. Gray, operations sergeant; Frederick B. Martin, supply sergeant; and Terry W. Rathert, administration sergeant—had their hands full. In the spring of 1974, Granville D. "G-II" Lasseter replaced Gibbs for the drafting of the second stage of the Minerva Plan.

This detailed study and timeline on how to "phase" women into the all-male Corps of Cadets became the pivotal document in allowing the process to be as orderly as possible. The Minerva Plan, in the years that followed, was studied by all the service academies as well as the other military universities as the benchmark on the subject. Drafted in 1973–74, the plan was broken into three primary phases: (1) the initial introduction of women on a day-student basis into the Corps and

TABLE 9-2
Corps of Cadets, 1974–75

Corps Staff: Rickey Gray '75, Commanding					
First Brigade—Army			Second Brigade—Navy/Marine		
1 Batt	2 Batt	3 Batt	4 Batt	5 Batt	6 Batt
A-1	D-1	R-1	B-2	C-2	E-2
B-1	M-1	Sq-13	D-2	N-2	K-2
C-1	F-1	E-1	F-2	H-2	S-2
I-1	K-1	W-1	L-2		
L-1					
	First Wing—AF			Second Wing—AF	
	1 Group	2 Group		3 Group	4 Group
	SQ-1	SQ-4		SQ-7	SQ-10
	SQ-2	SQ-5		SQ-8	SQ-11
	SQ-3	SQ-6		SQ-9	SQ-12
					SQ-15
		Combined Band			
	Maroon Band			White Band	

entry into the ROTC program; (2) the gradual assimilation of the women into an on-campus uniformed unit; and (3) final phase of adjustment within the Corps. There were no precedents for these actions other than a strong feeling that both time and strong leadership would be needed to make the transition. The cadets developed a broad five-year agenda.[38] In phase one of the plan the drafters of the Minerva Plan were candid as well as apprehensive regarding their concern over the impact on the Corps: "With the advent of ROTC for women at Texas A&M University we find ourselves confronted with the question of whether or not to extend this same outstanding training to women by allowing them membership in the Corps of Cadets. Admittedly, the consideration of such a question is directly contrary to the established tradition of the Corps and in the view of many individuals could, in effect, destroy the very system which we seek to extend to women."[39]

During the initial school year beginning in the fall of 1974, the unit to be called the Women's Detachment or W-1, was given "day duck" status (students living off campus and in the Corps but not actively assigned to a unit on the Quad) and attached to Corps staff. It was assumed that maintaining the more casual day duck status would ease the transition into the Corps. Training and drill was conducted after classes during the week and all were enrolled in basic ROTC. There were no uniforms. The new female cadets were identified only by name tags. Steven Don Roper '75 was selected to lead and train the detachment. Roper was assigned to Corps staff in a newly created position of personnel officer. In effect, Roper was the unit's commanding officer. A staff of three male junior cadets in the Class

of '76—Mark S. Machala, Michael Wollam, and David Dean—functioned as training instructors. To assist with the unit organization and transition, the Army assigned Lt. Theresa A. Holzmann, the first woman ROTC staff officer at Texas A&M, as faculty advisor to W-1. Roper's new unit started with fifty-one female cadets (two juniors, six sophomores, and forty-three fish). All were designated as "fish" regardless of their academic standing.[40]

Opposition to the women cadets, known as "Waggies" or "Maggies," in W-1 was often fierce and hostile and came not only from the male cadets, but also from professors, non-Corps coeds, civilian students, and alumni. Many A&M traditionalists were, according to General Simpson, "violently opposed to the idea of having women in the organization." They were vocal and bitter about the end of an era marking ninety-eight years of the all-male Corps. Students and former students were further irritated by the massive construction that had the campus in chaos and by the razing of one of the major icons of the campus—the old metal water tower with its distinctive greeting, "Welcome to Aggieland!"[41]

Officials in A&M's administration and Commandant's Office closely monitored the addition of women to the ranks of the Corps. "We recognized that there would be a great deal of resentment on the part of the male cadets," noted Gen. Ormond Simpson, who after retirement from the U.S. Marine Corps joined the Texas A&M staff as assistant vice president for student services and supervisor of military programs in 1974. The general was optimistic that the changes would be both smooth and swift: "Realistically, it will take four years for that resentment to fade away." However, many former cadets now think that the special treatment and very procedures that were instituted to protect the women Corps members from physical and verbal taunting actually helped prolong the resentment.[42]

Nonetheless, every effort was made to ease the transition of women into the cadet lifestyle. The new female fish were given a detailed list of privileges and duties. They were required to sign-in twice daily in Lounge B in the Quad. With no uniforms, they were allowed only to watch march-ins at Kyle Field during the fall football season. By Christmas, 1974, a number of the academic "upperclass" fish strongly requested in a detailed memo to Rickey Gray '75, the Corps Commander, that they be "allotted" new privileges and be treated as "frogs," on an equal basis with the male cadets. Gray as Deputy Corps Commander (DCC), in the early fall, 1974, had replaced Cadet Corps Commander John D. Chappell in late October, due to his withdrawal from the university. The surprise of Chappell's departure and the change of command did not drastically alter plans for W-1. Some modifications were made (such as the formation of the Women's Drill Team in 1975), but it was decided to follow the Minerva Plan through the 1974–75 school year. Only twenty-five women cadets completed the first year. The planning and timetable to bring the women into the mainstream of cadet life was altered only slightly in the fall of 1975 with the addition of on-campus housing. Dorm 2 was renovated to house W-1. Additionally, the unit was detached from Corps staff and placed on the organization chart in the Third Battalion of the First Brigade.[43]

Male cadet leadership of the unit continued through Final Review, 1976, when Ruth Ann Schumacher '77 of Gettysburg, Pennsylvania,

was named the first woman cadet commander for 1976–77. Furthermore, Schumacher went on to be the first woman officer commissioned a second lieutenant in the U.S. Army from an ROTC program at one of the major six military schools. In addition to Schumacher, May, 1977, commissions were also presented to Lt. Gayla J. Briles '77, U.S. Air Force, and Ensign Gretchan Ann Looman '77, U.S. Navy. The annual contingent of women cadets at A&M during their first decade in the Corps averaged sixty-five cadets. Of the original fifty-one women fish who entered in the fall of 1974, five graduated as cadets and four of the five were commissioned in the Army or Air Force.[44]

The next few years would set the tone and direction of the Corps for the balance of the decade. In 1975–76, Joe Chandler '76 became Corps Commander, and the Deputy Corps Commander was Bill Helwig. They, along with their very active staff of Mark Probst, Stewart Gregory (also 1975–76 Ross Volunteers Commander), John Hatridge, Mike Marchand, and Bill "Chili" Flores, worked to "reorient" the direction of the Corps. They had good reason to single out two primary areas for improvement—academics and leadership training. Interclass relations had deteriorated and needed vast improvement. Chandler noted that the practice of "whipping out [fish procedure for meeting upperclassmen] should become less belligerent," and second, civilian-Corps relations were at times fragile. With a loss of 29 percent of the freshman class during 1974–75, the Corps at Final Review, 1975, dropped below 2,000 members for the first time since World War II. To better inform the units, a weekly Corps newsletter, *The Quadrangle*, was published, with Bill Flores as editor.[45] Notwithstanding, national attention was focused on the Corps and Texas

Cadets wildcatting into Duncan Dining Hall. Courtesy Texas A&M Administration

A&M when 2nd Lt. Steve Eberhard '75 was awarded the Hughes Trophy in a Washington, D.C., ceremony as the outstanding Army ROTC graduate of 1975.[46]

CENTENNIAL CELEBRATION

Campus growth became very apparent by the mid-1970s. The A&M campus was the victim of its own success. The Quad received a long overdue facelift in 1975, adding the "Corps Arches" and Plaza at the north end and expanding the sidewalk drill areas. The cost of an education at Texas A&M remained extremely attractive. Full room and board, tuition, and fees totaled $959.20. The seven-day meal plan could be paid in three installments of $137.80 each.[47] At Northgate, a favorite evening locale, the "Dixie Chicken" was opened in 1975. In an address to the Student Senate, President Williams said that although the fall, 1975, enrollment jumped 3,784 for a total of 25,247, approximately 3,000 applicants were turned down for admission due to a lack of space, primarily housing. Enrollment management, campus dorm space, and office space for the increased number of students as well as the faculty needed to provide for an expanded number of classes became major concerns. With no new dorms or office space scheduled for availability until the fall of 1977, Dr. Charles E. McCandless, director of facilities and planning, noted that he hoped "there would be enough flexibility" to accommodate and address the growth. All available space was put to use. World War II–era barracks on the north side of campus were hastily converted into "temporary" classrooms and office space. McCandless recalled, "I was here in '56 and they were temporary then!" Williams pro-

jected enrollment would top 30,000 in the fall of 1976 and peak at this total number of students for the balance of the decade. As the campus mushroomed, the Corps housing remained unchanged and cadets filled most of the Quad area, except for the use of Dorms 10 and 12 for civilian students from 1972 to 1977.[48]

The shortage of space in no way dampened Texas A&M's 1976 centennial celebration. Celebrants could not help but draw parallels to earlier A&M growth periods when limited space and growth were also an issue. Notwithstanding, the 5,200-acre campus was a far cry from the barren post oak prairie that greeted the first six students a century earlier. In a series of events throughout the year, A&M remained in the media spotlight statewide. In addition to a loyal alumni numbering more than 70,000, Texas A&M ranked eighth nationally among public institutions of higher education in the volume of financial support from private sources. Furthermore, the university offered 209 degree programs in eighty-one fields of study. Gov. Dolph Briscoe hailed Texas A&M, the state's oldest institution of public higher education, as "one of the greatest universities in the world . . . Aggies have always proudly served their nation and state in time of war and time of peace."[49]

In addition to a twenty-six-minute Texas A&M documentary, *That Certain Spirit,* televised statewide on April 21, the university held a grand convocation on October 4 as the forum to mark the first hundred years and to usher in the new century. Speaking for the alumni were Association of Former Students President Mayo J. Thompson '41 and Congressman Olin Teague '32. Student Body President Frederick D. McClure '76 made remarks prior to the keynote address by former

Texas governor John B. Connally. Absent from the gala event was Dr. Williams, who was recovering from a mild heart attack.[50]

By the fall of 1976, the Corps had experienced a modest growth that resulted in the reorganization of a number of staffs and increased the number of units from thirty-eight to forty-one. The Aggie Band for the first time was considered a major unit, or the equivalent of a wing or brigade due in large part to the growing organizational complexity as it reached a strength of more than 350 cadets. The band's new organization consisted of two battalions—artillery and infantry, each composed of two outfits. In addition to creating a new Navy-Marine unit, Company M-2, the Fifth and Sixth Battalions were redesignated units of the newly created First Regiment, replacing the Second Brigade. For the balance of the century, the Corps would comprise four major unit commands: the Army Brigade, the Air Force Wing, the Navy-Marine Regiment, and the Fightin' Texas Aggie Band. Colonel Parsons noted, "Texas A&M's ROTC programs, which rely on the Corps for 'leadership lab' training, continue to provide more officers than any other school except the service academies." While the number of A&M commissioned officers for the armed services was down because of the lower demand, there remained a viable core of enthusiasm for membership in A&M's Corps.[51]

After six years as Commandant, Colonel Parsons retired from the Army in mid-1977. In addition to being appointed as an honorary Cadet Colonel of the Corps by the cadets, the U.S. Army awarded Parsons the Legion of Merit. Colonel Parsons remained in College Station as an employee of the university until 1983. On July 15, 1977, Col. James R.

Woodall '50 became Commandant. Much like Parsons, Woodall recalled, "I always aspired to the position; I was in Germany and wrote to both President Williams and Congressman Teague about the job. I flew to College Station from Europe and was interviewed by John Calhoun, Dr. Koldus, Colonel Parsons, and General Simpson. The next day I was notified that I was accepted."[52] As a cadet, Woodall was commander of Company D, a Ross Volunteer, a member of Saddles and Sirloin Club as well as co-editor of the *Aggieland*. He was selected as a distinguished military student in the fall of 1949 and commissioned in the Army in June, 1950. In addition to assignments in Europe, Woodall had been stationed in Korea and Vietnam and was the recipient of the Legion of Merit, Silver Star, and Bronze Star. Eager to begin, the forty-eight-year-old Commandant set as his goal for 1978 to increase the size of the Corps from 2,300 to 2,500 cadets.[53]

With the assistance of recently installed forty-eight-year-old A&M president Dr. Jarvis Miller '50 (and classmate in Company D Infantry), Woodall's first step in preparing for increased growth in the Corps was to launch a much-needed renovation of the twelve cadet dorms.[54] While the repairs to the facilities were long overdue, there was an increasing degree of interest in the military and ROTC. With the reduced number of commissioning slots, the Corps of Cadets began to appeal to those who wanted to be a part of the traditions of the Corps and the university. All freshman and sophomore cadets were required to take at least two years of ROTC training in one of the three armed service branches. This requirement was crucial to the ongoing allotment or "commutation" of federal funds earmarked for providing and maintaining the cadet uniform. Nationally, this uniform funding came under attack as an unnecessary expense. A&M was vocal in protecting the unique uniform. However, efforts to expand the Corps proved difficult.

In the fall of 1976, Corps Commander Robert W. Harvey '77 and Corps Sgt. Maj. Michael H. Gentry '78, in search of fresh ideas for the Corps, visited the U.S. Naval Academy, Virginia Tech, and VMI. Gentry recalled, "The idea was to improve morale during an otherwise uneventful time of the year." The solution was the modern-day reintroduction of the "March to the Brazos" in spring, 1977. The second concern was fish retention. Harvey delegated the development of a summer orientation program for incoming fish to the

juniors on Corps staff. In the late spring the Corps proposed the creation of "Fish Orientation Week" or FOW. Approved by the Commandant, Colonel Parsons, and championed by General Simpson, all university approvals were obtained to include a week of room and board at no additional cost for approximately 800 freshmen. Maj. Jake Betty '73, military advisor; Dale A. Lazo '78, commander of the first FOW; and the 1977–78 Corps staff completed the FOW training schedule after Army summer camp in July, 1977. The program has been a tremendous success for more than two decades.[55]

Thus, the Corps in August, 1977, began the first FOW to introduce new cadets to the

Boots, boot trousers (known as pinks), and midnight shirts make the Aggie uniform one of the most distinctive in the nation. Pictured (left to right) from the Class of '73 are Don Eglinton, David Young, Tim Tenant, Glen Cernik, John Adams, Bill Bambrick, Rusty McInturff, Joe Waltz, Carl Walker, and Mike Irwin. Photo from author's collection

TABLE 9-3
Corps of Cadets, 1979–80

				William Dugot '80, Cadet Corps Commander					
First Brigade				*First Regiment*		*First Wing*			
1 Batt	2 Batt	3 Batt	4 Batt	5 Batt	6 Batt	1 Grp	2 Grp	3 Grp	
A-1	B-1	E-1	B-2	C-2	E-2	Sq-1	Sq-6	Sq-10	
C-1	F-1	V-1	D-2	D-1	K-2	Sq-2	Sq-7	Sq-11	
I-1	K-1	W-1	F-2	H-2	M-2	Sq-3	Sq-8	Sq-12	
	M-1		L-2	N-1	S-2	Sq-4	Sq-9	Sq-14	
				P-2		Sq-5		Sq-15	

Combined Band

Artillery Batt	Infantry Batt
A Btry	A Company
B Btry	B Company

requirements of the Corps before the start of class. Gentry noted that sophomores were prohibited from being on the Quad during FOW. There was a continuing emphasis on academics, and the university was focused on growth. However, while traditionalists, both in and out of the Corps, wanted growth, they remained resistant to sweeping changes that would alter the cadet lifestyle or would impact traditions. The *Aggieland* put the dilemma brought on by change in the 1970s in perspective: "Tradition need not get in the way of progress, nor progress in the way of tradition." One area of Corps activity that did receive both local and nationwide attention was reaction to the image of women in the ranks.[56]

THE ZENTGRAF CASE

During the balance of the 1970s there arose a growing uneasiness toward women in the Corps. The women, averaging about fifty-five to sixty during this period, occupied two units, W-1 and Squadron Fourteen. Leadership in both units was provided by women cadets by late 1976. The women cadets gradually became more vocal in their concern on three primary issues: discrimination in obtaining Corps leadership positions, inequities in the way they felt they were treated by the male cadets, and concern over access to and membership in such special units as the Ross Volunteers, the Texas Aggie Band, Parsons Mounted Cavalry, Rudder's Rangers, and the Brigade Color Guard. To meet this concern, the Commandant appointed Cadet Colonel of the Corps Bob Kamensky '79 along with eight cadets (five men, three women) in January, 1979, to accept complaints and "carefully review the problems caused by both men and women" as well as provide recommendations on how to ease the tension within the Corps on these key issues.[57] The peer review committee established by the Commandant was the result of a November, 1978, internal report from Lt. Gen. Ormond R. Simpson '36, assistant vice president for student services, to President Jarvis Miller that pointed out the rising incidents of friction among cadets.[58]

General Simpson was candid in his assessment to the president, stating that problems had surfaced between the men and women cadets in the Corps that "we thought would dissolve by time and turnover of [cadet] personnel . . . as we found ways to accelerate the process."[59] However, there grew an ongoing contempt among male cadets and resentment among some women. Kamensky noted in a feature article in the *Battalion,* "Some men have acted unbecoming of a cadet toward the women. . . . it's mostly verbal abuse and the slighting of [class] privileges. For instance, some seniors have refused to meet ["whip out" to the women] or acknowledge the existence of the 'Waggies.'"[60] Colonel Woodall strongly supported the work of the review group and provided staff members to assist in the evaluation. And by February an initial draft of recommendations was forwarded to the Commandant.

However, prior to the conclusion and release of the final report of the special committee on women in the Corps, the situation became more contentious as a result of a junior cadet's public appeal for redress directly to the national media, elected officials in Washington, and to the Brazos Chapter of the American Civil Liberties Union. In March, nationally syndicated columnist Jack Anderson in Washington released a blistering article in support of cadet Melanie Zentgraf, highly critical of her reported mistreatment in the Corps. While it seems unlikely that Anderson was ever on campus to obtain what he reported as "first-hand" information, the story appeared in newspapers nationwide. Nonetheless, he reviewed how Zentgraf had

been barred from the color guard and was the victim of "childish harassment" by male cadets when she wore senior boots to the pre-Bonfire tradition Elephant Walk in mid-November, 1978. The unauthorized wearing of the senior boots by an underclass cadet in public prior to Final Review angered many male cadets. According to Anderson, "She was surrounded by 20 outraged male cadets and forced to take the boots off."[61] Her actions and the male cadet reaction was considered an internal cadet matter by the Corps. Yet her public protest and the Anderson article precluded any internal redress.

Texas A&M officials questioned both the accuracy and fairness of the column written by Anderson. It became apparent that, as one cadet noted, "she was dragging external forces into a Corps-type matter." More importantly, these events revealed a reactionary anti–Corps of Cadets element in the local community and on campus that encouraged capitalizing on the Zentgraf affair to undermine the image and credibility of the Corps. Groups and individuals that did not like the Corps of Cadets or the "military" regimen would, over the next two decades, use lawsuits and threats under Title IX—the 1972 law and federal government regulation prohibiting sex discrimination in institutions receiving federal funds—in an attempt to discredit the Corps and advance their own agenda. The allegations that Zentgraf and other women in the Corps had been discriminated against created a cause and purpose for the "external" anti-Corps involvement. The events in 1978–79 resulted in one of the first major cases outside of college athletics entered by the federal government to enforce Title IX.[62]

On May 11, 1979, a class action suit was filed in U.S. District Court in Houston, Texas, by Zentgraf and the Brazos County and Houston chapters of the American Civil Liberties Union on behalf of her and other female cadets in the Corps. The suit charged that "policies, practices and customs" of the Corps of Cadets violated the Fourteenth Amendment to the U.S. Constitution, the Texas Equal Rights Amendment, the Texas Constitution, and Title IX. The university attempted to negotiate a settlement, but talks broke down when the ACLU demanded a public admission by the university that there was discrimination against women in the Corps. The suit named as defendants Dr. Jarvis Miller, president; Dr. John J. Koldus, vice president for student services; Col. James Woodall, Commandant; and Robert Kamensky, Cadet Corps Commander. Responding to the allegations, the Texas A&M Board of Regents voted unanimously to "defend this suit with vigor."[63]

U.S. District Court Judge Ross Sterling presided over the case, which was not resolved until early 1985. During the early stages of the suit, the U.S. Justice Department sent a team of attorneys from Washington to campus to interview Colonel Woodall. The Justice Department would have ordinarily defended Colonel Woodall because he was a federal employee, but after obtaining all the information they could from him, he was dropped from the case in November, 1979, in order that the Justice Department would have a free hand to intervene solely on the behalf of Zentgraf. Information he had provided to *defend* his role in the case was applied instead to strengthen the government's case against A&M. In retrospect, Woodall noted, "I think they violated the attorney-client rights . . . they seemed to be a group of liberal lawyers looking for a case."[64] In June, 1980, Judge Sterling invalidated a request by the Justice

Department "to compel the University to comply with a federal statute requiring that 'any military college,' in order to maintain its designation, provide qualified female undergraduates the opportunity to receive military training." Furthermore, the Texas Constitution prohibited monetary damages in federal court against an agency of the state [Texas A&M], unless the state consents—which it did not do. Notwithstanding, Sterling noted that Zentgraf did have "cognizable" claims under the equal protection clause of the Fourteenth Amendment.[65]

Prior to the final resolution of the suit the Commandant's office and university administration were required to provide volumes of documents and information on the operation of the Corps of Cadets and the activities of the various organizations within the Corps. Texas A&M contended throughout the process that the suit was unwarranted. The seriousness of the case to the university and the Old Guard alumni was accented at the June, 1980, commencement ceremonies, when President Miller refused to shake hands with Zentgraf as she received her diploma. Lane Stephenson of the campus Public Information Office said Miller "was an extremely honest and straight forward individual, and if he shakes someone's hand, he means all that it implies."[66] On July 10, 1980, Miller was terminated as president, primarily because of a misunderstanding over the duties and role of the president of the university as well as administrative disagreements with A&M System Chancellor Frank Hubert. Dr. Charles H. Samson, Jr., was named acting president.[67]

The final disposition of the Zentgraf case was the issuing of a consent decree agreed to and signed by Texas Attorney General Jim Maddox and filed on January 24, 1985. The dozen-page document set forth the court's guidelines that allowed any cadet, regardless of the "the individual's sex," to gain membership in any activity or organization within the Corps. Any and all "references to any male-only programs and activities" were to be eliminated. Policies and guidelines in all campus publications were to reflect these changes.[68] Furthermore, the judge ordered that the Commandant of Cadets be responsible for monitoring all "selection procedures and activities . . . as well as insure corrective measures which he deems appropriate." The university was to maintain records on all action related to the decree, as well as annually submit a detailed report on the Corps of Cadets and the membership activities of women cadets to Judge Sterling, for a period of three years ending July 1, 1988.[69]

From the time the Zentgraf case was filed in mid-1979 until its final settlement in 1985, the Corps operated in much the manner as it had throughout the 1970s. The number of women cadets in the Corps did not dramatically increase in the early 1980s. The overall Corps averaged 2,100 cadets. As the Zentgraf case proceeded it became apparent that the Corps in the post-Vietnam, all-volunteer environment needed to reassess its approach to training as well as its public image. The Citadel, VMI, VPI, and North Georgia all experienced significant drops in cadet enrollment. The reduced number of available commissions in the armed forces along with lingering antimilitary sentiments had a significant impact during this period. In 1979 the Corps represented 8.5 percent of the undergraduate student body. To address both the image and change, the Corps was given a tremendous boost by a vocal group of alumni that thought the institution they loved was in jeopardy.[70]

THE CORPS DEVELOPMENT COUNCIL

The adjustment to modern demands on the conduct of on-campus military programs proved to be a significant challenge for the cadets, Texas A&M administrators, alumni, and programs sponsored within the Corps of Cadets. In May, 1982, Col. Jim Woodall retired and was replaced by Waco native Col. Donald L. Burton '56, effective August 1. Burton was a career field artillery officer with vast experience in the Army. As a cadet he was commander of the First Regiment and president of Cadet Court; in his new position he faced the direct implementation of new programs to address concerns in the early 1980s. Texas A&M maintained its high visibility and viable role of officer production, yet in ever smaller numbers. Commissioning slots for A&M cadets remained on the reduced post-Vietnam manpower model, in spite of the year-to-year excellence of cadets both on campus and at the yearly ROTC summer training camps.[71]

Retention of fish cadets continued to average only 70 percent per class during the early 1980s. Total Corps year-to-year losses during the period from 1980 to 1986 ranged between 14 and eighteen percent. Attrition was also a concern at West Point, Annapolis, and VMI. The six predominantly military universities with historically large cadet programs as well as the armed forces academies were assessing their roles and the level of commitment to adequately address the modern demands to recruit and retain cadets.[72]

A significant factor that began to challenge efforts at retention was the increased academic demands.[73] Due to the demands placed on the cadets, Burton concluded in early 1984 that the "overwhelming reason (for decreasing

TABLE 9-4
Corps Strength Statistics, 1980–86

	Sept	Dec	Jan	May	% change Sept–May
1980–81	2262	2133	2008	1953	14
1981–82	2374	2277	2060	1950	18
1982–83	2282	2182	2032	1952	14
1983–84	2154	2036	1870	1848	14
1984–85	2072	1947	1750	1702	18
1985–86	1955	1812	1688	1640	16

Source: Office of the Commandant

numbers) has been poor academic performance, or at least a perception that they cannot handle both the demands of the Corps and the classroom."[74]

However, further evaluation of changes in the Corps by cadet leaders and the Commandant's Office determined that there were "other contributing factors" that accounted for the lack of retention including poor leadership in the ranks. Six key areas of major concern "relative to the future of the Corps" were highlighted: (1) loss of respect for the authority of senior cadets; (2) the practice of inflicting degrading acts and requirements upon freshmen, and in some instances upperclassmen; (3) failure to accept and live by the Aggie Honor Code; (4) continued below average academic performance by the Corps body; (5) a lack of commitment to military excellence in some Corps units; and (6) weaknesses in ability to conduct exemplary drill and ceremony.[75]

There arose a realization in the early 1980s that the Corps needed a broader focus as well as active benefactors to ensure the institution's continued viability. Concerned with the continued attrition, a group of Texas A&M alumni led by Colonel Burton established an advisory body that would, as its main charge, focus on increasing funds in four primary areas: cadet scholarships, cadet recruiting, endowed funds for special programs, and funding to construct a significant edifice to house the heritage of the Corps of Cadets. Though the 1982 Texas A&M University Annual Report stated that A&M "remains unequivocally committed to a strong Corps of Cadets," it was apparent that new strategies should be developed to attract students from across the state. By 1980, A&M attracted more than one-third of the total enrollment from metropolitan areas in the counties of Harris, Dallas, Bexar, and Tarrant.[76]

Cadet recruiting and retention as well as improved academic performance were the primary focus of the Corps Development Council. The second area of priority was to enlarge the number of cadets in the Corps and endow its survivability into the future. The initial organizers for the CDC gathered at the Aggieland Inn (later named the Ramada Inn) in late 1983 and included Burton, Raul Fernandez '59, Harold Sellers '56, Ken Durham '54, Bill Heye '60, Mike Humphrey '78, Robert Ingram '80, Don Johnson '55, Dave Thompson, Glynn "Buddy" Williams '60, and ex-officio member Preston Abbott '84, 1983–84 Corps Commander. Planning meetings continued during the fall and spring, resulting in the formal drafting of bylaws and the selection of Sellers as the chairman of the board. The first official meeting of the CDC was held in Rudder Tower on March 31, 1984, at which time Harold Sellers provided the initial $100,000 gift to the CDC as a "challenge gift" to stimulate alumni giving to the Corps. The success of the organization has been substantially in the funding of endowed Sul Ross Scholarships first established in early 1984. The Sul Ross stipends as well as other cadet scholarships proved very critical to recruiting and numbered more than a thousand available scholarships by 1999–2000. Furthermore, the development of academic tutoring programs, recruiting, and the completion of the $5 million Sam Houston Sanders Corps of Cadets Center located west of the Quad enhanced the image of the Corps. By the turn of the century a capital campaign spearheaded by former Yell Leader Col. Richard M. Biondi '60 and Maj. Gen. Tom Darling '54, had amassed in two decades a total committed endowment for the Corps of Cadets from all sources that exceeded $60 million.[77]

CADET GOODRICH

Tragedy struck in the early fall of 1984 with the heat stroke death of cadet Bruce W. Goodrich. The "frog" transfer student (a Corps term for those who enter the organization after their freshman year) was subjected to an extensive early morning session of calisthenics and running, in what upperclassmen deemed a "motivational exercise," as part of an outfit initiation ritual. Such so-called rituals were not a part of any Corps-wide policy or procedure. The four cadets involved were indicted on charges of hazing and negligent homicide. Three were suspended and one was expelled from the university. These events drew unfavorable nationwide attention to Texas A&M and the Corps.[78]

The Corps was in a state of shock. The tragic death of cadet Goodrich, in an exercise conducted by a few individuals outside the norms of the Corps, could not be tolerated. All authorized Corps activities such as outfit runs and drills were subject to approval and to be conducted only at times of low heat and low humidity. Colonel Burton and his staff, along with General Simpson, moved swiftly to reaffirm that such rituals were not authorized, nor were they in any way "traditions."[79]

Furthermore, the A&M Board of Regents on August 31, 1984, directed Cadet Corps Commander Charles H. Rollins III '85 to conduct a full internal assessment of any unauthorized activities in the Corps and those events or actions that could be termed hazing. Rollins, working with the twelve-member Cadet Court, looked into a three-year period from the fall of 1981 through late 1984. His report to the A&M regents concluded that there had been an abuse of the letter and spirit of rules against unauthorized exercise activity in a number of outfits. He noted that the cadet leadership and the Trigon were vigilant in exacting punishment against all offenders and that some unit commanders had been relieved of command and lost their ROTC contracts. Further, the Commandant had directed a thorough review of all university regulations, policies, and procedures, including the cadet *Standard* as well as Corps policies to ensure full compliance. The intent was to confirm to all cadets that actions considered "good bull" or carrying on so-called traditions in the name of "Old Army" would not be tolerated. The Corps Commander's primary concern was that "there will probably always be some people who will try to beat the system, but they now know that the odds of getting caught are high and the penalty will probably be dismissal from Texas A&M."[80] Rollins was to be commended for his attention to duty. There was no question among cadets about the seriousness of the Goodrich incident, Rollins noted: "The Goodrich tragedy stunned the Corps. Nothing like it had ever happened before. The Corps was bewildered and confused. Duncan Dining Hall was completely quiet for two days. The entire Corps seemed to realize without being told that its reputation was very severely tarnished and that drastic action was necessary to restore its image and reputation."[81]

Though threatened, no investigation was undertaken by the state legislature, but new penalties for hazing were passed in Austin.[82] There also was greater concern for the legal ramifications of hazing incidents not only at A&M but on campuses nationwide. College administrators as well as ROTC officials were charged with the responsibility of guarding against such extreme circumstances.[83]

New procedures for entering freshman cadets were instituted in August, 1985. Each cadet was required to undergo a physical exam and the Fish Orientation Week conference, and indoctrination was enhanced with a new guide book called "The Fish Buddy."[84] The summer Commanders Conference for outfit commanders and first sergeants also received an extensive briefing on the policies and procedures. Expanding membership to women in all cadet activities was a prime priority. For the first time women were admitted to the Aggie Band. Of the first three coed fish, Andrea Abat '89, would persevere to become the first female senior member of the band. University-wide enrollment for the fall of 1985 was 35,701, down 1,100 students from 1984–85.[85]

SELECT COMMITTEE ON THE CORPS

During mid-1985 President Frank E. Vandiver initiated the Select Committee, a review panel on the Corps comprised of members from inside and outside the university. Vandiver, a military historian who was previously at Rice University and North Texas State University, became president of Texas A&M in August, 1981. While amicably disposed to the Corps, the cadets and their supporters proved to be a real challenge to the president. At issue was the future direction of the Corps and, most importantly, the future staffing of the Commandant's Office as well as an assessment of the role to be played by the ROTC personnel. Vandiver, in extensive correspondence to the committee members, targeted five primary topics: (1) Should the Commandant of Cadets be an ROTC professor or a university employee? (2) Is there a need for the position of the head of the School of Military Science to serve as liaison between the ROTC detachments and the university? (3) Should ROTC personnel serve as unit advisors only, or as tactical officers as well? (4) Should the Commandant of Cadets have a staff of university employees to supervise the Corps? and (5) Should the university pay the ROTC personnel a stipend for performance of advisory or supervisory duties?[86]

President Vandiver convened a final session in December, 1985, to review recommendations on the above topics as well as to assess any further actions in light of the Goodrich incident and Zentgraf consent decree. Concern was also directed to the drop in Corps numbers from 2,088 in 1984–85 to 1,977 in fall, 1985. The decline in overall Corps strength marked the fifth consecutive year of lower numbers. In addition to a 17 percent drop in Corps strength, ROTC enrollment dropped 422 to 1,495 cadets, down from a fifteen-year high of 1,917. D&C cadets averaged 470 junior and seniors between 1980 to 1986.[87]

The special committee formed by President Vandiver was composed of military and academic experts from around the nation and provided insight into their opinions of the changing role of the Commandant. Marine Corps Commandant Gen. P. X. Kelly chaired the committee. Meeting and corresponding throughout 1983–84, the panel recommended that the president and Board, first, continue to request that each branch of armed service at A&M maintain and enhance its detachment staffing and, second, proceed to develop a method to identify and employ a "civilian" Commandant—preferably an A&M retired flag or general officer. This second proposal closely followed the scheme at VMI and The Citadel. Upon completion of Colonel Burton's tour of duty in mid-1986, a search was begun to identify a new Commandant to fill the role envisioned by Vandiver and the review panel. In the meantime, Dr. J. Malon Southerland '65 was chosen by Dr. John Koldus to serve as the interim Commandant. Southerland maintained his position as assistant vice president for student services while assigned as Commandant. Within the Office of Student Services oversight for the military programs remained under Gen. Ormond Simpson until

his retirement in 1985, at which time Howard Perry assumed oversight duties.

The image and role of the Corps of Cadets within the changing university environment was a key issue. The Corps of Cadets, while reduced in size compared to the overall student body, remained highly visible. Even though the number of available slots for active duty commissions diminished during the 1980s, the Corps continued to attract a large number of entering male freshmen. Support from alumni and the Board of Regents confirmed that the Corps would maintain a prominent role at the university. Yet a growing non-Aggie faculty and many civilian students, all unfamiliar with A&M traditions, questioned the emphasis placed on the cadets. Thus, the prime concern was how best to prepare and position the Corps for the future.

In terms of facilities, the Board of Regents in early 1986 authorized and contracted what would become the beginning of a $36 million renovation of the Corps dorms and a substantial upgrade of Duncan Dining Hall.[88] However, facilities improvements were only a small element facing the future of the Corps. The challenges were many, but two aspects would dominate the future course of the Corps: first, cadet attitudes toward the demands and nature of the academic community; and second, the need for full acceptance of women cadets in the ranks. Furthermore, detractors of the organization would find fertile grounds to attack the Corps if these items were not addressed. At the year-end meeting, a draft of a revised "Mission Statement for the Corps of Cadets" was reviewed. Its tenor was to maintain the emphasis on the prime objectives of the Corps to "provide exemplary citizens and qualified officer candidates for the armed

Cadet Todd Krugel of Company H-2 "on-the-wall" for an inspection by Thomas L. Farmer, Jr. '85, executive officer, 5th Battalion. Courtesy Texas A&M University Archives, Cushing Library

services," while amending the mission statement to incorporate the "higher order" of academic pursuits deemed appropriate to the changing demands on the student-cadets. There is no record of this being formally adopted, yet it accented the scope of the Corps:

> The Corps of Cadets has chosen an essentially military model as a vehicle for developing its members along these dimensions. While clearly consistent with its desire to identify and develop candidates for the nation's officer corps, the University views the military model to be an excellent and appropriate vehicle for developing the character and leadership traits required of the citizen soldier.
>
> The military model is relaxed where necessary to accommodate the higher-order interest of academic pursuit and to respect the rights of non-Corps members of the University community to establish an environment consistent with their own wishes and needs.[89]

In response, the cadet bulletin, *The Quadrangle,* noted:

> *We are proud of our past and confident of our future . . .*
> *We are guardians of A&M tradition, and the*
> *Spirit of Aggieland is alive and well in our ranks . . .*
> *To our university, our state, our nation,*
> *And all who join our Corps, we can and do offer . . .*
> SOMETHING EXTRA.

10

SO HERE
WE STAND

The Corps of Cadets will continue to play a significant role in student life as a crucible for the development of leadership, repository of many of the University's oldest traditions, and a source of commissioned officers for the Armed Forces of the United States.

Texas A&M Board of Regents, January 28, 1986

The Corps of Cadets remains a vital and relevant part of the overall University community today, both as the "keeper" of many of the University's cherished traditions and as a repository and champion of values that make Aggies and Texas A&M truly unique.

Blue Ribbon Committee on the Corps, December, 1993

THE FINAL DECADE of the twentieth century began to unfold with an increased emphasis on a truly more progressive Corps of Cadets. Internally the Corps set as its priorities retention, leadership training, and improved academic standing within the university. Externally, the Corps would become the focus of intense scrutiny, due in large part to a shift in national public policy toward military organizations. Gradually during the late 1980s and into the 1990s, these changes manifested themselves in our society at large first as the armed forces and a sense of citizen obligation toward military service were de-emphasized with the end of the Cold War; second, the military was downsized with the elimination of the draft; and third, the armed services were often criticized for being out of step with the broader changes within the nation—oddly reminiscent of the anti-ROTC-military tones of the pre–World War II era in the late 1920s and 1930s. The precipitous decline of the "all-voluntary" American armed forces greatly impacted morale in the ranks and proved an affront to the legacy of the citizen-soldier. Furthermore, these influences led to change in the Corps and Texas A&M in a number of ways.[1]

The Select Committee on the Corps of Cadets established by President Frank Vandiver in 1985 and a number of studies by university officials and alumni over the next decade set in motion a re-evaluation of the policies and procedures that shaped the direction and image of the Corps. The influence and aftermath of the Zentgraf and Goodrich cases would have implications for the Corps and the Corps's actions through the balance of the century. The internal issues of the Corps and court-mandated changes were elevated to national attention. Given an air of legitimacy, a small group of anti-Corps detractors of campus military programs in the early 1990s were quick to attack all elements of the organization as well as the traditions that singled out the uniqueness of Texas A&M.

For its part, the high-profile Corps of Cadets, with its enthusiasm, confidence, and sometimes arrogance—along with its rich heritage—often became, as an institution, a victim of isolated individual actions that detractors used to claim that the entire organization mirrored and condoned such actions. Reaction by administration officials, alumni, and cadets was swift, yet often muted in press accounts that suggested deep problems in the Corps and Texas A&M. It is interest-ing to note that this blossomed at a time in the early 1990s when the armed forces were viewed with disdain by politically motivated antimilitary factions in Washington, D.C., and in the press.[2]

TRANSITION AND TRADITION

Following the Goodrich incident in August, 1984, and the final Zentgraf decision in January, 1985, it became apparent that campus issues involving nearly any level of dispute or an allegation of harassment could at any time be referred to the judicial system. Pursuing legal action, in advance of an internal attempt to resolve the issue in question, was to become a pervasive tactic after the mid-1980s. This litigious trend affected not only A&M, but also other universities and society as a whole. Vandiver, a respected military historian and academician—biographer of Generals Thomas "Stonewall" Jackson and John J. "Blackjack" Pershing—yet never a member of the armed forces, was concerned with the impact on, as well as future of, the Corps of Cadets at A&M.[3] As president, Vandiver had a profound aversion to conflict—especially if it involved the Corps of Cadets. Early in his administration an incident at the 1981 Texas A&M–Southern Methodist University football game in which a senior cadet OD (Officer of the Day) drew a saber on the field during an altercation with SMU cheerleaders clouded the president's image and impression of the Corps. Angered by the incident, Vandiver demanded immediate action, and the Commandant ordered sabers off limits in Kyle Field while the episode was under investigation.[4]

The Corps was once again a target. Opposition began to surface from a number of

Senior Cadet Greg Hood drew national attention during a confrontation with an SMU cheerleader at Kyle Field in the fall of 1981. Photo copyright 1981 Gregory L. Gammon

anti-Corps factions. Ironically, the A&M Faculty Senate that Vandiver established in 1983 was often at the vanguard of the attacks on the Corps and A&M traditions.[5] This body in the late 1980s and early 1990s progressively became more liberal. In his final state of the university address to the Faculty Senate, Vandiver was quick to address the role accorded A&M traditions:

I'm proud of the recent statistic that indicates our general student retention is around 86%, considerably higher than any other public institution in the state. That we have

done so well can, I believe, be attributed to many things; but the faculty deserves a larger measure of the credit. . . . You are all to be congratulated.

But there is another factor that helps, I think, weld our many and diverse students into a student body that stays with us. That factor is tradition—the stuff of A&M. Although we are quick to defend our academic traditions, some of us seem to find institutional traditions trivial, even juvenile in some instances. I think that blasé view obscures a vital quality of A&M. The Twelfth-Man kickoff squad has helped draw our football team and the student body closer together. Midnight Yell Practice does more than provide support for the team. It, like Bonfire, provides an outlet for high spirits, but welds together that special Spirit of Aggieland. Silver Taps needs no special praise, nor does Muster. Their value is self-evident to those with soul. My Aggie daughter, Nancy, explained Silver Taps to me—"Daddy, they will call my name when it's time—and I know it!" Our Fish Camps go a long way toward bringing new freshmen into the feeling of belonging—which is a quality that sets A&M decently apart.

Of course we have some [so-called] "traditions" that are less than wholesome. On occasion, hazing masquerades as a "rite of passage" in some fraternities or as "Good Bull" and "old Army." Those have no place here, despite their deep roots, and I would ask the help of the whole community in eliminating them from A&M.[6]

Vandiver concluded that to enhance the growth of the Corps and manage the cadets better, a civilian Commandant and staff enhancement would be needed.[7] Staff members in the Commandant's Office visited The Citadel, VMI, VPI, and Norwich, all of which had larger staffs and a cadre of tactical officers. The commandants at these institutions were retired general officers. Administrations of these military institutions believed that a university employee had both better control and better oversight of the operations of campus military programs. Given the fact that the armed services were considering de-emphasizing campus ROTC programs due to the reduction in the active duty armed forces, this change seemed inevitable.[8]

During the transition, an extensive effort was undertaken to identify a general officer and A&M former student to become Commandant. In the 1986–87 conversion period, Dr. J. Malon Southerland '65 was named interim Commandant. A member of the Fish Drill Team as a cadet in the early 1960s, he served in the U.S. Army in Germany and returned to A&M in 1968, filling a variety of university positions ranging from Corps advisor to assistant to the president, 1979–81. Occupying the Commandant's Office in the fall of 1986, he proved a timely catalyst to bridge the transition of leadership in the Trigon. President Vandiver recalled, "Southerland held the Corps steady—he was not easily fooled and he had a tremendous rapport with the students." During this period there was very strong cadet leadership. Cadet Corps Commander Garland Wilkinson '87 and Deputy Corps Commander Mandy Schubert '87, the first female Deputy Corps Commander, proved very effective. Wilkinson noted in 1986, "The new [civilian] commandant will be involved and provide longevity to the position . . . this will help in opening up the Corps to the student body."[9]

Southerland's priorities were to maintain normal operations while placing tremendous emphasis on leadership and improving the living and dining facilities in the Quad. Fully familiar with operations at A&M, Southerland proved valuable in both reminding the Corps of the broader scope of involvement on the campus as well as the importance of the transition to a civilian Commandant. In 1986, the A&M Board of Regents earmarked $5 million to upgrade Duncan Dining Hall, and in 1987 a $30 million project to address cadet housing (fifteen times the original cost of the twelve dorms) was begun to renovate all the Corps-style dorms. Improved fish class retention and leadership at all levels would be needed to fill the dorms. The renovation of Duncan Dining Hall had a significant impact on the Corps. Within the Corps, one key element of the lifestyle and cadet training that was altered in the name of progress was meal service in Duncan. Citing better efficiency, "family-style" dining with eight cadets per table, served by cadet waiters, was eliminated in 1987. The traditional dining scheme was replaced by "fast-food" lines, plastic trays, and the promise of "hot" food. The cadet newsletter, the *Saber,* in 1987 recounts the passing of the century-old custom:

We marched to the doors and in we went. After locating our tables, we stood behind our chairs until all the holes [chairs] were filled. We then spoke to all the upperclassmen at the table and proceeded to stand until the command to "sit!" was given. And when we asked for food, we were actually asking for it to be passed to us. There wasn't any of this look-on-the-tray-and-call-out-what-you-have stuff. When it was passed it was often cold. But the dish had to be emptied before we could get seconds which were usually warmer and fresher. The aisles were patrolled by cadet-types in clean, white jackets. These people were known as Duncan Volunteers, D.V.'s, or simply waiters. They waited on the tables, removing and refilling empty dishes. Duncan rarely had enough help to suffice, however, and we fish often had to get up and get refills.

Eating family style was not as bad as it sounds. Life that way was different, interesting, *and Corps,* through and through. And that was the best part.[10]

In addition to facility improvements, there was an ongoing emphasis on academic excellence. The Corps Development Council provided funding to establish Fish Academic Survival Training (FAST) in conjunction with the Student Counseling Service. Cadet leaders and the Commandant's Office worked to enhance both the Fish Orientation Week program as well as the upperclass program for "Cadre Leadership Training." After one year as Commandant, Southerland returned to the staff of Dr. John Koldus. Upon Koldus's retirement in August, 1993, Southerland was named the vice president of student affairs.[11]

In the fall of 1986, the search for the new Commandant was narrowed to three candidates, and, in December, U.S. Air Force Maj. Gen. Thomas Darling '54 made an extensive visit to the campus. He was selected in early 1987. Due to his military obligation, he would not report until August, 1987. A decorated command pilot with more than 500 combat hours in the B-52, Darling had served a tour as commandant at the Armed Forces Staff College in Norfolk, Virginia, and as chief of staff of the U.S. Atlantic Command just prior to retirement from the Air Force. While a cadet, he played on the A&M varsity basketball team and was selected as a Distinguished Military Graduate.[12]

The new Commandant and the Corps were ever mindful of the scrutiny they faced. Incoming Cadet Corps Commander Patrick W. Thomasson '88 and the senior class endeavored to stress grades, retention of the fish class, and leadership fundamentals throughout the outfits. The Corps in 1987–88 comprised less than 6 percent of the student body. Yet the year proved very successful for the Corps: 200 more cadets enrolled, compared to the previous year; the Aggie Band for the first time in years numbered more than 300 marching members; women in the Corps exceeded 100 for the first time; and 9 cadets were selected to Who's Who in American Universities. During early 1988, the Commandant's Office—made up of Col. Donald J. Johnson '55, Lt. Col. Buck Henderson '62, and Maj. Jake Betty '73—was moved to Lounge A in the Quad to allow the Trigon to be renovated. But the big story was that the spring, 1988, grade point average (GPA) for the Corps fish—Class of '91—was 2.534 overall—outpacing their civilian university counterparts. This began a trend that continued throughout the next decade.[13]

Optimism ran high as the Commandant and the Corps Development Council, chaired by Bill Heye '60, set a goal to fill all the dorms on the Quad—2,600 cadets by year 2000. In 1988, 208 Sul Ross Scholarships were awarded to freshman and sophomore cadets, and 179 senior cadets received officer's commissions in the armed services. The goal to "fill the dorms" would be the focus of recruiting and

retention efforts for the balance of the century. To assist with the broadening activities of the Commandant's Office, five new positions were approved by the university—a recruiting coordinator and assistant, a staff assistant, and three training officers. One popular program that was started to attract new cadets was the "Spend the Night with the Corps" program, which allowed prospective high school junior and senior students a first-hand look at day-to-day activities in the Corps. Since 1987, more than 12,000 potential cadets have participated in this event.[14]

ENROLLMENT MANAGEMENT

By fall, 1987, with a total student enrollment of 39,079 students, attention focused on the problems associated with the dramatic growth of the university. The 1987–88 A&M budget was $456 million, a slight gain over 1986–87. Complaints of overcrowding in the dorms, lack of adequate classroom space, and shortage of

TABLE 10-1
Corps of Cadets Fall Enrollment, 1990–2000

Year	Fish	Soph	Juniors	Seniors	Corps
1990	635	519	465	417	2116
1991	596	401	469	414	1880
1992	540	405	400	459	1804
1993	708	388	407	402	1905
1994	706	512	368	367	1953
1995	762	530	501	353	2146
1996	704	552	502	485	2243
1997	646	498	518	463	2125
1998	776	484	460	477	2197
1999	564	529	438	431	1962
2000	668	389	486	404	1947

Source: Office of the Commandant

classes in the basic subjects resulted in the implementation of an enrollment management policy by the board of regents. The targeted enrollment for the fall of 1988 was set at 41,000 students based on the annual admission of 6,600 new freshmen. Admissions standards were upgraded to include more emphasis on SAT scores and high school class standing. In response to faculty insistence on more research and advanced degree studies, the regents planned to reduce the number of freshmen per entering class from 6,600 in 1988 to 5,600 by 2002, while during the same period increasing the graduate student population from 6,300 to 10,150. The goal was to have the post-baccalaureate students comprise 25 percent of the student body by the turn of the century. Concerned supporters of the Corps predicted that such policies and enrollment "caps" could impact the Corps's efforts to recruit (by reducing the pool of potential new Corps fish) and retain cadets. Others contended that numerous qualified applicants were denied admissions because they expressed interest in the Corps of Cadets. Furthermore, the university was hesitant to grant "extra credit" for applicants who wanted to join the Corps, in spite of directives approved by the A&M Board of Regents. These concerns were calmed when the board named members Douglas R. DeCluitt '57, Joe H. Reynolds, and Royce E. Wisenbaker '39 to the "Ad Hoc Corps Enhancement Committee" to monitor any items detrimental to the Corps. One issue that persisted and received board attention was overcrowding in the Cadet dorms each fall semester.[15]

In January, 1988, after a seven-year tenure as president, Dr. Vandiver announced his resignation effective in September. Vandiver was named to head the Mosher Institute for Defense Studies funded by Houston steel magnate Edward J. Mosher '28 with a contribution of $1.5 million over a ten-year period. This, along with the establishment of the Military Studies Institute created by the Board in early 1984 and housed in the Department of History established the only two specific centers of research at Texas A&M focused toward the study of military affairs. Following the departure of Vandiver, Dr. William H. Mobley was named president on August 1, 1988.[16]

ROTC: NATIONAL DOWNSIZING

It was most ironic that as official ceremonies were being planned to celebrate the fiftieth anniversary of Reserve Officers Training Corps in Washington, D.C., steps were under way to "down-size" ROTC programs nationwide. In correspondence to President Mobley, the Commanding General of the Cadet Command (ROTC) noted, "a balance must be struck between societal requirements and officer requirements. Too precipitous a reduction in the size of Cadet Command will jeopardize the 'value-added' peacetime societal return as well as mortgage the Army's future leadership."[17] In response Mobley expressed surprise at the shortsighted efforts of the manpower planners in the Pentagon to diminish years of successful officer production via ROTC, a "unique system of education for military leaders." He further responded, "ROTC has contributed to our past in planning the future. The national loss in scaling down ROTC will be immeasurable. The cost to this nation underwriting the ROTC programs will pale in comparison to the return in value of these citizen-soldiers."[18]

Dr. William Mobley, Student Body President Stephen Ruth '92, and Sen. Lloyd Bentsen. Courtesy Texas A&M University Archives, Cushing Library

In the four-year period from 1986 to 1990, the combined national enrollment in ROTC—Army, Navy, and Air Force—declined 10 percent, to 86,000. The Army closed 50 of its 416 college and university campus units. Furthermore, during approximately the same period the Army decided to close the Third ROTC Regional Headquarters at Fort Riley, Kansas, based on "projections of a smaller Army and a shrinking national defense budget"; and there once again surfaced an anti-ROTC notion among the faculty at many universities to "expel" campus ROTC. However, the A&M campus, as in previous antimilitary periods, was generally a vast contrast to other universities. At all levels, emphasis at Texas A&M was placed on enhancing the Corps of Cadets. National attention was constantly directed on

the Corps as well as the Texas Aggie Band, under the seventeen-year direction of Col. Joe T. Haney '48 until August, 1989, and then under Col. Ray Toler for the balance of the century.

While some antagonists on campus did not like the high profile of the Corps and emphasis on the cadets (especially the tremendous investments in upgrading the Quad and Duncan Hall), the commitment to the Corps was clear and unmistakable. These actions coupled with significant efforts to provide the Corps with a permanent endowment were augmented with the establishment of the "Corps of Cadets Center." Located in Spence Park west of the Quad, this 19,300-square-foot multi-use facility was made possible by a substantial donation by

The Corps Visitor Center and Museum opened in January, 1992, and was named in honor of former cadet Sam Houston Sanders '22. Courtesy Texas A&M Photographic Services

Dr. Sam H. Sanders '22 of Memphis as well as gifts from thousands of former cadets.[19] The Sanders Corps Center—a combination visitor center and museum—provides a daily link with the heritage of the Corps as well as a base for recruiting and tutoring programs. At dedication ceremonies in September, 1992, Raul Fernandez '59, chairman of the Corps Development Council, elaborated on the values of the Corps: "The lessons that I learned in the Corps—the teamwork, problem solving, the camaraderie—all of these lessons have stayed with me for a lifetime. I learned that individuals must follow before they can lead. I learned that each member of a team must do excellent work if the team is to achieve true excellence. I learned to strive for that extra something because my buddies depended on me and I didn't want to let them down. I learned that in keeping the traditions of the past, we can set our course for the future."[20]

TABLE 10-2

Texas A&M Officer Production, 1990–2000

Year	Army	Air Force	Marine	Navy	Other*	Total
1990	69	48	28	27		172
1991	31	50	30	37	5	153
1992	57	47	22	31		157
1993	53	40	32	18	3	146
1994	49	31	20	26		126
1995	58	26	21	16	3	124
1996	58	30	13	17		118
1997	57	31	25	13	2	128
1998	38	50	23	18		129
1999	53	48	24	27		152
2000	57	47	31	21		156
					TOTAL	1,561

*Commissions in the Texas State Guard or foreign military service

Source: Office of the Commandant

DESERT STORM

The impact of ROTC reductions on Texas A&M had been gradual, yet came as a direct result of the end of the Cold War. An ominous sign of the times was President George Bush's executive order on September 27, 1991, to take B-52 bombers off alert status.[21] Notwithstanding, A&M continued to lobby in Washington to be allowed the maximum number of commissioning slots available. Within the Corps, the number of D&C (non-contract) cadets rose to nearly 60 percent. The Corps in the fall of 1991 numbered more than 1,800 cadets in an overall student body of 41,171. As the debate of words and policy makers continued over the man-power needs of the armed forces and the future of ROTC, war erupted in the Persian Gulf region. Operation Desert Shield turned to hostilities under Operation Desert Storm as allied forces began a blitzkrieg against Iraqi forces. Shortly after midnight on January 17, 1991, Col. George W. "John Boy" Walton '71, commander of an F-4G Wild Weasel Squadron, was ordered to knock out enemy radar systems. He was credited as the first American and one of the first of many Aggies to see action in the Persian Gulf War.[22] In all, more than 300 Aggies saw service in the war, including 21 cadets called to active duty in activated reserve units, who on very short notice moved from "student" to "soldier" status. Three former cadets were lost in action: former Company E-2 Reveille Mascot corporal, Lt. Daniel V. Hull '81; Maj. Richard M. Price '74, a former member of the Fightin' Texas Aggie Band; and 1985–86 Squadron One executive officer, Lt. Thomas C. "Cliff" Bland '86. While the "100-day war" marked a milestone for the post-Vietnam U.S.

military and its capabilities, it did little to convince policy makers in Washington to restore the military strength to adequate levels.[23] On campus, a fight of a different sort erupted in the fall of 1991.

SO HERE WE STAND . . .

In 1989–90, there had surfaced a nationwide focus on harassment and hazing in colleges and universities. The military colleges and service academies were singled out for special attention. A combination of court-ordered mandates, the rapid introduction of women into previously all-male military environs, antimilitary sentiments, and a trend toward higher attrition in institutions with uniformed students promoted further inquiry into this phenomenon. Reports and investigations into harassment and hazing incidents at West Point, the Naval Academy, the Air Force Academy, Norwich, The Citadel, VMI, North Georgia College, and Texas A&M were partly focused on the declining number of students willing to be a part of such programs and an increased likelihood that students displeased with such a regimented program might withdraw. Concerned with reports of "gender assimilation problems," the service academies were subjected to a series of extensive investigations by a U.S. Congressional inquiry conducted by the General Accounting Office (GAO).[24] Many times claims and charges of "hazing" were invoked for any perceived disciplinary action, thus triggering an alarm and creating the impression that such a claim could be used by cadets wanting to separate from the academies. More importantly, the courts had become so involved in attempting to enforce Title IX and antihazing measures

that the *Chronicle for Higher Education* quoted officials of military institutions as saying, "It is sometimes difficult to decide what constitutes hazing."[25]

In the case of the service academies at West Point, Annapolis, and Colorado Springs there were reports that 10 percent attrition of the freshman class was not unusual. In the case of the Naval Academy, about 22 percent of the cadet midshipmen left the academy in the first two years. "The class system" at The Citadel, VMI, and Texas A&M was deeply embedded and contributed to yearly reinforcement of underclass hazing every year. There also was a shift in the types and scope of claims of harassment and hazing—still not fully defined, and, if defined, hard to enforce and monitor. Furthermore, nearly all the acts of hazing were carried out by individuals violating the established policies and rules. Incidents and claims of sexual harassment increased due to the growing numbers of coeds in military schools. Although the practice of hazing is officially banned at all military institutions, initiation rituals assumed to be acceptable practice or tradition and "carried out under a veil of secrecy" resulted in a passive acceptance, because upperclassmen partaking in such events had endured the same treatment as freshmen.[26]

The Corps of Cadets at A&M became the target of the hazing-harassment allegations. As the first military-oriented institution in the nation to bring women into an all-male Corps in 1974, Texas A&M had by 1990–91 become a barometer on gender issues. Interest groups outside the Corps lobbied to intercede in the management of how women should be integrated into the Corps. It was decided by the Commandant's Office and cadets that the all-women units, W-1 and Squadron Fourteen, would be eliminated in the fall of 1990 and the women cadets integrated into seven of the thirty-nine cadet units. General Darling noted, "Reports we received noted that the armed services were having reasonable success with integration and reported that men and women usually performed better when integrated."[27] A pilot program on integration of female cadets in 1989–90 proved successful. This initial strategy involved assigning no fewer than eight women to an all-male unit—two from each class. Furthermore, the Commandant established the "Advisory Panel for a Discrimination-Free Corps of Cadets" composed of cadets, faculty, and staff. However, due to the small number of women in the Corps—less than 5 percent of the total cadets—and because of the need for female support in integrated units, it was thus impractical to instantly integrate women in every Corps unit. Vocal members of the A&M Faculty Senate demanded faster integration and threatened to form a special committee to monitor and oversee integration. Their prime argument was that without such a committee the "progress of integration would continue to be slow."[28] General Darling received a tremendous amount of pressure to move at once to solve this problem. However, he noted those making the demands wanted "total integration even if it meant only one female cadet per unit. I knew this would not work."[29]

The integration of women into selected Corps units proceeded during the fall of 1990 and into 1991. Darling, his staff, and the cadet leadership worked to identify new methods and programs for recruiting, training, retention, and information to improve all aspects of the Corps.[30] Women, totaling seventy-six cadets in 1991, were given access to all the special units of the Corps and all activities.

In March, 1991, Deputy Corps Commander Kurt Sauer spent a week at VMI to observe the VMI Cadet Corps, aspects of its Honor Code, new cadet orientation, peer evaluation methods, and the demerit system.[31] VMI had not yet admitted women to their program and would not do so until the late 1990s. Notwithstanding, isolated individual incidents of verbal abuse, catcalls, and harassment continued at A&M, but students were swiftly disciplined when reported. The Aggie Bonfire, a traditional all-male bastion, was a particularly difficult venue. The Corps and the administration worked to balance the integration of women in the face of pockets of objections.[32]

In September, 1991, a female cadet alleged she had been assaulted by sophomore members of the Parsons Mounted Cavalry. Anti-Corps forces at the university and in the media at once jumped to conclusions, claiming that certainly the entire Corps of Cadets was at fault and guilty as charged. A statement by the board of regents noted such efforts to malign the Corps were the work of those "trying to capitalize on recent events to attack the Corps of Cadets."[33] At once the Commandant suspended operations of the Cav, except for the care and feeding of the horses. He canceled all public appearances until further notice, and he began a full investigation and review of the unit's operational procedures, training methods, and qualification for membership. On the evening of September 25, President William Mobley addressed the entire Corps of 1,850 in Duncan Hall, stating he would not tolerate any form of intimidation or harassment. A special eight-member panel was formed by Mobley to investigate the allegations. However, in advance of any findings, the "news media wasted no time

Cadets and caisson of the Parsons Mounted Cavalry, shown in 1997. Courtesy Texas A&M Photographic Services

from coast-to-coast spreading the story that Texas A&M University and particularly its Corps of Cadets had serious problems."[34] In response to the coverage in the press, the administration was deluged by mail and calls on the "allegations" against the Corps. Ty Clevenger '91, 1990–91 student body president and a feature writer for the *Battalion* in the fall of 1991, noted in a detailed article in the *Texas Aggie* that it appeared that "someone was manipulating the matter for reasons other than helping the state of women in the Corps."[35]

By mid-October, evidence surfaced that there was an orchestrated effort by factions both on and off campus to manipulate the media in order to undermine and destroy the Corps of Cadets. Mail and calls flooded the

president's office as well as the regents' office in support of the Corps and demanding the rapid resolution of the allegations against the Cav. There was a growing demand that anti-Corps activities and organizations be challenged and held accountable. Former students, the Association of Former Students, parents, A&M Clubs, and A&M Mothers Clubs statewide, although denied a meeting with the president, took statewide action on October 21. The A&M Mothers Clubs were mobilized by a candid letter to Dr. Mobley by Linda Armstrong and Margaret R. Hinton. Their letter of indignation on the treatment of the Cav, along with presentations to A&M Clubs and civic organizations across the state expressed concern with the treatment of the cadets:

It is increasingly apparent that an organized plan is being executed on the campus of Texas A&M to discredit, dishonor, and eventually dispose of the Corps of Cadets. It is our opinion that the political influences sponsoring this effort are the Faculty Senate and two on-campus student organizations: the National Organization of Women and the Gay Liberation and Sexual Freedom organization. There are procedures in place at the University for these people to express their concerns and opinions; however, they have apparently chosen a subversive route to draw unfavorable publicity to the Corps.

We have been patient and quiet, trying to let the TAMU Administration handle this matter, believing in the fairness and justice of the system. A month has passed since the first alleged assault was reported. During this month, our young people have had their privileges revoked without being proven guilty; and they have been publicly slandered and humiliated.

It is imperative that one point be perfectly clear: these incidents have been mishandled; our children are being made to suffer and possibly to be scapegoats to satisfy public concerns, and we cannot continue to sit idly by and watch this happen.[36]

On October 22, the media reported that the female cadet that had leveled the charges against the Cav admitted in a handwritten statement that she had fabricated certain facts and that the events reported "had never happened."[37] Notwithstanding, the Commandant proceeded with his ongoing investigation, resulting in the punishment of more than thirty cadets in the Cav for harassment incidents unrelated to the alleged physical assault charges leveled in mid-September.

Activities of the Cav remained suspended. In spite of revelations that the charges had been recanted, the special presidential panel appointed by Mobley continued its deliberations, urged on by the Faculty Senate. Interpreting a much broader charge for the committee, some faculty members felt that the investigation should proceed and steps should be taken to curb the influence of the Corps.[38]

The ongoing manipulation of the facts and the blatant discrediting of the Corps was forcefully addressed on October 25 by Association of Former Students President Bill Youngkin '69 at a press conference covered by the news media from across the state. Present at this meeting were Ty Clevenger and Liz Tisch, also a reporter for the *Battalion*. Noting that the Corps had been the target of special interest groups on campus, Tisch candidly told the reporters, "I thought the story was about discrimination and sexual harassment . . . it had nothing to do with sexual harassment or discrimination. It had to do with lies, manipulation, and disrespect, and the Corps of Cadets was the target."[39] The plot by anti-Corps advocates to further their agenda at the expense of the century-old Corps was fully unveiled and discredited. Reflecting on the events in the early 1990s, Dr. E. Dean Gage '65, former executive vice president and provost (1990–93), sensed broader implications: "There is no question that the conservative long term values of the *entire university* were under attack . . . those with an agenda wanted to tear down A&M."[40]

Exposure of the detractors of the Corps did not lessen the challenge to integrate women. The pattern of transition of women into all-male military enclaves proved difficult in organizations across the United States in the late 1980s and 1990s. The all or nothing approach (i.e., "disband the Corps") precipitated by new laws, political agendas, and court orders mandating the immediate integration of women proved a difficult and inexact task among the U.S. armed forces as well as at all military schools. Dictating that the process should be accomplished with haste and no reverberations failed to take into account the dynamics of the process. VMI and the Citadel, for example, delayed the admission of women until the mid-1990s via the courts—taking their case before the U.S. Supreme Court— yet failing in the end to block admission of women.[41] They, too, experienced transitional problems, attacks from antimilitary groups and massive media scrutiny upon the admission of women.[42] Furthermore, at all military schools the process was agitated by lone individuals engaged in unauthorized actions which were given disproportionate attention. Such actions provided antimilitary factions which abhorred all aspects of the military with a "cause" and a targeted means to discredit the Corps, while at the same time promoting their own agendas. "Due process" was often overlooked by zealous antimilitary antagonists. Although harassment and misconduct existed and will probably never be fully eradicated from any such organization. In no way is this the policy or norm, nor condoned as the accepted practice. Reflecting on the events of September–October, 1991, Cadet Colonel of the Corps John B. Sherman '92 noted:

It [this press conference] is in order to clarify a point that has been grossly distorted in the media. In an organization consisting of 1,800 people, there are unfortunate implications of having such a cross section of society, even if the training environment is beyond reproach to the training of ethical behavior. A few bad apples are going to slip through the system, since there has not been a perfect training system devised by anyone. I challenge any one of the Corps's critics to find any historical precedent of an organization free of members who demonstrate some forms of misconduct.

With this in mind, it should be clear that a very small percentage of misfits are going to slip through an organization such as the Corps. No amount of education will change these peoples' attitudes, nor will any amount of punishment. . . .

So here we stand, with the Corps of Cadets being convicted as being a discriminatory organization without any due process to clear its name . . . there is a clear difference between incriminating an organization as a whole or incriminating a few misfits whose attitude never adjusted to fit the best interest of the organization.[43]

During the next three years the Corps of Cadets and its programs came under continued scrutiny. The Commandant increased efforts to bring women into the mainstream of activity, yet Corps opponents were there to rehash old events and link any misstep of the Corps to some trend of harassment. To permit the staff in the Commandant's Office to better direct cadet activities, four new officers were added, including Maj. Rebecca L. Ray, USAFR, who was given the direct responsibility to monitor the activities of women in the Corps and serve as a role model. In addition to Major Ray routinely interviewing each female cadet, tremendous efforts were taken to track their progress as well as ensure they had fair access to all Corps programs and organizations. At the end of her first round of interviews, Major Ray noted, "To my surprise, a few even expressed the opinion that our program is not tough enough."[44]

In the meantime, the Parsons Mounted Cavalry was reorganized, staff members in the Commandant's Office were assigned to the dorm area for oversight, and an extensive training program was established to address ethics and nondiscrimination, as well as positive leadership and management skills. The *Standard* was updated to further define and review conduct that was deemed unacceptable, and an "intensified" effort by cadets included rejuvenated internal oversight by the Cadet Honor Board. Many expressed concern that the day-to-day "fun" was being eliminated. Any spontaneous event by the cadets that contributed to *esprit* or "good bull"—such as April Fool's Day antics, fish details in the mess hall, or a fish class "Rain Dance" on the Quad "to solicit rain for the purpose of canceling the Fall Review scheduled for the next day"—was brought into question by the Bulls in the Trigon. There was heightened awareness among the cadets. For example, Military Weekend in late February, 1992, which included invited cadets and officers from other military schools, featured a roundtable forum moderated by Lt. Col. Robert McDannell from West Point on "Sexual Harassment in the Military."[45]

While the alumni and media were routinely advised of these ongoing events, many former students continued to express grave concern about the treatment the Corps had received. Irritation increased with the release of negative press coverage and highly derogatory articles such as the feature story on A&M and the Corps—"Love and Hate at A&M"—in *Texas Monthly* magazine, referred to by President Mobley as "particularly repulsive."[46] This coverage, along with an article that appeared

in the *New York Times,* "Harassment Cited In Cadet Program," in late November, 1991, on the "interim" report of the special committee set up by Mobley to investigate the Corps, resulted in the U.S. Department of Justice making a formal request on the status of "possible harassment in the Corps of Cadets, including complaint resolution procedures."[47] This inquiry, followed by a complaint on an already resolved allegation to the office of Lloyd Bentsen, Texas senator, resulted in a U.S. Justice Department investigation of the Corps that lasted more than two years. General Darling's calm resolve in handling the continuing investigations and allegations helped settle the issues in question.[48]

The president's special committee on the Corps concluded its investigation on April 30, 1992. The report, resulting from interviews with fifty-three witnesses, cautioned that the Corps "risks becoming anachronistic" if the body does not change its attitude and insensitivity to harassment.[49] A list of recommendations was provided, most of which had already been enacted and accepted by the Corps and the Office of the Commandant. The committee noted that societal changes would impact the demographics of the Corps: "As traditional social institutions become less stable, the value of the Corps increases as an environment which provides a solid structure for young people who want and need to acquire self-discipline and good habits in a close-knit community. However, the widespread belief that bonding is produced by inflicting and enduring physical and psychological pain has on occasion contributed to unacceptable and indeed, illegal behavior by a number of cadets."[50] Following the report, President Mobley established a second committee, the "Ad Hoc Advisory

Reveille V, mascot of Texas A&M from 1984 to 1993, surrounded by cotton. During Rev V's tenure the Aggies won six Southwest Conference Football Championships and two Cotton Bowl titles. Courtesy Texas A&M Photographic Services

Group on the Corps of Cadets," to monitor the Corps's efforts to eliminate gender discrimination and harassment within its ranks.[51] And shortly after the dedication of the Sam Houston Sanders Corps of Cadets Center in late September, 1992, President Mobley asked Dr. John Koldus to contact members of the A&M Board of Regents to explore the possibility of setting up yet a third committee to provide an "external" assessment into "Corps recruitment and vision of the Corps for the 21st century." Little notice was given to the fact that the cadets of the A&M Army ROTC detachment were recognized as having the top training program among the sixteen largest

institutions nationwide.[52] Notwithstanding, for the first time in months the focus shifted toward the future of the Corps.

CADET LIFE

The Corps endeavored to continue with its broad scope of activities and events, yet the crush of observation and scrutiny eliminated any activity or event that could be construed to be in any manner harassing. The Ross Volunteers served as honor guard at the 1993 and 1997 Texas gubernatorial inaugurations, the Cav expanded its number of appearances on campus and statewide, the Fish Drill Team, with the absence of a national drill competition, concentrated on regional events and expositions such as the Fiesta Flambeau Parade in San Antonio. The profiles of the entering Corps fish indicated that recruiting efforts were attracting a broad cross section of the state's best students. For example, the Corps of Cadets Class of '95 that entered A&M in the fall of 1991 reflected a changing make-up of the Corps: 15 percent were minorities, 55 percent lettered in one or more high school sports, 26 percent captained an athletic team, 33 percent graduated in the top 10 percent of their graduating class, and 46 percent were members of the National Honor Society. With academic excellence stated as the number one priority, the new fish were well suited for the academic rigors of the university. Part of the success in recruiting was directly attributable to an increased number of Sul Ross Scholarships, the effective activities of the Corps Leadership Outreach program (i.e., former student volunteers) organized by Don Crawford '64 in communities across the state, as well as high school students' visits under the "Spend the Night with the Corps" program.[53]

Fall semester for the Corps continued to be dominated by the annual gridiron season. Home football weekends were marked by Friday midnight Yell Practice in Kyle Field, the traditional march-in of the Corps prior to kickoff, and the victory plunge for the five Yell Leaders at the post-game Yell Practice at the fish pond and on the steps of the YMCA. Students maintained the Twelfth Man tradition of standing throughout the game and at half-time seniors formed the traditional "Boot Line" to welcome the team back on the field. When the Aggies score, the cannon crew at the south end of the stadium accented the event by firing the Spirit of '02. Company C-2 enlivened each Halloween with their flight of the "Great Pumpkin." The ritual of a junior wearing a huge pumpkin on his head while members of the Aggie Band lay in wait to deter the annual "flight" or journey through the Quad was halted in the late 1990s. Each fall, by mid-November, attention turned toward the construction of the bonfire on the northeast corner of the campus. While its height had been restricted to about 55 feet, the event generated a tremendous amount of student (Corps and civilian) involvement. And in October, 1999, shortly after the burial of Reveille V, the A&M gridiron program reached a milestone, winning its six hundredth game with a victory over Oklahoma State.[54]

Spring continues to be marked by a number of social events ranging from the Military Weekend, the Senior Ring Dance, and the Combat Ball to the Boot Dance for the junior class. Parents Weekend marks the change of command for cadets and features ceremonies in which individual and unit-level award recipients are recognized. The active spring schedule includes the annual fourteen-mile March to the Brazos for the March of Dimes. The largest collegiate fund-raiser for March of Dimes in the nation, it has raised in twenty-four years more than $900,000 for charity. The formal end of the school year culminates with Final Review on the Simpson Drill Field.[55]

With the goal to enhance retention and develop leadership skills, an intensive effort was begun in the late 1980s to bolster academic preparation and prepare drill and ceremony (D&C) cadets for the civilian employment market. General Darling, whom many have termed the "academic commandant," placed a tremendous emphasis on academic excellence. With the assistance of funding from the Corps Development Council, an extensive tutoring program, an academic advisor program, and a cadet consultation program were established at the Sanders Corps Center under the direction of Laura Arth '75, Corps Academic Coordinator. These efforts, along with opportunities for recognition in the Ormond R. Simpson Honor Society established in 1980, enhanced the academic performance of each class. The second phase of cadet enrichment was the development of career programs for cadets that were not entering the armed forces. By the late 1990s, more than 70 percent of the juniors and seniors were noncontract drill and ceremonies cadets. To address the changing composition of graduates from the Corps, Maj. Rebecca Ray developed an extensive leadership and career-preparation course for D&C seniors.[56]

THE BLUE RIBBON COMMITTEE

At the year-end 1992 Board of Regents meeting, Regent (and Chairman of the Corps Development Council) Raul Fernandez '59 expressed concern about the future of the Corps as well as the apparent decline in the number of cadets in the Corps—he said, "1,804 is low enough!" A variety of ideas surfaced in conversations within the administration. General Darling and Provost E. Dean Gage '65 initially concentrated on scholarship opportunities as well as a review of enrollment procedures to "assure maximum admissions to the Corps."[57] Gage challenged the premise of the age old recruiting tactics and questioned the overall strategy of the Corps. "Rather than market the Corps of Cadets as the 'engine that fires

the Aggie spirit,'" he noted, "a more effective strategy nationally is to sell the unique opportunity which our Corps of cadets offers to young people who are interested in the ROTC experience but who wish to attend a major university offering a wide variety of academic programs and student leadership development activities."[58]

A&M Regents contacted the board of the Association of Former Students and urged the formation of an alumni committee to review current trends and policies within the Corps as well as provide recommendations on the future state of the Corps. Bill E. Carter '69, association past president, was appointed to chair the panel officially known as the Committee on Corps Enrollment and Strength—yet commonly referred to as the "Blue Ribbon Committee." The alumni group focused on two primary areas: first, the internal nature and operations of the Corps, chaired by Weldon Kruger '53; and second, external issues that affect the image and growth of the organization, chaired by Royce H. Hickman '64. Four subcommittees dealt with the "fish" experience (and retention), academics, leadership, and the organization and operation of Corps units. In advance of the Blue Ribbon study two reports provided detailed background material; first, an independent research firm conducted extensive interviews and focus groups comprised of cadets, non-cadets, parents, staff, and faculty. Second was an extensive report compiled by members of the 1991–92 Corps staff and edited by Scott Phelan '92—"Marching to the Future: A Proposal for Revitalizing the Corps of Cadets." Noting the concern with which the report was drafted, Phelan stated, "Change, however drastic[,] is most effective and acceptable if it comes from within . . . if not, we risk being changed by outsiders." This data formed the initial basis of the deliberation of the Blue Ribbon Committee.[59]

As the alumni committee proceeded throughout 1993 to assess the Corps, the administration—with good cause due to the ongoing inquiry by the Justice Department—had mixed feelings on the progress of change as well as the manner in which discipline in the Corps should be handled. It became apparent that many alumni and the regents thought that the Commandant should handle all discipline problems, while Koldus stressed caution about the due process needed to fully handle cadet discipline cases. The prime concern was that there existed a perception in the media as well as among detractors of the Corps that the Corps protects its own and would tend to mitigate its discipline problems. The Commandant, his staff, and the cadets received notice in June, 1993, that the president's Ad Hoc Advisory Group had concluded their oversight, stating, "Overall, we feel that the Office of the Commandant is committed to continuing the progress that has been made so far in addressing the issues of harassment and discrimination in the Corps so that it can remain a viable and important part of the University in the years to come."[60] However, in the meantime a debate evolved between the Commandant and Vice President Koldus over the discipline policy. In the face of repeated regent comments that the Commandant controlled discipline, the eventual resolution was the use of the Center for Conflict Resolution, where all alleged violations of university regulations by A&M students—Corps and civilian—would be heard and handled equally. A representative of the Commandant's Office would participate in hearings involving cadets.[61]

Turmoil within the administration became evident in September, 1993, with a wholesale shake-up in both the Texas A&M System and the university. Dr. Mobley was promoted to replace System Chancellor Herbert Richardson, and Provost E. Dean Gage vaulted to interim president of the university. A number of other moves were made, including Ed Davis '67, executive deputy chancellor of the university system becoming director (later president) of the Texas A&M Foundation, which in 1993 had an endowment of $1.6 billion. On campus the Blue Ribbon Committee prepared to release its final report. The Corps of Cadets by December numbered 1,905 cadets, 4.5 percent of the university's 41,700 enrollment. Cadet recruiting for 1993–94 involved a network of 113 Corps Leadership Outreach representatives in forty-four communities across Texas.[62]

During the December, 1993, holiday break a group of University of Texas students "dognapped" the four-month-old A&M mascot, Reveille VI, from the Dallas backyard of the mascot corporal, Jim Lively '96. Exhorting Aggies to "take the high road," both Dr. Gage and Dr. Southerland urged calm. Fortunately, Little Rev was returned without harm in time to make a highly publicized appearance on nationwide television at the 1994 Cotton Bowl.[63]

Cadets during the spring of 1994 intensified recruiting efforts. Funding from the Corps Development Council and an expansion of the Corps Leadership Outreach program helped in targeting high school students across the state. One key to success was the Spend the Night with the Corps Program. The recruiting program invites high school juniors and seniors to the campus for an advanced look at the cadet lifestyle and programs offered by the Corps. This recruiting program increased new student contacts from 151 in 1988–89 to 1,911 in 1994–95. The result was 764 new Corps fish in the fall of 1994 (total Corps strength 2,010), up 43 freshmen over the previous year. The Texas Aggie Band totaled 352 members (130 fish and 222 upperclassmen)—one of the largest Aggie Bands ever assembled. In February, 1995, there were 79 women in the Corps.

The overall goal during the decade was to fill all twelve dorms in the Quad with 2,600 cadets. This bold target proved difficult in spite of increased recruiting and scholarships for all who joined the Corps.[64] Enrollment management, increased numbers of women, and a lower emphasis on active duty military service stymied real growth of the Corps. However, the "Take Back the Quad" motto remained the primary recruiting and retention challenge. One key to sustained growth rested with the level of attrition, a long-time nemesis of the Corps, of each new fish class and their assimilation into the rigors of both the cadet lifestyle and the academic environment at the university. Fish retention has a tremendous impact on the efforts to expand the Corps to a strength of 2,600. Thus, fish attrition of 27 percent in 1994–95 was of a great concern.[65]

In the fall of 1995, Dr. Ray M. Bowen '58, 1957–58 Deputy Corps Commander, was named the twenty-second president of Texas A&M. Prior to coming to A&M he was provost and interim president at Oklahoma State University and department chair of mechanical engineering at Rice University. Bowen recalled, "Neither of my two Aggie brothers, Herbert '48 and Jerry '55, thought I'd make it through my fish year in Aggieland let alone graduate . . . but I persevered." The

new president found a far different place than the one he remembered from the mid-1950s. By the mid-1990s, the growth and recognition of the diversified programs, the growing endowment, and research expenditures ranked the university among the top ten academic institutions nationwide. The Corps of Cadets realized its viability and growth was allied with academic performance. Awards, leadership positions, and image clearly focused on the academic results of the cadets.[66]

General Darling stepped down after nine years as Commandant at Final Review 1996, ending a sometimes tumultuous yet stabilizing period. Concerned with the future of the Corps, Darling was able to deal with a broad cross section of issues and transitional changes within the Corps that left the organization in good shape. His focus on academics, retention, and leadership brought the Corps into the mainstream of the university at a time when the cadets became a smaller percentage of the overall student body every year. One key to the success in fostering the Corps programs has been to harness alumni efforts. To better address projects, the Corps of Cadets Council was formed in early 1996—merging the Corps Development Council (fund raising), Corps Leadership Outreach Program (recruiting), and the Corps of Cadets Association (data base of former Corps members). Upon retirement, General Darling chaired the Corps Endowment Campaign from 1996 to 2000. His unique focus on the Corps proved invaluable. Although the initial campaign goal was $16 million, more than $33 million was raised by early 2000.[67]

Col. Donald J. Johnson '55, a thirty-year veteran of the Commandant's Office, was interim Commandant from May to August, 1996. The legacy of the Darling period firmly established the return of the non–active duty "civilian Commandant."[68] With the growth of personnel and support elements in the Commandant's Office, the role of the active duty ROTC officers cadre as military advisors also changed. The U.S. Army, Navy/Marine, and Air Force ROTC officers, except for classroom instruction and weekend branch orientation training, had fewer duties with the day-to-day management of the Corps of Cadets.[69] Each of the three service branches emphasized the establishment of parallel specific programs that address training objectives for summer camp and pre-commissioning requirements. In contrast, the Office of the Commandant grew from a staff of six in the late 1980s to fifty-four personnel by the time of General Darling's retirement in the mid-1996. These personnel were added to address the expanding role of the Commandant's Office to oversee the operations of the Corps. Furthermore, the changing role of the Commandant and his staff of university employees revived the "tac officer" function, newly labeled Cadet Training Officers (CTOs), that daily monitored cadet activities on the Quad.[70]

By the late 1990s, positive changes within the Corps of Cadets included inexhaustible alumni support and an improved image. The downsizing of the armed forces (the U.S. Army by late 1994 had shrunk by 25 percent and defense spending dropped to its lowest point as a percentage of GDP since Pearl Harbor) created a severe readiness problem and the decline was mirrored in the yearly number of A&M cadets commissioned.[71] Campus ROTC detachments once again came under increased pressure to fulfill their mission and quotas with fewer resources.[72] Concerned with growing anti-ROTC action

on university campuses, Congress passed sweeping new laws in early 1996 that prohibited federal agencies from providing funds to any institution deemed to have "antimilitary" policies.[73] Furthermore, attempts were made to deny the six military colleges and universities any "privileged" status afforded their graduates. Sen. Phil Gramm of Texas along with Chairman of the Senate Armed Services Committee Sen. Strom Thurmond of South Carolina (home of The Citadel) killed these attacks and facilitated passage of a new law that protected the active duty status of graduates from the senior military colleges—Texas A&M, Norwich, VMI, The Citadel, VPI, and North Georgia. By the late 1990s, the goal at Texas A&M was to commission a minimum of 125 officers each calendar year. And on campus, shortly before the formal opening of Reed Arena, the December graduation and commissioning ceremony (the last of three ceremonies held annually) for 47 new Aggie officers was conducted in old G. Rollie White Coliseum in December, 1997, more than four decades after Aggies first graduated in the coliseum in 1954.[74]

THE FISH DRILL TEAM

Upon the arrival of Maj. Gen. M. T. "Ted" Hopgood, Jr., USMC (Ret.)—Class of '65, former Aggie Yell Leader, a Vietnam veteran, and past president of the Marine Corps University at Quantico, Virginia—it was apparent that academic achievement had become the number one priority of the Texas A&M Corps. Working closely with the administration and the Blue Ribbon Committee, Hopgood reinforced the goal to "take back the Quad" with the goal of recruiting and

Three decades of Texas A&M Commandants pictured on the plaza at the Sanders Corps Center (left to right): *Maj. Gen. Ted Hopgood '65, Col. Tom Parsons '49, Maj. Gen. Tom Darling '54, Col. Jim Woodall '50, Dr. J. Malon Southerland '65, Brig. Gen. Don Johnson '55, and Col. Donald L. Burton. Courtesy Office of the Commandant*

retaining 2,600 cadets. In spite of the fact that the committee issued its findings in late 1993, it remained active to monitor the implementation of changes and determine future needs of the Corps. Their charge was essentially complete by 1994–95, with the exception of one primary goal to create a senior level "Cadet Leadership Institute." In line with the Commandant's goals to "recruit, retain, and graduate" outstanding cadets, the recruiting

staff and Corps Leadership Outreach volunteers set a goal of 900 new fish each year. Between 1993 and 1996 the entering fish class had averaged 720 new cadets. In 1997, as The Citadel and VMI grappled with the challenges of admitting women, seventeen of the Corps's thirty-one units had been integrated. In a briefing paper to cadets, the Commandant noted, "In our Corps, we have only Aggies. The only color we should see is maroon, the

TABLE 10-3

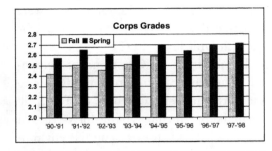

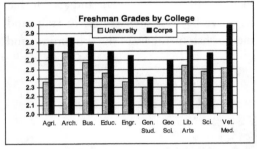

Beginning in the early 1990s, the Corps set as a goal to improve all areas of academic performance. By the 1997–1998, overall Corps grades as well as the results of the freshman class indicated tremendous improvement university-wide.

Source: Corps Academic Center

only gender we should see is Aggie, with the only creed required for membership a belief in the Spirit of Aggieland."[75]

The Corps of Cadets grew at a 5 percent rate beginning in 1991 (to a strength of 1,804 cadets) for five years to 2,243 in uniform by the fall of 1996. In 1996, the Corps added 704 fish, 75 of them women. Yearly, additional resources were added to the recruiting program; however, the pool of potential male freshmen admitted to the university due to the enrollment management policy remained stagnant. Thus, real growth via the fish class would be an ongoing challenge. While the number of coed freshman admissions were up slightly, new women cadets averaged only 60 each year. Between 1994 and 1997 the total student enrollment averaged 41,700 each fall and remained close to the university's enrollment management guidelines, while graduate students averaged 7,500 and the freshman class averaged 6,200 during the same period. During the late 1990s, with the consolidation and

TABLE 10-4
Corps of Cadets, 1999–2000

Corps Staff Forrest Lane, Corps Commander					
1st Brigade		2nd Brigade		Regiment	
1st Batt	2nd Batt	3rd Batt	4th Batt	5th Batt	7th Batt
A-1	D-1	I-1	A-2	H-1	G-2
B-1	E-1	K-1	D-2	N-1	K-2
C-1	G-1	L-1	F-2	C-2	P-2

Wing		Band	
1st Group	2nd Group	A Company	
Sq-1	Sq-12	B Company	
Sq-3	Sq-16	A Battery	
Sq-8	Sq-17	B Battery	

disbanding of companies V-1 and G-3, Squadron Thirteen, and Squadron One, the 1999–2000 Corps numbered 27 units.[76]

Streamlining the Corps's organizational structure was intended to create better accountability as well as larger units. Unit size and retention were critical elements in order to be considered for annual awards. Juniors and seniors were expected to be more responsible not just to their units, but also to overall Corps policy. These measures, except for the elimination of units that had traditional roots with numerous classes of cadets, were well received in the ranks, yet tensions were heightened with reports of misconduct by the Fish Drill Team (FDT).

In early 1997, the Commandant "temporarily suspended" nine student advisors to the Corps's precision drill unit, the Fish Drill Team. Former members of the forty-year-old organization had created a cult-like environment of routine harassment and assault that went far beyond their duties to train and mentor fish members of the drill unit. Amid

The 1999–2000 recipients of the General George F. Moore Award, B Company of the Fightin' Texas Aggie Band. Standing from left: *Chris Migl '00, Commander; David Seith '02, Flag Bearer; Daniel Lilly '02, Guidon Bearer.* Seated from left: *Commandant MG Ted Hopgood '65; Cadet Ryan Zeitler '01, Company 1st Sergeant; and Unit Advisor Keith Stephens '71. Courtesy Texas A&M Photographic Services*

Women have assumed a major role in the Corps. Clockwise from left: *Cadets Adrienne Bender, Monica Strye, Megan Kinne, Erica Smith, and Kelly Garrity, all Class of '99. Courtesy Office of the Commandant*

allegations of hazing and assault, the Commandant declared that the drill unit would not compete in 1997–98. A pattern of extensive harassment and complaints by team members resulted in the suspension of all future teams. Reaction was swift. In addition to discipline charges by the university, the judicial system became involved in the case due to the 1987 Texas Hazing Law. The university and Corps had maintained a strong commitment to zero tolerance toward hazing. However, the hazing law, considered by many to be ambiguous, was subject to varying interpretations that resulted in a court ruling that those hazed could not be forced to report hazing incidents.[77]

Negative media coverage on the drill team incident was partly to blame for the drop in the size of the fall, 1997, fish class that totaled 646, down from 704 in 1996.[78] Legal action and litigation involving various members of the Fish Drill Team continued into the year 2000. Notwithstanding, 776 fish were enrolled in the fall 1998, bolstering total Corps strength to nearly 2,200 cadets. The continued emphasis on academics dominated cadet activities. During calendar year 2000 A&M commissioned 156 new officers into the armed services, the most since 1992. The 1999–2000 Corps numbered 149 women in the ranks and Corps strength in the fall of 2000 totaled

1,947.[79] The benefits of the Corps of Cadets were viewed as the key to growth and retention. By the turn of the century the recruiting focus was centered around the Corps "experience" as stressed by Lovell W. Aldrich '65, chairman of the Corps Development Council:

Many people across the nation have seen firsthand what the Corps of Cadets is about. They have seen that Aggies stand shoulder to shoulder in good times or bad. They have seen that our system works and the quality of its product.

Since 1876, Texas A&M has produced some of the finest military officers and great leaders of this nation in all fields . . . there is something else of which we must make the world aware. It begins with the adventure that is the essence of the Corps and is something that cannot be found without participating in the Corps at A&M.

Thus, the direction of the Corps of Cadets at Texas A&M today is to bring good students into a good academic environment and to help them achieve their goals as students in the academic world. The Corps will also help to develop a student into a well-rounded person who can handle a problem, put the distractions aside, and find the answer.[80]

BONFIRE TRAGEDY

Ever mindful of the uniqueness of A&M's time-honored traditions, the Corps continued its lead role in fostering the heritage and lore of the institution. In spite of the care and attention to safety, tragedy struck during the early morning hours of November 18, 1999, with the sudden collapse of the traditional Aggie Bonfire stack. As more than 70 climbers and stackers worked on positioning of logs, the logs shifted, rolling the top three layers over on the students, trapping and killing 12 and injuring 27. Never in the history of the institution had a tragedy of this magnitude occurred on the campus. Bonfire was first built in 1907 and, since the record 109-foot-tall Bonfire in 1969, the height of the structure had been restricted to 55 feet. In the aftermath of the collapse, the 1999 Bonfire was canceled for the first time since 1963, after the assassination of President John F. Kennedy. The outpouring of grief for the lost students, eight of whom were current or former members of the Corps of Cadets, was overwhelming.[81]

Around the world, people watched live television coverage and prayed as rescue crews scrambled to assist the injured. Students, alumni, parents, friends, and foes came to the aid of Texas A&M. Following campus memorial services at A&M and on the campus of the University of Texas, there followed an emotional Candlelight Ceremony at the Bonfire site, a much muted pregame Yell Practice at Kyle Field, and the traditional game with the Longhorns the day after Thanksgiving. President Ray Bowen '58 summed up the response: "I have never been more proud to call myself a Texas Aggie. We will never forget the concern and sympathy demonstrated by our friends and neighbors at the University of Texas and other college campuses throughout the nation. Their kindness has been a touching tribute to the special camaraderie we share. Although I'm sure our rivalries will continue to be spirited, I believe the core of our relationship with these schools will be changed forever."

KEEPERS OF THE SPIRIT

In the months that followed the Bonfire tragedy, an extensive investigation was conducted to evaluate all aspects of Bonfire. The final report concluded that design modifications and the structural dynamics of building such a large vertical project resulted in the collapse. While some few clamored for a Bonfire in the fall of 2000, it was agreed among the students and administration that the event would be postponed until 2002.

No single event in the history of the university, save the tremendous loss of Aggie life in the armed forces, has had such a profound impact on the students, faculty, alumni, and friends of A&M. The aftermath of the tragedy gave the larger world a glimpse of the unique nature and meaning of being a Texas Aggie and a part of the A&M family.

The foundation of the Aggie spirit, shaped over decades of dedication and adversity, lies in the values and commitment of A&M cadets and former students worldwide. The enthusiasm and loyalty have long been an enigma to many, yet no mystery to the keepers of the spirit.

For Aggies know as few men do,
There are no other ties,
So tight and true
To help us through,
As the Spirit that never dies.
Anonymous

EPILOGUE

A prolonged shrill whistle—the train stops—"College Station!" sounds in the ear of the young man who is seated in one of the coaches and looking out the window, eager to catch a glimpse of his port of destination, the A. & M. College of Texas, of which institution he intends to become a student.

He alights, and soon sees himself surrounded by a crowd of boys dressed in gray, who conduct him to the college building. From this instant he ceases to be boy so and so . . . he is—what is he? Well, he is simply a "fish." What will be thy fate?

fish Charles Prokisch, *The Journal*, December, 1889

THE CORPS OF CADETS at Texas A&M has persevered and prospered over the past 125 years. The historic bedrock of Texas' first institution of public higher learning, the Corps has provided the traditions and legacy that mark the rich heritage of one of the nation's largest universities. The mandate of the Morrill Land Grant Act of 1862 to include "military tactics" within the broader curriculum of the "agricultural and mechanic arts" was adhered to and pursued with vigor at the A&M College of Texas. The Corps, in time, not only shaped the students who entered the institution but also engendered lifelong dedication and loyalty among alumni. The agrarian roots of the nation, the frontier spirit of Texas, and the Southern military tradition were pivotal in molding a solid foundation for the citizen-soldiers who passed through the ranks of the Corps of Cadets. Furthermore, the environment of shared experiences, time-honored traditions, and exceedingly loyal alumni has proven to be critical to the longevity of the organization.

At times when similar land-grant institutions either discontinued or scaled back their commitment to a uniformed student body, Texas A&M held firm in its belief and heritage. By the early 1960s, the greater good of the university dictated that a broader student body and enhanced educational and research programs were needed. However, broadening the scope of Texas A&M did not imply elimination of the Corps. In time, growth and diversification of the university relegated the Corps to a position of fewer numbers when compared to the overall student body but did nothing to undermine its influence as the keeper of Aggie traditions. In large part, it is this legacy of traditions and the uniqueness of the Corps that eventually disarms those who, over the years, have endeavored to diminish the role of the Corps in exchange for programs and agendas that have no foundation in the fundamental values of Texas A&M nor in the hearts of the students and former students.

The shared experiences of those cadets who have walked the halls and drill fields of A&M have linked generations of Aggies in loyalty to each other and their alma mater. From the earliest days of the fledgling college, being a Texas Aggie has been a formidable experience. The fraternal regimen and esprit de corps shape an uncommon bond. The buddy system, whipping out, fish spurs, air outs, whistle jock, hot corner and cush, come-bys,

quadding, Corps trips, campusology, hitting the wall, your fish old lady, rams, the bull ring, the block "T," Midnight Yell Practice, Bonfire, boot-line, Silver Taps, Muster, the Twelfth Man, Elephant Walk, and Final Review each year immerse class after class of cadets in the ways and lore of Aggieland. The shared environment that all enter on an equal basis leaves a vivid image and encourages solid reflection on those who came before—remembered by the former cadets and alumni as the way it "really" was in Old Army days, for most just a moment in time in the mind's eye. Thus, with each new freshman class, a new generation of cadets—in the spirit of the times—passes the traditions forward to yet another generation of Aggies.

In war and peace the cadets and former students of A&M have fulfilled their citizen-soldier role in service to the state and nation. When called upon, the warrior spirit and valor of the Texas Aggies have soared to fill the annals of our nation's wars. Beginning with the Spanish-American War, Aggies have been among the first to answer the call to duty, providing a vanguard of leadership and gallantry. In the two great world conflicts of the twentieth century—World Wars I and II—Texas A&M former students played prominent roles. More than 16,000 officers and 6,000 enlisted men bolstered the ranks in these two conflicts—more than from any other institution in the nation. And in like fashion, they answered the call in the wind-swept hills of Korea, the rice paddies of Vietnam, and the sands of the Arabian peninsula. In total, since 1876 Texas A&M has produced more than 43,000 commissioned officers for the armed services, of whom more than 225 have reached the rank of general or flag officer. From these periods of armed conflict have come strength, loyalty, pride, and remembrance—something many who have not been a part of the Aggie experience cannot fully appreciate or understand.

The brave and the bold are not forgotten. Annually, recognition is bestowed on former cadets in ceremonies inducting them into the Corps of Cadets Hall of Honor. Numerous memorials, plaques, and monuments on campus recall the sacrifices and dedication of generations of former students. The uniqueness of purpose and common bond among Aggies is most dramatically displayed by students at the monthly Silver Taps observance on campus and the annual alumni answer to the Roll Call at hundreds of worldwide Aggie Muster ceremonies conducted each San Jacinto Day.

This response to duty is further exemplified by countless Aggies' tremendous peacetime endeavors and contributions to their chosen civilian vocations. Whether graduates serve in the armed forces or in civilian life in industry and public service, the Corps of Cadets has long provided the preparation and framework of leadership training for success in all walks of life. The downsizing of the armed forces and the peace dividend in the wake of the end of the Cold War have resulted in fewer Aggie military commissions, yet there has been no letup in the dedication to excellence and duty. The Corps remains one of the largest uniformed military cadet corps in the nation, behind only the federal service academies. Offering commissions in all branches of the armed services, the Corps has fulfilled the mandate of the Morrill Act to produce citizen-soldiers and leaders. Thus, the Corps of Cadets is constantly challenged to maintain high-level leadership training, retention within the ranks, and an atmosphere conducive to academic excellence.

From a bald post oak prairie with six students and as many professors in 1876, to today's dynamic university, past and current members of the Aggie Corps of Cadets remain the keepers of the spirit. The introduction in the cadets' 1947 little maroon book, *The Cadence,* captured the legacy and essence of the citizen-soldier and the A&M Corps experience: "The Aggie Spirit helps to mold a man's character. The things he does and the manner in which he lives with his fellow students make a permanent impression on him. . . . you will feel that you are a part of something greater and larger than yourself, something noble and moving . . . for there are others who depend upon you."

Per Unitatem Vis

APPENDIX I

PRESIDENTS, COMMANDANTS,

AND CADET CORPS COMMANDERS, 1876–2001

Date	President	Commandant	Corps Commander/Hometown
1876–77	Thomas Gathright July 7, 1876–Nov. 21, 1879 VMI, 1862	R. P. W. Morris Prof. & Maj.	Alva P. Smyth Mexia
1877–78		Geo. T. "Pomp" Olmsted, Jr. Capt. USA 2081	Charles Rogan Giddings
1878–79			Charles Rogan
1879–80	John G. James Nov. 22, 1879–Apr. 1, 1883 VMI, 1866		William Harrison Brown Navasota
1880–81			Salis Albert Hare, Jr. Sherman
1881–82		Jan., 1882–84 C. J. Crane Capt. USA 2684	David Rice Houston
1882–83			William Edwin Mosley Jefferson
1883–84	James R. Cole Apr. 1, 1883–June 26, 1883 (acting) June 26, 1883–July 19, 1883		R. E. Pennington

Date	President	Commandant	Corps Commander/Hometown
	H. H. Dinwiddie VMI, 1867 chairman of the faculty July 23, 1883–Dec 11, 1887		
1884–85		1884–Oct., 1886 John S. Mallory 2d Lt. USA 2815	Robert M. Rutherford Seagoville
1885–86			Frederick E. Giesecke New Braunfels
1886–87		Oct., 1886–Sept. 5, 1889 Guy Carleton 2895 Lt. USA 2nd Cav.	John B. Hereford Dallas
1887–88	Luis L. McInnis Jan. 24, 1888–Jul. 1, 1890		Walter H. Allen Martin
1888–89		William S. Scott Lt. USA 1st Cav. 2852 Sept. 6, 1889–Aug. 21, 1890 F. E. Giesecke '86 Aug.–Oct., 1890 (acting)	E. W. Hutchinson Denton
1889–90			Sam Houston Hopkins Waelder
1890–91	W. L. Bringhurst (acting) July 1, 1890–Jan. 20, 1891	Benjamin C. Morse Lt. USA 3054 Oct., 1890–Sept. 1894	Clifford R. Morrill Austin

Date	President	Commandant	Corps Commander/Hometown
	Lawrence S. Ross Jan. 20, 1891–Jan. 3, 1898		
1891–92			Edgar Wright Paris
1892–93			Bert C. Parsons Kerrville
1893–94			Frank N. Houston Holland
1894–95		Geo. T. Bartlett Lt. USA 3rd Art. 2888 Sept., 1894–Apr., 1898	A. U. Smith Huntsville
1895–96			C. M. Park Dallas
1896–97			Charles C. Todd Jefferson
1897–98	Ross death: Jan. 2, 1898 Roger H. Whitlock (acting) Jan. 17, 1898–July 1, 1898		G. Newton Milano
		Lt. C. C. Todd '97 Apr.–Aug., 1898 (acting)	
1898–99	L. L. Foster July 1, 1898–Dec. 2, 1901	H. B. Martin Aug., 1898–Mar. 16, 1899	E. J. Kyle Kyle
		Cadet E. J. Kyle '99 (acting) Mar.–May, 1899	
1899–1900		J. C. Edmonds Col. USA May, 1899–1901	W. I. Bryan Chambersville

Date	President	Commandant	Corps Commander/Hometown
1900–1901			W. T. Garbade Witting
1901–1902	Roger H. Whitlock (acting) Dec. 10, 1901–July 1, 1902	Frank P. Avery Capt. USA 2741	Victor M. Foy Corsicana
1902–1903	David F. Houston July 1, 1902–Sept. 1, 1905		L. W. Wallace Garfield
1903–1904		Herbert H. Sargent Capt. USA 2991 Sept. 23, 1903–1907	James E. Pirie San Antonio
1904–1905			Marion S. Church* *Cadet Maj.: first designated CC
1905–1906	Henry H. Harrington Sept. 8, 1905–Aug. 7, 1908		Ed C. Arnold Graham
1906–1907			Walter G. Moore
1907–1908		Andrew M. Moses Capt. USA 3797 1907–11	Richard H. Standifer
1908–09	Robert T. Milner Sept. 1, 1908–Oct. 1, 1913		David Shearer** **Cadet Lt. Col.
1909–10			Arland L. Ward Houston
1910–11			H. M. Pool Clifton
1911–12		Chauncy L. Fenton Lt. USA 4229 1911–12	T. George Huth San Antonio

Date	President	Commandant	Corps Commander/Hometown
1912–13		Levi G. Brown Lt. USA 4129 1912–14	Robert E. Baylor Montell
1913–14	Charles Puryear (acting) Oct. 1, 1913–Aug. 24, 1914		Virgil V. Parr Waelder
1914–15	William Bizzell Aug. 25, 1914–Sept. 1, 1925	James R. Hill Lt. USA 4806 1914–16	Ernest N. Hogue Paris
1915–16			Thomas F. Keasler Mineral Wells
1916–17		C. H. Muller Capt. USA 4045 1916–17	Jack Shelton Brownwood
1917–18		C. J. Crane Col. USA 1917–18	H. C. Knickerbocker Houston
		Fred W. Zeller Maj. USA 1918–Nov., 1918	
		Grant M. Miles Maj. USA Nov., 1918–Jan., 1919	
1918–19		C. H. Muller Col. USA Jan. 16, 1919–Aug., 1919	Douglas W. Howell Bryan
1919–20		Ike Ashburn Lt. Col. USA (Ret.) Sept. 1, 1919–Jan. 1, 1924	Waller T. Burns Houston

Date	President	Commandant	Corps Commander/Hometown
1920–21			Bonner H. Barnes Coleman
1921–22			Paul C. Franke El Campo
1922–23			John C. Mayfield Huntsville
1923–24		Charles C. Todd '97 Col. USA Jan., 1924–1925	Herman L. Roberts Corsicana
1924–25			Frank M. Stubbs Robstown
1925–26	Thomas Walton Sept. 3, 1925–Aug. 7, 1944	F. H. Turner Lt. Col. USA 1925–Aug., 1927	William M. Pinson Forney
1926–27			Robert L. Edgar Cleburne
1927–28		Charles Nelson Col. USA Sept. 1, 1927–1932	Lacy N. Bourland Clarendon
1928–29			William M. Patton Lockhart
1929–30			Joseph H. Taylor Dallas
1930–31			F. Eddie Bortle Longview

Date	President	Commandant	Corps Commander/Hometown
1931–32		John E. Mitchell Lt. Col. Tex. St. Guard 1932–35	A. C. Moser Dallas
1932–33			Jimmie W. Aston Farmersville
1933–34			George W. Holmes Gonzales
1934–35		Frank G. Anderson Col. USAR Field Artillery 1935–37	Joe C. McHaney San Antonio
1935–36			Earle D. Button Houston
1936–37			Louis E. Lee Houston
1937–38		George F. Moore '08 Col. USA Aug., 1937–May, 1940	A. D. Justice Post
1938–39			D. B. Thrift San Antonio
1939–40			Durward B. Varner Cottonwood
1940–41		James A. Watson Lt. Col. USA Aug. 1, 1940–Aug. 1, 1941	William A. Becker Kaufman
1941–42			Tom S. Gillis Fort Worth
1942–43		Maurice D. Welty Col. USA Inf. Nov. 3, 1941–1946	W. W. Cardwell Luling

Date	President	Commandant	Corps Commander/Hometown
	Frank Bolton (acting) Sept. 9, 1943–May 27, 1944		
1943–44	Gibb Gilchrist May 27, 1944–Sept. 1, 1948		John Mullins Carrizo Springs
1944–45			George Strickhausen Galveston
1945–46			Charles West unknown
1946–47		Guy S. Meloy Col. USA 1946–48	Edward Brandt Houston
1947–48	Frank Bolton Sept. 1, 1948–Jun. 3, 1950		William Brown Cleburne
1948–49		Haydon L. Boatner Col. USA 1948–51	Robert McClure Texarkana
1949–50	M. T. Harrington June 3, 1950–Sept. 1, 1953		Doyle R. Avant, Jr. Laredo
1950–51		Joe E. Davis '29 Col. USA Res. 1951–63	DeLoach Martin Dallas
1951–52			Eric Carlson Elgin
1952–53			Weldon Kruger Brenham

Date	President	Commandant	Corps Commander/Hometown
1953–54	David H. Morgan Sept. 1, 1953–Dec. 21, 1956		Fred Mitchell Galveston
1954–55			Frank Ford Lubbock
1955–56	David W. Williams (acting) Dec. 22, 1956–Sept. 1, 1957		Larry Kennedy Houston
1956–57	M. T. Harrington Sept. 1, 1957–July 1, 1959		Jack Lundford Houston
1957–58			Jon Hagler La Grange
1958–59	James E. Rudder July 1, 1959–Mar. 23, 1970		Don Cloud Kerens
1959–60			William Heye San Antonio
1960–61			Sydney Heaton Tyler
1961–62			James Cardwell Luling
1962–63		Denzil L. Baker Col. USA Aug., 1963–Aug., 1967	William Nix Canadian
1963–64			Paul Dresser Corsicana
1964–65			Neil Keltner Lansing, Mich.

Date	President	Commandant	Corps Commander/Hometown
1965–66			Ralph Filburn San Angelo
1966–67			Eddie Joe Davis Henrietta
1967–68		James H. McCoy '33 Col. USA Aug., 1967–Aug., 1971	Lonnie Minze Houston
1968–69			Hector Gutierrez Laredo
1969–70			Matt Carroll Annandale, Va.
1970–71	Alvin R. Luedecke '32 (acting) Mar. 30, 1970–Nov. 1, 1970		Van Taylor Temple
1971–72	Jack K. Williams Nov. 1, 1970–May 24, 1977	Thomas R. Parsons '49 Col. USA July, 1971–July, 1977	Tom Stanley Mount Pleasant
1972–73			Ron Krnavek Corpus Christi
1973–74			Scott Eberhard Dallas
1974–75			John D. Chappelle Aug.–Oct.,
1974			
			Rickey Gray Oct., 1974–June, 1975
1975–76			Joseph Chandler San Antonio

Date	President	Commandant	Corps Commander/Hometown
1976–77	Jarvis E. Miller '50 July 29, 1977–July 10, 1980		Robert W. Harvey Houston
1977–78		James R. Woodall '50 Col. USA July, 1977–July, 1982	Michael H. Gentry Huntsville
1978–79			Robert J. Kamensky San Antonio
1979–80	Charles H. Samson (acting) July 10, 1980–Aug. 26, 1981		William Dugat Weslaco
1980–81			Ken Richardson
1981–82	Frank E. Vandiver 26 Aug. 1981–31 July 1988		Kelly Castleberry Lake Jackson
1982–83		Donald L. Burton Col. USA July 1982–Sept., 1986	Mike Holmes Grand Prairie
1983–84			Preston Abbott Longview
1984–85			Charles H. Rollins III Pace, Fla.
1985–86			Matthew J. Michaels Houston
1986–87		J. Malon Southerland Interim Cmdt. Sept., 1986–July, 1987	Garland Wilkinson Littlefield
1987–88		Thomas G. Darling Maj. Gen. USAF (Ret.) July, 1987–June, 1996	Patrick W. Thomasson Athens, Ga.

Date	President	Commandant	Corps Commander/Hometown
1988–89	William H. Mobley Aug. 1, 1988–Aug. 31, 1993		Todd M. Reichert Arlington
1989–90			Matthew C. Poling San Antonio
1990–91			Jonathan D. Wittles Cloverdale, Ore.
1991–92			John B. Sherman Victoria
1992–93			Matthew J. Michaels Houston
1993–94	E. Dean Gage Interim Pres. Sept. 1, 1993–May 30, 1994		William J. Haraway Harlingen
		Donald J. Johnson '55 Col. USA (Ret.)—Interim Cmdt.* Sept.–Dec., 1994	
1994–95	Ray M. Bowen '58 June 1, 1994–July 31, 2002		Matthew P. Segrest Bryan
1995–96		Col. Donald Johnson Interim Cmdt. June–Aug., 1996	Tyson T. Voelkel Brenham
1996–97		M. "Ted" Hopgood Jr. '65 Maj. Gen. USMC (Ret.) Aug., 1996–May 2002	Stephen Foster Arp

*Note: Colonel Johnson was interim commandant during a medical leave of absence by General Darling.

Date	President	Commandant	Corps Commander/Hometown
1997–98			Danny Feather Menard
1998–99			Tase Bailey Plano
1999–2000			Forrest Lane Dallas
2000–2001			Mark Welsh IV Colorado Springs, Colo.
2001–2002		Donald J. Johnson '55 Brig., Gen., USA (Ret.)—Interim Cmdt.* May—Oct. 2002	Joe R. Dickerson Andrews
2002–2003	Robert M. Gates Aug., 1, 2002–Dec., 16, 2006		Spence Pennington Raymondville
		John A. Van Alstyne '66 Lt. Gen. USA (Ret.) Oct., 2, 2002–Jan., 15, 2010	
2003–2004			Will McAdams Colorado City
2004–2005		John Huffman (Fall Semester) College Station	Kyle Tobin (Spring Semester) Richmond
2005–2006			Matt Ockwood Plano
2006–2007	Eddie Joe Davis '67 Dec., 19, 2006– Jan., 3, 2008		Jackson Dashiell Austin
2007–2008			Nicholas Guillemete Copperas Cove

Date	President	Commandant	Corps Commander/Hometown
2008–2009	Elsa Murano '61 Jan., 3, 2008– June 15, 2009		Jordan Reid White House
2009–2010	R. Bowen Loftin June 16, 2009–Feb., 11, 2010 Interim Feb., 12, 2010– Present		Brent Lanier Katy
		Gerald R. "Jake" Betty '73 Col. USAR (Ret.) (Interim) Jan., 15, 2010– Present	
2010– 2011			David Keim College Station

APPENDIX II

COMMANDANTS, 1876–2001

Commandant/Rank	Dates	Alma Mater/Class
R. P. W. Morris Maj. (civilian)	July, 1876–Aug., 1877	VMI 1872
George T. Olmsted, Jr. Capt. USA	Sept., 1877–July, 1881	USMA 1865
Charles J. Crane Capt. USA	Jan., 1882–Oct., 1883	USMA 1877
John S. Mallory 2nd Lt. USA	Oct., 1883–June, 1886	USMA 1879
Guy Carleton Lt. USA	Oct., 1886–Aug., 1889	USMA 1881
William S. Scott Lt. USA	Sept., 1889–Aug., 1890	USMA 1880
Frederick E. Giesecke acting (civilian)	Aug.–Oct., 1890	A&M 1886
Benjamin C. Morse Lt. USA	Oct. 9, 1890–Sept. 30, 1894	USMA 1884
George T. Bartlett Lt. USA	Sept. 31, 1894–Apr., 1898	USMA 1881
Charles C. Todd Lt. USA (acting)	Apr., 1898–July, 1898	A&M 1897

Commandant/Rank	Dates	Alma Mater/Class
Harry B. Martin (civilian)	July, 1898–Mar. 16, 1899	unknown
E. J. Kyle (cadet acting) Cadet Corps Comm.	Mar. 16, 1899–May 1, 1899	A&M 1899
Frank P. Avery Capt. USA	July, 1899–Aug., 1903	USMA 1878
Herbert H. Sargent Capt. USA	Sept., 1903–Aug., 1907	USMA 1883
Andrew M. Moses Capt. USA	Sept., 1907–1911	USMA 1897
Chancey L. Fenton Lt. USA	1911–12	USMA 1904
Levi G. Brown 1st Lt. USA	June, 1912–14	USMA 1903
James R. Hill Lt. USA	1914–16	USMA 1909
Carl H. Muller Capt. USA	1916–17	USMA 1901
Charles J. Crane Col. USA (Ret.)	1917–Sept., 1918	USMA 1877
Fred W. Zeller Maj. USA	Sept.–Nov., 1918	unknown
Grant M. Miles Maj. USA	Nov., 1918–Jan., 1919	unknown
Carl H. Muller Col. USA	Jan., 1919–Aug., 1919	USMA 1901

Commandant/Rank	Dates	Alma Mater/Class
Ike Ashburn Lt. Col. USA (Ret.) (civilian)	Sept., 1919–Dec., 1923	Poly College (SMU)
Charles C. Todd Col. USA	Jan., 1924–1925	A&M 1897
F. H. Turner Lt. Col. USA	1925–27	unknown
Charles J. Nelson Col. USA	1927–32	Auburn 1897
John E. Mitchell Lt. Col. USA	Sept., 1932–1935	unknown
Frank Anderson Col. USAR	1935–37	unknown
George F. Moore Col. USA	1937–40	A&M 1908
James A. Watson Lt. Col. USA	1940–41	unknown
Maurice D. Welty Col. USA	1941–46	USMA 1910
Guy S. Meloy Col. USA	1946–48	USMA 1927
Haydon L. Boatner Col. USA	1948–51	USMA 1925
Joe E. Davis Col. USAR	1951–63	A&M 1929
Denzil L. Baker Col. USA	1963–May, 1967	A&M 1933

Commandant/Rank	Dates	Alma Mater/Class
James H. McCoy Col. USA	May, 1967–June, 1971	A&M 1940
Thomas R. Parsons Col. USA	July, 1971–July, 1977	A&M 1949
James R. Woodall Col. USA	July, 1977–July, 1983	A&M 1950
Donald L. Burton Col. USA	July, 1983–Sept., 1986	A&M 1956
J. Malon Southerland Interim Cmdt.	Sept., 1986–July, 1987	A&M 1965
Thomas Darling Maj. Gen. USAF (Ret.)	July, 1987–June, 1996	A&M 1954
Donald J. Johnson Brig. Gen. USA/TSG Interim Cmdt.	June–Aug., 1996, and May–Oct., 2002	A&M 1955
M. T. "Ted" Hopgood, Jr. Maj. Gen. USMC (Ret.)	Aug., 1996–May 2002	A&M 1965
John A. Van Alstyne Lt. Gen. USA (Ret.)	Oct., 2002– Jan., 2010	A&M 1966
Gerald R. "Jake" Betty Col. USAR (Ret.) (Interim)	Jan., 2010– Present	A&M 1973

Abbreviations: TSG, Texas State Guard; USA, U.S. Army; USAF, U.S. Air Force; USAR, U.S. Army Reserves; USMA, U.S. Military Academy

Appendix III

RECIPIENTS OF THE

GENERAL GEORGE F. MOORE AWARD, 1946–2000

Year	Unit	Commander	
1946	"F" Infantry	Robert B. MacCallum	'47 *
1947	"A" Signal	Dean M. Denton, Jr.	'48 *
1948	"E" Field Infantry	A. N. Hartman	'49 *
1949	"A" Army Security	J. M. Wallace	'49
1950	"B" Engineers	Monroe A. Landry	'50
1951	"H" Air Force	Douglas D. Hearne	'51
1952	"H" Air Force	Steve M. Vaught	'52
1953	Squadron 10 "H"	Charles R. Little	'53
1954	Squadron 10	Burt G. Holsworth	'54
1955	Squadron 10	Bobby E. Carpenter	'55
1956	White Band	Weldon C. Steward	'56
1957	"B" Field Artillery	J. M. Dellinger, Jr.	'57
1958	"B" AA Artillery	Ray O. McClung	'58
1959	Squadron 4	Roy E. Davis	'59
1960	Company F-1	William M. Strough	'60
1961	Company C-2	Jerry I. Gilliland	'61
1962	Company G-1	John F. Imle, Jr.	'62
1963	Squadron 8	Robert H. Hackett	'63
1964	Company F-1	Thomas R. Ransdell	'64
1965	Squadron 6	Henry H. Norman	'65
1966	Company A-1	Jerry L. Lummus	'66
1967	Company C-2	Charles T. Jones	'67
1968	Squadron 11	Robert L. French	'68
1969	Company F-1	Melvin D. Sanders	'69
1970	Company F-1	Richard J. Oates	'70
1971	Squadron 4	Randall T. Schulze	'71
1972	Company D-2	Michael A. Thompson	'72

Year	Unit	Commander	
1973	Company D-2	William A. White	'73
1974	Company D-2	Gary L. Gouch	'74
1975	Squadron 9	David R. Tiller	'75
1976	Company M-1	Gregory Knape	'76
1977	Company D-1	Patrick K. Harrigan	'77
1978	Company F-2	Joseph K. Marshall	'78
1979	Company B-2	John A. Vertegen	'79
1980	Company D-2	Michael S. Hanley	'80
1981	Company D-1	Timothy S. Lyda	'81
1982	Company D-1	Scott Jordan	'82
1983	Company D-1	Thad Hill	'83
1984	Squadron 10	Michael G. Walker	'84
1985	Company B-2	Gilbert Huron	'85
1986	Company K-2	Charles Hall	'86
1987	Company K-2	Jeffrey P. Davis	'87
1988	Company K-2	Dale Daniel	'88
1989	Squadron 16	Brian Yates	'89
1990	Company K-2	Dave Carr	'90
1991	Company G-2	Jay Cothem	'91
1992	Squadron 2	James Sukenik	'92
1993	Company G-2	Tim Sultenfuss	'93
1994	Company D-2	Brian Pedder	'94
1995	B Battery	Paul Reininger	'95
1996	B Company	David Childers	'96
1997	Squadron 2	Daniel Yurasek	'97
1998	Company D-2	Chris Halpin	'98
1999	Company B-1	Martin Ramos	'99
2000	B-Company (Band)	Chris Migl	'00

*Due to the special three-semester accelerated program and wartime adjustments, commanders and class standing varied.

APPENDIX IV

GENERAL AND FLAG OFFICERS, TEXAS A&M

ABBREVIATIONS

AirNG	Air National Guard
ANG	Army National Guard
AUS	Army of the United States
TSG	Texas State Guard
USA	United States Army
USAF	United States Air Force
USAFR	United States Air Force Reserve
USAR	United States Army Reserve
USN	United States Navy
USNR	United States Naval Reserve
USMC	United States Marine Corps
USPHS	United States Public Health Service

★★★★ GENERAL ADMIRAL

Otto P. Weyland '23 USAF
Bernard A. Schriever '31 USAF
Jerome L. Johnson '56 USN
Joseph W. Ashy '62 USAF
Patrick K. Gamble '67 USAF
Hal M. Hornburg '68 USAF

★★★ LIEUTENANT GENERAL/VICE ADMIRAL

Andrew D. Bruce '16 USA
John T. Walker '17 USMC

Ian M. Bethel '25 USMC
Robert W. Colglazier '25 USAR
Harry H. Critz '34 USA
Ormond R. Simpson '36 USMC
Woodrow W. Vaughan '39 USA
James F. Hollingsworth '40 USA
Herron M. Maples '40 USA
Jay T. Robbins '40 USAF
John H. Miller '46 USMC
Eivind H. Johansen '50 USA
Herman O. Thomson '51 USAF
Richard T. Gaskill '52 USN
Kenneth E. Lewi '52 USA
James S. Cassity, Jr. '57 USAF
Melvin F. Chubb, Jr. '62 USAF
David B. Robinson '62 USN
James T. Scott '64 USA
Herbert A. Browne '65 USN
Theodore G. Stroup '65 USA
Leonard D. Holder, Jr. '66 USA
Donald L. Peterson '66 USAF
Randolph W. House '67 USA
Patrick P. Caruana '72 USAF

★★ MAJOR GENERAL/ REAR ADMIRAL

Bennett Puryear '06 USA
George F. Moore '08 USA
Howard C. Davidson '11 USAF

William E. Farthing '14 USAF
Roderick R. Allen '15 USA
Percy W. Clarkson '15 USA
Edmond H. Leavey '15 USA
Ralph H. Wooten '16 USAF
Albert M. Bledsoe '17 USN
Harry H. Johnson '17 USA
Herman M. Ainsworth '19 USA
J. D. Hill '21 USA
Gerald Bogle '23 USN
Robert B. Williams '23 USAF
William D. Old '24 USAF
Manning E. Tillery '26 USAF
William L. Kennedy '28 USAF
George P. Munson, Jr. '28 USAR
Stuart S. Hoff '29 USA
Benjamin H. Pochyla '29 USA
Frederick H. Weston '29 TexasANG
Alvin R. Luedecke '32 USAF
James Earl Rudder '32 USAR
John H. White '32 USAF
Robert F. Worden '33 USAF
Bruno A. Hochmuth '35 USMC
Raymond A. Moore '35 USN
Raymond L. Murray '35 USMC
Wood B. Kyle '36 USMC
Robert L. Pou, Jr. '37 TexasANG
Andrew P. Rollins, Jr. '39 USA
John H. Buckner '40 USAF
Homer S. Hill '40 USMC
William A. Becker '41 USA
George L. Cassell '41 USN
Otto E. Scherz '42 TexasANG
Harold C. Teubner '42 USAF
Harold B. Gibson '43 USA
Tom E. Marchbanks '43 USAF
Homer D. Smith '43 USA
Merton D. Van Orden '43 USN
Wesley E. Peel '46 USA
William L. Webb, Jr. '46 USA
Guy H. Goddard '47 USAF

Charles R. Bond, Jr. '49 USAF
Robert E. Crosser '49 USAR
Charles I. McGinnis '49 USA
James L. Brown '50 USAF
Glenn H. Kothmann '50 TexasANG
Waymond C. Nutt '51 USAF
Howard H. Haynes '52 USN
Charles H. Kone '52 TexasANG
John H. Storrie '52 USAF
James W. Taylor '52 USAF
William R. Wray '52 USA
Dionel E. Aviles '53 USAR
Charles R. Cargill '53 USAFR
Harry V. Steel, Jr. '53 TexasANG
G. J. Wilson, Jr. '53 USAR
Thomas G. Darling '54 USAF
George H. Akin '56 USA
Don O. Daniel '56 TexasANG
Thomas R. Olsen '56 USAF
Paul L. Greenberg '58 USA
Sam C. Turk '58 TexasANG
Ira E. Scott '59 TSG
James R. Taylor '59 USA
Robert Smith III '61 USNR
Darrel P. Baker '62 TexasANG
John J. Closner III '62 USAFR
Jay T. Edwards III '62 USAF
Charles W. McClain, Jr. '62 USA
Walter B. Moore '62 USA
Gerald H. Putman '62 USA
Jay D. Blume, Jr. '63 USAF
Billy G. McCoy '63 USAF
Hiram H. Burr, Jr. '65 USAF
Marvin Ted Hopgood '65 USMC
James M. Hurley '65 USAF
John E. Simek '65 USAR
Alvin W. Jones '66 USAR
John A. Van Alstyne '66 USA
Frank D. Watson '66 USAFR
Joe M. Ernst '67 USAR
John B. Sylvester '67 USA

Richard L. Engel '68 USAF
Michael C. Kostelnik '68 USAF
William W. Pickavance, Jr. '68 USN
Kenneth W. Hess '69 USAF
Wilbert D. Pearson, Jr. '69 USAF
Gerald F. Perryman, Jr. '70 USAF
Ted M. Moseley '71 USAF
James W. Robinson '75 TSG

★ BRIGADIER GENERAL/REAR ADMIRAL (LOWER)

Douglas B. Netherwood '08 USA
John A. Warden '08 USA
William C. Crane, Jr. '10 USA
John F. Davis '11 USA
Oscar B. Abbott '13 USA
Eugene A. Eversberg '13 USAR
Jerome J. Waters '13 USA
Robert R. Neyland, Jr. '14 USA
Victor A. Barraco '15 USMC
Claudius M. Easley '16 USA
Durant S. Buchanan '17 USMC
Walter T. H. Galliford '17 USMC
Nat S. Perrine '17 USA
George H. Beverly '19 USAF
Paul L. Neal '19 USA
John T. Pierce '19 USA
Arthur B. Knickerbocker '21 TexasANG
Cranford C. Warden '21 USA
Aubry L. Moore '23 USAF
Spencer J. Buchanan '25 USAR
William R. Frederick '25 USA
Richard J. Werner '25 USA
William L. Lee '27 USAF
James P. Newberry '27 USAF
Charles S. Hays '32 USA
Travis M. Hertherington '32 USAF
John A. Hilger '32 USAF
Graber Kidwell '32 USAR

John M. Kenderdine '34 USA
Odell M. Conoley '35 USMC
Kay Halsell II '35 TexasANG
Clifford M. Simmang '36 USAR
Carter C. Speed '36 USA
Jack T. Brown '37 TexasANG
Kyle L. Riddle '37 USAF
Theodore H. Andrews '38 USA
David L. Hill '38 TexasANG
Robert M. Williams '38 USA
Clarence A. Wilson '38 USAR
O. D. Butler '39 USAR
Andrew W. Rogers '39 AUS
Joe G. Hanover '40 AUS
Thomas F. McCord '40 USA
George P. Cole '41 USAF
Hubert O. Johnson '41 USAF
Seaborn J. Buckalew, Jr. '42 USA
Charles M. Taylor, Jr. '42 CaliforniaANG
Victor H. Thompson, Jr. '42 USAF
Mike P. Cokinos '43 USAR
Charles V. L. Elia '43 USA
Jack N. Kraras '43 USA
Guy M. Townsend '43 USAF
George W. Connell '45 USA
Irby B. Jarvis, Jr. '45 USAF
Joseph E. Wesp '45 USAF
Allen D. Rooke, Jr. '46 USA
David O. Williams '46 USAF
Thomas G. Murnane '47 USA
Carl D. McIntosh '48 USAR
Robert M. Mullens '48 USA
John D. Roper '48 USAF
Billy M. Vaughn '49 USA
Walter O. Bachus '50 USA
Wilman D. Barnes '51 USA
Keith L. Hargrove '51 USAR
Frank A. Ramsey '51 USA
Walter J. Dingler '52 TexasANG
Louis L. Stuart, Jr. '52 USAR
Robert C. Beyer '53 USA

George R. Harper '53 TexasANG
Robert O. Petty '53 USAF
Charles M. Scott '53 ArizonaANG
Donald J. Johnson '55 TSG
Woodrow A. Free '56 USAR
Dennis A. Wilkie '56 USA
Donald L. Moore '57 USAF
Paul L. Carroll, Jr. '58 Calif.ANG
James E. Freytag '59 USAF
John Serur '59 USAF
Charles R. Weaver '59 TSG
Kenneth F. Keller '60 USAF
Edmond S. Solymosy '60 USA
Malcolm Bolton '61 USAF
Richard A. Box '61 TSG
Don M. Ogg '61 TexasANG

Jimmy L. Cash '62 USAF
Michael M. Schneider '62 USA
John A. Hedrick '63 USA
Ronald D. Gray '64 USAF
George E. Chapman '65 USAF
Lee V. Greer '67 USAF
Stephen D. Korenek '68 AlaskaANG
James M. Richards III '69 USAF
Robert T. Howard '70 USA
Michael H. Taylor '70 TexasANG
Joseph F. Weber '72 USMC
Robert L. Herndon '73 USA
William M. Fraser III '75 USAF
Loyd S. Utterback '75 USAF
Robert C. Williams '76 USPHS

NOTES

INTRODUCTION

1. Text of the Morrill Land Grant Act of July 2, 1862, with amendments, in Texas A&M *Catalogue*, 1887–88, pp. 62–65. See also Rod Andrew, Jr., "Soldiers, Christians, and Patriots: The Lost Cause and Southern Military Schools, 1865–1915," *Journal of Southern History* (Nov., 1998): 677–710; and Edward D. Eddy, Jr., *College for Our Land and Time: The Land-Grant Idea in American Education*, pp. 1–66.

2. H. P. N. Gammel, ed., *The Laws of Texas, 1822–1897*, Vol. 2, pp. 134–36; Thomas Lloyd Miller, *The Public Lands of Texas 1519–1970*, pp. 120–25; Frederick Eby, *The Development of Education in Texas*, p. 80; *Journal of the House of Representatives of the Republic of Texas*, 2nd Congress, Adjourned Session, p. 31; Lewis B. Cooper, *The Permanent School Fund of Texas*, pp. 1–41; Patsy McDonald Spaw, ed., *The Texas Senate*, College Station: Texas A&M University Press, 1990, Vol. 1, p. 39.

3. Clarence Ousley, *History of the Agricultural and Mechanical College of Texas, Bulletin of the Agricultural and Mechanical College of Texas*, 4th ser., vol. 6, no. 8 (Dec. 1, 1935), pp. 1–18; Edward D. Eddy, Jr., *The Land-Grant Movement: A Capsule History of the Educational Revolution Which Established Colleges for All the People*, pp. 4–5; H. G. Good, *A History of American Education*, pp. 291–3; Marie Guy Tomlinson, "The State Agricultural and Mechanical College of Texas, 1871–1879: The Personalities, Politics, and Uncertainties," May, 1976, M.A. thesis, Texas A&M University, pp. 15–25; Fred A. Shannon, *The Farmer's Last Frontier*, pp. 272–80.

4. Frank E. Vandiver, *Blood Brothers*, p. 18.

5. Ibid.; "Agricultural Colleges," *The Congressional Globe*, Washington, D.C.: June 19, 1862, p. 2770; Governor's Messages, Coke to Ross, 1874–91, p. 137; Land Scrip No. 1 in the George Pfeuffer Papers, Texas A&M University Archives; United States, *United States Code Annotated*, Title 7 Agriculture, sec. pp. 301–304; Texas A&M *Catalogue*, 1887–88, p. 66. See also Robert F. Smith, "A Brief Sketch of the Agricultural and Mechanical College of Texas," manuscript, College Station, 1914 [*sic*, actually written ca. 1904], pp. 1–3, in the Texas A&M Archives. The grant of 180,000 acres was based on the fact that Texas had two senators and four representatives in Congress.

6. *Journal of the Senate of the Twelfth Legislature of the State of Texas*, 1st Reg. Sess., p. 490; Tomlinson, *The State Agricultural and Mechanical College of Texas, 1871–1879*, pp. 34–44; United States, *United States Code Annotated*, Title 7, Agriculture, sec. 301–308, pp. 321–31.

7. Eddy, *Colleges for Our Land and Time*, pp. 41–42.

8. Land-Grant Colleges: Act of 1862 Donating Lands for Colleges of Agriculture and Mechanic Arts in Ousley, *History of the A&M College*, pp. 8–10; Tomlinson, *The State Agricultural and Mechanical College of Texas, 1871–1879*, pp. 10–13.

9. J. M. Moon, secretary of state, letter to Louis L. McInnis, Mar. 23, 1888, L. L. McInnis Papers, Texas A&M Archives.

10. Ousley, *History of the A&M College*, p. 39; Gammel, ed., *Laws of Texas*, Vol. 6, pp. 1186–87; John Henry Brown, *Indian Wars and Pioneers of Texas*, p. 595; Robert F. Smith, "An Historical Sketch of the Texas Agricultural and Mechanical College" *Longhorn*, 1904, p. 168; *Bryan Eagle*, Feb. 14, 1901, and Apr. 26, 1928. Smith in *A Brief Sketch* notes that other locations for the college included San Antonio, Austin, San Marcos, Waco, and Tehuacana Hills.

11. Ernest Langford, *Getting the College Under Way*, pp. 1–7; Deed Records of Brazos County, Vol. M, p. 142ff.; *Galveston Daily News*, June 23 and 29, 1871; George Sessions Perry, *The Story of Texas A and M*, pp. 52–56; Ousley, *History of the A&M College*, pp. 38–40; J. M. Moore, secretary of state, to Louis L. McInnis, Mar. 23, 1888, McInnis Papers. See also Joe M. Carson '45, letter to Association of Former Students, seen in File HL 0032, Texas A&M Archives.

12. Smith, *A Brief Sketch*, p. 1.

13. Elmer Grady Marshall, "The History of Brazos County, Texas," Master's thesis in history, University of Texas, 1937, pp. 90–92, 104–106; Austin *Tri-Weekly Statesman*, May 28, 1872; Tomlinson, *The State Agricultural and Mechanical College of Texas, 1871–1879*, pp. 64–69; J. E. Fee, "Early Student [William M. Sleeper '79] of A. & M. Tells School's Story," *Dallas Morning News*, Oct. 15, 1933. Five-sixths of the population of Texas lived in the eastern third of the state.

14. *Galveston Daily News*, Aug. 24, 1872, Oct. 18, 1873, May 16, 1874, June 2 and 3, 1875; Governor's Messages, Coke to Ross, 1874–91, p. 138; Langford, *Here We'll Build the College*, pp. 1–30a; Tomlinson, *The State Agricultural and Mechanical College of Texas, 1871–1879*, pp. 78–117. See also Texas A&M College, *The Olio*, p. 25–29.

15. John A. Adams, Jr., *We Are the Aggies: The Texas A&M University Association of Former Students*, pp. 5–6.

16. Andrew, "Soldiers, Christians, and Patriots," *Journal of Southern History*, Nov., 1998, p. 682. See also Robert D. Meade, "The Military Spirit of the South," *Current History* 30 (Apr., 1929): 55–60, and Charles Reagan Wilson, *Baptized in Blood: The Religion of the Lost Cause, 1865–1920*.

17. Ousley, *History of the A&M College*, pp. 42–44; Perry, *The Story of Texas A and M*, pp. 58–59; John E. Weems, *To Conquer a Peace*, pp. 213, 306–309.

18. Eddy, *Colleges for Our Land and Time*, pp. 1–59.

19. Ousley, *History of the A&M College*, pp. 42–43. Also see Henry C. Dethloff, *Centennial History of Texas A&M University 1876–1976*, Vol. 1, pp. 30–31.

20. Ousley, *History of the A&M College*, pp. 44–45, 48; Smith, *A Brief Sketch*, p. 2; Dethloff, *Centennial History*, Vol. 1, pp. 31–33; Virginia Military Institute [VMI] *Register of Former Cadets*, p. 73.

21. Texas A&M *Annual Catalogue*, 1876–77, pp. 25–26.

22. Board of Directors, Announcement and Circular of the State Agricultural and Mechanical College, 1876, p. 31; The Seventy-fifth Anniversary Committee, "Opened Oct. 4, 1876: The Agricultural and Mechanical College of Texas," College Station: 1950, pp. 1–6; Young Men's Christian Association (YMCA), *Students' Handbook of A. & M. College of Texas for 1914–15*, College Station: 1914, pp. 1–40; *The Cadence: A Handbook for Freshmen, 1947–48*, p. 137; *Articles of the Cadet Corps*, Sept., 1961, p. 62; *The Standard* (1968), p. 47; J. E. Loupot, *Aggie Facts and Figures*, p. 39.

23. Ousley, *History of the A&M College*, p. 46; Texas A&M *Annual Catalogue, 1877–78*, College Station: 1878, pp. 37–38; Smith, *A Brief Sketch*, p. 3. See also Octavia F. Rogan, *Land Commissioner Charles Rogan*, pp. 73–75.

24. "First Student of A. & M. Dies," *Battalion*, Jan. 22, 1930, p. 1; Perry, *The Story of Texas A and M*, p. 60; Tomlinson, *The State Agricultural and Mechanical College of Texas, 1871–1879*, pp. 2–21; *Galveston Daily News*, July 12, 1890. See also "Lieutenant Colonel G. W. Hardy," *Galveston Daily News*, May 6, 1884, p. 2.

25. *Galveston Daily News*, July 12, 1890; Texas A&M *Catalogue, 1878–79*, p. 30; "Cushing Writes of First Days at College," *Reveille*, Dec. 7, 1918, p. 1.

26. *Texas Senate Journal, 1879*, pp. 205–10; *Texas Aggie*, Oct. 29, 1926, and Nov. 1, 1824; Ousley, *History of the A&M College*, p. 46. See also "W. A. Trenckmann '78, Relates History of Early Days," *Texas Aggie*, Dec. 15, 1928, p. 1.

27. *Galveston Daily News*, Oct. 5, 1876. A *Galveston Daily News* reporter wired the paper on October 4 that "though the attendance of students is not large, there being not more than 50 cadets present, yet a start has been made and this number will rapidly swell up to 300 or 400 during the session," in Perry, *The Story of Texas A and M*, pp. 60–61. Further confusion was added at the semi-centennial celebration of the college in October, 1926, when the inaccuracy gained

more credence. Dean Charles E. Friley of the School of Arts and Science delivered the keynote address at Guion Hall to the Corps of 2,350 cadets, stressing the fact that the September 17 date was aborted for lack of interest and the formal opening was reset for early October, 1876. Friley's remarks were printed and widely distributed in the *Texas Aggie* magazine. And Clarence Ousley, in *History of the Agricultural and Mechanical College of Texas* in 1935, repeated the 1928 Rogan story. Thus the inaccuracy was again printed and became campusology legend, and the legend became accepted as fact.

28. Inauguration of the State Agricultural and Mechanical College of Texas: Bryan, Oct. 4, 1876. Address of His Excellency Gov. Richard Coke, and Thos. S. Gathright, President of the College, 1876; *Galveston Daily News,* July 17, 1890.

29. Address of His Excellency Gov. Richard Coke, 1876.

CHAPTER I

1. Eddy, *Colleges for Our Land and Time,* pp. 64–65.

2. Ibid.; Texas A&M *Annual Catalogue,* 1876–77, pp. 4–5, and 1877–78, pp. 1–15; College Station *Texas Collegian,* Jan., 1880; Thomas Gathright to George Pfeuffer, Nov. 2, 1879, Gathright Papers, Texas A&M Archives; Robert W. Jeffrey, VMI Public Relations Director, to B. A. Hardaway, College Station, Mar. 18, 1958, Texas A&M Archives; Lester Austin Webb, *"The Origin of Military Schools in the United States Founded in the Nineteenth Century,"* Ph.D. dissertation, University of North Carolina, 1958, pp. 148–58. See also Arthur J. Klein, "Survey of Land-Grant Colleges and Universities," U.S. Department of Interior, *Education Bulletin,* 1930. Washington, D.C.: GPO, 1930.

3. "Original 'fish' Guyler Relates Old Experiences," *Texas Aggie,* May 20, 1926. See also "W. A. Trenckmann, '78 Relates History of Early A. & M. Days," *Texas Aggie,* Dec. 15, 1928, and David Chapman, "Gathright Hall a Magnificent Afterthought," *Texas Aggie,* Mar., 1995, p. 24.

4. Smith, *A Brief Sketch,* p. 3; Austin E. Burges, *A Local History of A. & M. College 1876–1915,* p.

5; David B. Cofer, ed., *Early History of Texas A&M College through Letters and Papers,* pp. 130–31.

5. "College in 1876," general correspondence files Texas A&M Archives. Professor Banks and his wife were first at A&M from 1876 to 1879, departed the college to set up a private school in Salado, and returned to A&M in 1893. During the first couple of years the size of the Corps did vary as "cadets were coming in every day," yet the number never exceeded 200. Duncan Martin was the son of Dr. Carlisle P. B. Martin, professor of agriculture and science. Gen. Hamilton P. Bee was steward and superintendent of the farm. See introduction to Smith, *Aggies, Moms, and Apple Pie,* pp. xiii–xvii.

6. *Rules and Regulations of the Agricultural and Mechanical College of Texas, 1876,* pp. 1–28; Texas A&M *Annual Catalogue,* 1876–77, p. 27; William A. Banks to Jos. Richardson, Jan. 9, 1877, Historical Files, Box 1, and "First A&M Student Dies," n.p., n.d., Clipping File in Texas A&M Archives.

7. Texas A&M *Annual Catalogue,* 1876–77, pp. 21–26, and 1877–78, pp. 25–26; Walter Bradford '68, interview, Sept. 3, 1998, U.S. Army History Center, Washington, D.C.

8. "Cushing Writes of First Days at College," *Reveille,* Dec. 7, 1918, p. 1; "Lieutenant Colonel G. W. Hardy," *Galveston Daily News,* May 6, 1884, p. 2.

9. C. B. Perry to Mama, Oct. 3, and Nov. 1, 1877, letters by Charles Bryan Perry in James Franklin Perry Papers, Center for American History, University of Texas, Austin (hereafter Perry Papers).

10. *Rules and Regulations, 1876,* pp. 5–16.

11. Texas A&M, *1877 Annual Catalogue,* pp. 2–41.

12. Ousley, *History of the A&M College,* p. 46.

13. Ibid., pp. 49–50.

14. "Early Days of A&M Recalled by Judge Rogan," *Texas Aggie,* Nov. 1, 1924, pp. 1–2.

15. "Christmas at the Agricultural and Mechanical College of Texas," *Galveston Daily News,* Jan. 2, 1877, p. 2.

16. "Early Days of A&M Recalled by Judge Rogan," *Texas Aggie,* Nov. 1, 1924, p. 2; Ousley, *History of the A&M College,* pp. 45–46.

17. Thomas S. Gathright, "Agricultural College Report," *Journal of the Senate of the Sixteenth Legislature of the State of Texas,* Austin: June 10, 1877, pp. 46–50.

18. U.S. War Department, Adjutant General's Office, Report of the Adjutant General, Washington, D.C.: 1984, p. 322. Nationwide in 1881, thirty colleges, universities and institutions had been authorized to have an assigned regular army officer. One of the primary qualifications to have an officer detailed was to have no fewer than 150 male students.

19. C. B. Perry to Brother, Jan. 25, 1878, Perry Papers; Texas A&M *Catalogue 1877–78,* p. 16; "The Cadet Lynching Scrape," *Galveston Daily News,* Nov. 6, 1978, p. 2.

20. "A&M's First Commandant," *Texas Aggie,* June 16, 1924, and "Judge Page Morris Is Hoping to Meet Great Gathering of Men of Seventies," *Texas Aggie,* May 16, 1924.

21. William Rufus Nash to mother, Oct. 3, 1877, W. R. Nash Papers, Texas A&M Archives; C. B. Perry to Emmett, Jan. 31, 1878, Perry Papers; College Station, *The Texas Collegian,* Mar., 1979, pp. 5–7.

22. Nash to mother, Oct. 3, 1877, Nash Papers. See also C. B. Perry to Emmett, Jan. 31, 1878, and C. B. Perry to Mama, Oct. 3, 1877, both in Perry Papers.

23. Nash to mother, Oct. 23, 1877, Nov. 14 and 18, 1877, and Dec. 14, 1877, Nash Papers; "Early Days of A&M Recalled by Judge Rogan," *Texas Aggie,* Nov. 1, 1924, p. 2; "Cushing Writes of First Days at College," *Reveille,* Dec. 7, 1918, p. 1; C. B. Perry to Mama, Oct. 10, 1877, Perry Papers.

24. *The Texas Collegian,* May, 1879.

25. Ibid., Apr., 1879. See also W. R. Nash Papers; Smith, *A Brief Sketch,* p. 4.

26. Thomas Gathright to George Pfeuffer, Oct. 15, 1879, Gathright Papers; "Military Staff," *The Texas Collegian,* Mar., 1879, p. 7.

27. R. P. W. Morris, William Banks, D. B. Martin, and Alexander Hogg to Board of Directors of the State A. & M. College, Aug. 28, 1879, Pfeuffer Papers; *Galveston Daily News,* July 12, 1890. In an interview with Hogg, he passed off the November–December, 1879, affair as one in which the faculty got at "loggerheads" with each other and the president. Hogg felt that Governor Roberts's dismissal action was "justified by the situation as it then existed" (*Galveston Daily News*).

28. Thomas Gathright to George Pfeuffer, Nov. 13, 1879, Gathright Papers; Thomas Gathright to James R. Cole, Nov. 4, 1879, James R. Cole Papers, Texas A&M Archives.

29. Cadet John C. Crisp to Gov. O. M. Roberts, Nov. 21, 1879, Gathright Papers; Hamilton P. Bee, San Antonio, Tex., to George Pfeuffer, Bryan, Tex., Nov. 14, 1879, Pfeuffer Papers.

30. Thomas Gathright to George Pfeuffer, Sept. 13 and Nov. 2, 1879; Cadet Petition to Governor O. M. Roberts, Nov. 16, 1879, Gathright Papers; Resolution of the Cadets of the A. & M. C. to Secretary of the Board of Directors, via Thos. L. Gathright, Nov. 20, 1879, Gathright Papers; *Galveston Daily News,* July 12, 1890.

31. Rufus C. Burleson to O. M. Roberts, Nov. 20, 1879, Historical Letter File, Box 1, Texas A&M Archives; George Pfeuffer to L. L. McInnis, Nov. 27, 1879, McInnis Papers; *The Texas Collegian,* Jan., 1880.

32. Rod Andrew, Jr., *Long Gray Lines: The Southern Military School Tradition, 1839–1915,* Chapel Hill: University of North Carolina Press, 2000 p. 4.

33. George T. Olmsted, captain, "Commandant's Report," *Annual Report of the A. & M. College of Texas,* College Station, June 23, 1880, pp. 32–33.

34. Adams, *We Are the Aggies,* pp. 7–13.

35. John G. James to F. H. Smith, Superintendent VMI, Aug. 21, 1867, John G. James Papers, Texas A&M Archives; Albert Z. Conner, "VMI at War: VMI's Civil War Soldiers [James]," *VMI Alumni Review,* spring, 1998, p. 16. TMI was first opened in San Marcos, Texas, in the fall, 1867, and moved to Bastrop (1868–70) and then to Austin (1870–79).

36. Texas A&M Board Resolution "To the People of Texas," Nov. 24, 1879, Pfeuffer Papers. See also VMI *Register of Former Cadets,* p. 54, and William Couper, *One Hundred Years at V.M.I.,* pp. 110, 130–31.

37. J. R. Cole to Mrs. Cole, Nov. 23, 1879, Cole Papers.

38. *Galveston Daily News,* July 12, 1890. For a complete list of the new 1880 faculty and staff see Smith, *A Brief Sketch,* p. 4.

39. Guyler to Poindexter, May 21, 1880, Robert W. Guyler Papers, Texas A&M Archives.

40. David B. Cofer, ed., *Fragments of Early History of Texas A&M,* pp. 5–7; *A&M College Record* 11, no. 3 (Apr., 1902): 7; Texas A&M *Fourth Annual Catalogue* (1880–81), pp. 17–19.

41. *Texas Legislative Record,* 17th sess., Vol. 2, no. 43, Mar. 3, 1881, pp. 1–3.

42. Texas A&M *Fourth Annual Report, 1880,* pp. 32–33; *Kaufman Sun,* Aug. 11, 1881, in James Cole File, Texas A&M Archives; Charles J. Crane, *The Experiences of a Colonel of Infantry,* pp. 128–29, 139.

43. J. G. James to Mr. [P. H.] Hayne, Oct. 2, 1882, in Daniel M. McKeithan, ed., *Selected Letters of John Garland James to Paul H. Hayne and Mary M. M. Hayne,* pp. 68–69.

44. John G. James, College Station, to George Pfeuffer, Austin, Feb. 14, 1883, Pfeuffer Papers; Texas A&M, *Annual Report, 1882–1883,* pp. 1–3, 144–45; *Texas Senate Journal, 1883,* p. 145; Texas A&M *Catalogue,* 1883–84, p. 6; "Educational Facilities," *Brazos Pilot,* Sept. 1, 1882, p. 4.

45. Crane, *Experiences of a Colonel of Infantry,* p. 131; C. J. Crane, 1st lieutenant, "Department of Military Science," *Report of the Acting President of the A. & M. College of Texas,* June, 1883; Texas A&M *Catalogue,* 1883–84, p. 20.

46. Crane, *Experiences of a Colonel of Infantry,* pp. 130–32.

47. Dethloff, *Centennial History,* pp. 85–87. Colonel James attended the semi-centennial celebration on the A&M campus in 1951.

48. John G. James, Wichita Falls, Tex., to Louis L. McInnis, College Station, Dec. 14, 1885, McInnis Papers. See also William Couper, *The V.M.I. New Market Cadets,* pp. 103–104.

49. Smith, *A Brief Sketch,* p. 4.

50. *Report of the Agricultural and Mechanical College of Texas,* 1885, pp. 9–10; Crane, *Experiences of a Colonel of Infantry,* pp. 130–33.

51. "Texas News Items," *Galveston Daily News,* May 6, 1884, p. 5.

52. *Bryan Enterprise,* July 4 and Oct. 3, 1883, in McInnis Scrapbook, McInnis Papers; *Galveston Daily News,* Sept. 15 and 26, 1883; *Houston Post,* July 27 and Aug. 4, 1883; *Annual Report of A&M, 1884,* n.p.

53. *Galveston Daily News,* June 25, 26, and 27, 1883; Texas A&M *Catalogue,* 1883–84, p. 6; "Report of the Acting President of the College," June, 1883, Cole Papers; James R. Cole, *Seven Decades of My Life,* pp. 95–97; Couper, *The V.M.I. New Market Cadets,* pp. 59–60; Clement A. Evans, *Confederate Military History,* pp. 350–51; William Couper to Dr. S. W. Geiser, Nov. 4, 1940, in Dinwiddie Bio File, Texas A&M Archives. James R. Cole had been a representative in the Twelfth Legislature of Texas, which chartered Texas A&M on April 17, 1871.

54. Dethloff, *Centennial History,* pp. 96–97; H. H. Dinwiddie to George Pfeuffer, Apr. 27, 1885, Pfeuffer Papers.

55. Louis Mackensen '85 to David B. Cofer, Oct., 1952, Letter File HL 0096, Texas A&M Archives.

56. J. G. Garrison to George Pfeuffer, Nov. 12, 1885, Pfeuffer Papers; Minutes of the Board of Directors of Texas A&M, May 31, 1886; Lt. John S. Mallory, "Report of Department of Military Science and Tactics," *Report of the Agricultural and Mechanical College of Texas,* pp. 41–43; Minutes of the Board of Directors, May 31, 1886, p. 8. See also "Agriculture and Mechanical College," *Texas Review,* Mar., 1886, pp. 433–36, and H. H. Dinwiddie, "Industrial Education in Texas," *Texas Review,* Mar., 1886, pp. 407–16.

57. Lucius Holman to Mother, Apr. 25 and Oct. 31, 1888, Historical Letters File, Box 2; John E. Hill to Mama, Mar. 27 and Apr. 3, 1887, both in Texas A&M Archives; Minutes of the Board of Directors, May 31, 1886, p. 8.

58. *Report of the Agriculture and Mechanical College of Texas,* 1885, p. 10. Also see W. R. Nash to Mother, Oct. 3, 1877, Nash Papers.

59. John E. Hill, College Station, to Mother, Mar. 13, 1887, John E. Hill, letters, Texas A&M Archives; Minutes of the Board of Texas A&M, Jan. 2, Mar. 7, and Oct. 12, 1887, and May 30, 1888; Cofer, *Fragments of Early History of Texas A&M College,* pp. 23–27; *Austin Daily Statesman,* June 7, 1888; Texas A&M College, *The Olio,* 1895, pp. 65–66. Lieutenant Carleton was provided board in exchange for teaching two mathematics classes per day.

60. Minutes of the Board of Directors of Texas A&M, Jan. 24, 1888; Texas A&M *Annual Report, 1889,* p. iv.

61. Minutes of the Board of Directors of Texas A&M, May 30, 1889, and Dec. 13, 1888; Eddy, *Colleges for Our Land and Time,* p. 94.

62. Willie R. Ratchford to Mother, Dec. 12, 1889, Ratchford Papers, Texas A&M Archives.

63. John S. Mallory, 2nd Lieutenant, "Department of Military Science," *Annual Report of the A. & M. College of Texas, 1883–84;* Lucius Holmon, letter to mother [P. G. Weeks], Apr. 25, 1888, in Texas A&M Archives.

64. Guy Carleton, 1st Lieutenant, "Report of the Department of Military Science and Tactics," *Report of the Agricultural and Mechanical College of Texas,* 1889, pp. iii–iv.

65. Minutes of the Board of Directors at Texas A&M, May 30, 1889; Texas A&M *Annual Report, 1889,* p. iv; W. R. Ratchford to father, Feb. 26, 1890, Ratchford Papers.

66. W. R. Ratchford to parents, Mar. 14, 1890, Ratchford Papers.

67. Willie R. Ratchford to mother, Dec. 12, 1889, Ratchford Papers; Andrew, "Soldiers, Christians, and Patriots," *Journal of Southern History,* Nov., 1998 pp. 677–86.

68. Ratchford letters, to father, Feb. 26, 1890, and to mother, Oct. 24, 1890; Minutes of the Board of Directors, Dec. 13, 1888.

69. Ratchford, letter to parents, Mar. 14, 1890.

70. W. R. Ratchford, letter to J. H. Ratchford, June, 1890, Ratchford Papers.

71. A. J. Rose, letter to W. S. Scott, Aug. 21, 1890, McInnis Papers.

72. Letters, Rose to Scott, Aug. 24 and 29, 1890; Letter, Scott to the adjutant general, U.S. Army, Washington, D.C., Aug. 31, 1890; Scott, letters to L. L. McInnis, Oct. 7, 1890, Nov. 4, 10, 1890, in McInnis Papers. See also "Story of 'Old Army' Fish Recalls Past," *Battalion,* Sept. 10, 1964, p. 2.

73. Ratchford, letter to parents, Sept. 9, 1890, Ratchford Papers; *Report of the Agricultural and Mechanical College of Texas,* 1891, pp. 54–58.

CHAPTER 2

1. Lawrence S. Ross Clarke, interview with R. Henderson Shuffler, Jan. 4, 1963, L. S. Ross Papers, Texas A&M Archives; *Galveston Daily News,* Jan. 21, 1887. Also see Rosalind Langton, "Life of Colonel R. T. Milner," *Southwestern Historical Quarterly* 44 (Apr., 1941): 439, and John A. Adams, Jr., "Lawrence Sullivan Ross," *Texas Aggie,* July, 1979. There are many versions of the Shapley Ross inauguration story which put the location of the event either in Austin or San Antonio.

2. Adam Rankin Johnson, *The Parson Rangers of the Confederate States Army,* p. 29; *Dallas Herald,* Oct. 10, 1858; Judith Ann Benner, *Sul Ross: Soldier, Statesman, Educator,* pp. 1–33. Also see Thomas E. Turner, Sr., "A Career That Bred Legends," *Baylor Magazine,* Apr.–May, 1986, pp. 19–21; *Waco Tribune-Herald,* Oct. 30, 1949, and "Texas Swinging Guardian Angel," *Houston Post—Texas Magazine,* Sept. 11, 1966. See also Margaret Schmidt Hacker, *Cynthia Ann Parker: The Life and the Legend,* pp. 21–30. In addition to Ross, four of the officers of the Second Cavalry involved in the Wichita campaign of 1858 became Confederate general officers during the Civil War: Major Van Dorn, Capt. Nathan George "Shanks" Evans, Lt. James Patrick Major, and Lt. Charles W. Phifer.

3. William B. Philpott, ed., *The Sponsor Souvenir Album and History of the United Confederate Veterans' Reunion,* pp. 98–100; L. S. Ross to The President of the United States, Aug. 4, 1865, Case File of Application from Confederates for Presidential Pardons ("Amnesty Papers") 1865–1867, Poll 54 in the Ross Papers; Homer L. Kerr, *Fighting with Ross' Texas Cavalry Brigade, C.S.A.,* pp. vii–viii, 38–39, 48, 61; S. B. Barron, *The Lone Star Defenders: A Chronicle of the Third Texas Cavalry, Ross' Brigade,* p. 125; Benner, *Sul Ross,* pp. 35–114. See also Shelly Morrison, ed., *Personal Civil War Letters of General Lawrence Sullivan Ross,* pp. 1–70, and Victor M. Rose, *Ross' Texas Brigade: Being a Narrative of Events Connected with Its Service in the Late War between the States.* Victor Rose, editor of the *Victoria Advocate,* was an active supporter of Ross for governor.

4. Benner, *Sul Ross*, p. 113.

5. Ibid., pp. 115–59; R. Henderson Shuffler, *Son, Remember . . .* , n.p.; *Galveston Daily News*, Feb. 15, 1886. See also Alwyn Barr, *Reconstruction to Reform: Texas Politics, 1876–1906*, and Turner, "A Career That Bred Legends," *Baylor Magazine*, Apr./May, 1986, pp. 19–21. L. S. Ross received his federal pardon on October 22, 1866, which cleared the way for him to run for elected office. In the 1886 gubernatorial bid Ross received 228,776 votes out of 313,300 cast statewide; Texas' population was 2.1 million.

6. Dethloff, *Centennial History*, pp. 122–25; *Galveston Daily News*, Feb. 25, 1886; Texas Governors, *Governor's Messages*, pp. 638–39, 668; "Sul Ross Career Full of Thrills from Time He Was Born in a Frontier Cabin," *Waco Tribune-Herald*, Oct. 30, 1949, p. 16.

7. Shuffler, *Son, Remember . . .* , n.p. See also Frank X. Tolbert, "Sully Saved A&M from 'Lunatic Role,'" *Dallas Morning News*, Oct. 26, 1965.

8. A. J. Rose et al., to L. S. Ross, July 5, 1890; L. S. Ross to A. J. Rose, Aug. 8, 1890, both in Ross Papers; *Galveston News*, July 6 and 7, 1890; Benner, *Sul Ross*, pp. 199–204. See also "Governor's Message to the Texas Senate and House of Representatives," Jan. 18, 1889, pp. 1–20, and Texas A&M College, *The Olio*, 1895, pp. 32–33.

9. L. S. Ross to A&M College Board of Directors, Aug. 8, 1890, in Ross Collection, Texas Collection, Center for American History, Austin.

10. L. S. Ross to H. M. Holmes, Feb. 20, 1891, Ross Papers. See also John Henry Brown, *History of Texas*, Vol. 2, pp. 494–96.

11. L. S. Ross to H. M. Holmes, Feb. 13 and 20, 1891, Ross Papers; David Chapman, "The President's Home," *Texas Aggie*, Dec., 1997, p. 13.

12. Ousley, *History of the A&M College*, p. 60; L. S. Ross to H. M. Holmes, Aug. 9, 1891, Ross Papers.

13. U.S. War Department, Headquarters of the Army, General Orders, No. 15, Feb. 12, 1890, pp. 1–3; Minutes of the Texas A&M Board, Dec. 12, 1890.

14. Lt. B. C. Morse, commandant, to the President [Ross], July, 1891, Texas A&M, *Report of the A&M College*, 1891, pp. 24–25, 55–58; "The Col-lege Is Successful," *Galveston Daily News*, Feb. 27, 1892, p. 2, and Sept. 25, 1893, p. 5.

15. Walter D. Adams to D. B. Cofer, Nov. 2, 1950, File 0002, Texas A&M Archives; L. S. Ross to Holmes Aug. 9, 1891, Dec. 20, 1891, Apr. 20, 1892, and July 22, 1892, Ross Papers; "La grippe," *The Journal* (College Station), Mar., 1982, p. 15.

16. Minutes of the Association of Former Students of Texas A&M University, June 6, 1892, Vol. 1, p. 2; F. E. Giesecke to All Alumni, Oct. 30, 1892, F. E. Giesecke Papers, Texas A&M Archives; *College Journal* (College Station), 4 (Feb., 1893): 13. See also Adams, *We Are the Aggies*, pp. 15–46.

17. "The College Is Successful," *Galveston Daily News*, Feb. 27, 1892, p. 2; Texas A&M, *Report of the A&M College*, 1891, pp. 55–58. See also Sixteenth *Annual Catalogue*, 1891–92, A&M College of Texas, 1892, pp. 34–35.

18. "A&M College: An Institution of Practical Education for Young Men," *Bryan Eagle*, Apr. 14, 1892; L. S. Ross to Holmes, July 22, 1892, Ross Papers.

19. *Texas Farmer* quoted in "A&M Colleges: To Teach the Boys How Not to Farm?" *Bryan Eagle*, Jan. 28, 1893; "Much Ado about Nothing" *Bryan Eagle*, May 3, 1893. See also Robert C. Cotner, *James Stephen Hogg: A Biography*, pp. 580–81.

20. L. S. Ross to H. M. Holmes, Dec. 20, 1891, Ross Papers; "La grippe," *The Journal*, Mar., 1892, p. 15.

21. L. S. Ross to Holmes, Apr. 20, 1892, Dec. 30, 1893, Ross Papers.

22. *Battalion*, May–June, 1894, pp. 2–3.

23. "Major Vroom's Report," *Battalion*, Oct. 15, 1893.

24. Minutes of the Board of Texas A&M, Nov. 7, 1893. See also "A. & M. College," *Galveston Daily News*, Sept. 25, 1993, p. 5.

25. A. G. (byline), "College Improvements," *Battalion*, 1893, in the clippings file at Texas A&M Archives; Texas A&M *Catalogue of the A&M College of Texas*, 1893–94, pp. 67–68.

26. L. S. Ross to H. M. Holmes, July 22, 1892, and June 22, 1891, Ross Papers; *Battalion*, June, 1897, pp. 10–11.

27. Vick Lindley, *The Battalion: Seventy Years of Student Publications at the A&M College of Texas* (College Station: N.p., 1948), pp. 1–7; Adams, *We Are the Aggies,* pp. 22–28, 30, 32; "Early Aggie Football Days Reviewed by Hal Mosley," *Battalion,* Jan. 22, 1924, p. 1; David L. Chapman, "Well, Not Exactly!" *Texas Aggie,* Dec., 1994, p. 21; John A. Adams, Jr., "Sul Ross Defined a Young Texas A&M," *Bryan Eagle,* Jan. 11, 1998, p. A10.

28. Nemo (byline), "A Day as a Cadet," *Battalion,* Dec. 1, 1893, pp. 7–8, and Feb. 1, 1894, p. 12. See also Adams, *We Are the Aggies,* pp. 3–66.

29. Minutes of the Board of Texas A&M, Dec. 12, 1890. In September, 1867, the secretary of war was authorized to allow the army to have a peak strength of 56,815. During the Indian Wars and through the late 1890s the paper strength of the army was 25,000–28,000. See Robert M. Utley, *Frontier Regulars: The United States Army and the Indian, 1866–1891,* pp. 13–25.

30. Texas A&M *Annual Catalogue* 1894–95, pp. 24–26; Minutes of the Board of Texas A&M, June 2 and 27, 1894.

31. "The Varsity Team Wins," *Galveston Daily News,* Oct. 21, 1894; "Hot Game of Football," *San Antonio Daily Express,* Oct. 22, 1894; "Varsity vs. A. and M.," *Fort Worth Gazette,* Oct. 23, 1894; "Visiting College Station," and "Was Football Day," *Galveston Daily News,* Nov. 29, 30, 1894. See also David L. Chapman, "Well, Not Exactly!" *Texas Aggie,* Dec., 1994, p. 21.

32. Texas A&M College, *The Olio,* 1895, pp. 37, 55; Association of [West Point] Graduates, *Assembly* 9, no. 2 (July, 1950): 10; Minutes of the Board of Texas A&M, June 2, 1894.

33. *Battalion,* June, 1897, p. 52.

34. Ibid., p. 53.

35. "The Agricultural and Mechanical College," *Galveston Daily News,* Oct. 24, 1897, p. 19.

36. Jos. E. Abrahams, letter to D. B. Cofer, Apr. 30, 1952, File HL 0001, Texas A&M Archives; Minutes of the Board of Texas A&M, Jan. 17, 1898.

37. Austin E. Burges, *A Local History of A. & M. College 1876–1915,* p. 14; *Battalion,* June, 1897, pp. 48–50; Minutes of the Board of Texas A&M, May 7, 1895.

38. *Battalion,* June, 1897, p. 5.

39. Ibid., pp. 33–47; Ida Wipprecht Kernodle [daughter of Walter Wipprecht '84], interview, Jan. 6, 1976, Bryan, Tex.

40. U.S. Army inspection report quoted in *Battalion,* June, 1897, pp. 35–40; David Chapman, "Texas A&M's Castle: The Old Mess Hall—1897–1911," *Texas Aggie,* Oct., 1995, p. 9; Chapman, "Bernard Sbisa (1843–1928): Culinary Artist on a Grand Scale," *Texas Aggie,* Nov., 1995, p. 11.

41. *Battalion,* June, 1897, pp. 56–57.

42. War Department, Adjutant General's Office, *Annual Report,* 1898, pp. 232–36.

43. C. C. Todd, "Valedictory Address," *Battalion,* June, 1897, pp. 10–11.

44. *Bryan Eagle,* Mar. 24, 1898; Texas A&M *Catalogue 1898–1899,* p. 87, and *1899–1900,* pp. 93–94; G. W. Hardy '79, 4th Texas Volunteer Infantry, to E. E. McQuillen, Executive Secretary of the Association of Former Students, Aug. 19, Sept. 18, and Oct. 5, 1946, and Bonney Youngblood '02, Co. I, First Texas Volunteer Infantry, to Col. Hardy, Sept. 18, 1946, in the Spanish-American War Files, Association of Former Students.

45. Letter, To whom it may concern from George T. Bartlett re: W. C. Martin '98, May 28, 1898, File HL 0015, Texas A&M Archives.

46. War Department, Adjutant General's Office, *Annual Report,* 1901, p. 24.

47. Texas A&M *Biennial Report,* Dec., 1898, p. 31; Minutes of the Board of Texas A&M, June 6, 1898.

48. *Assembly,* July, 1950, p. 55.

49. *Longhorn,* 1903, p. 80, and 1904, pp. 106–108.

50. Texas A&M College, *The Olio,* 1895, pp. 65–66; Elizabeth W. Cook, "A Brief History of the Ross Volunteers," College Station, manuscript, Oct. 10, 1966; Office of the Commandant, *A History of the Ross Volunteers,* Apr. 10, 1963; *Longhorn,* 1915, p. 133; "The Honor of Being a Volunteer—A Ross Volunteer," *TAMU Today,* Mar.–Apr., 1975, p. 3; Bob Oliver, "Men in White," *Battalion,* Aug. 13, 1938, p. 9, and June, 1897, p. 8; J. Wayne Stark, director MSC, letter to Oliver O'Bar, Feb. 6, 1957, author's collection.

51. Ross Volunteer documents and history, author's collection; Gov. Allan Shivers, letter to Cadet

Capt. Richard A. Ingels, Feb. 11, 1952, and Livingston (King Rex Organization, New Orleans), letter to Richard Ingels, RV commander, Mar. 17, 1952.

52. *Alumni Quarterly* articles: W. B. Bizzell, "Memorial to Governor Ross," Feb., 1918, p. 15; "Memorial Service," Feb., 1919, p. 3; and "Lawrence Sullivan Ross Statue Unveiled,"
pp. 11–2; W. B. Philpott, "L. S. Ross Monument," *Battalion,* Jan., 1899, p. 36; "Unveiling Seen by Large Crowd," *Reveille,* May 6, 1919, p. 1.

53. L. S. Ross Clarke, interview by R. Henderson Shuffler, Jan. 4, 1963, Ross Papers.

CHAPTER 3

1. "Mourning Ross' Death: All Revere His Memory," *Galveston Daily News,* Jan. 5, 1898; Adams, *We Are the Aggies,* p. 45; Robert Eugene Byrns, "Lafayette Lumpkin Foster: A Biography," M.A. thesis, Texas A&M University, 1964, pp. 153–57; Minutes of the Board of Directors, June 7, 1898, Vol. 1, p. 207. Pallbearers for the Ross funeral were F. E. Giesecke, A. L. Banks,
C. C. Todd, Joseph Spencer, J. C. Nagle, and A. M. Soule. See also David L. Chapman, "A Tragic Loss: Lafayette Lumpkin Foster, 1851–1901," *Texas Aggie,* Oct., 1997, p. 7.

2. Texas A&M, Biennial Report, Dec., 1898, p. 31; *Battalion,* Jan., 1899, pp. 60–64; Minutes of the Board of Texas A&M, June 11, 1901.

3. Ousley, *History of the A&M College,* p. 62.

4. *Bryan Eagle,* Jan. 22, 1899, and Sept. 28, 1899; Dethloff, *Centennial History,* p. 181; *Longhorn,* 1904, p. 164.

5. "Col. J. C. Edmonds Killed," *Battalion,* Feb. 6, 1907, p. 1; "Military," *Battalion,* Apr.–May, 1900, pp. 18–19.

6. *Bryan Eagle,* Mar. 23, 1899; Byrns, "Lafayette Lumpkin Foster," pp. 196–97, 200; Texas A&M *Catalogue,* 1899–1900, p. 79; Minutes of the Board of Texas A&M, Dec. 10, 1901.

7. *Battalion,* Apr.–May, 1900, p. 18.

8. "L. L. Foster Is Dead," *Galveston Daily News,* Dec. 3, 1901, p. 1; Minutes of the Board of Texas A&M, Dec. 10, 1901; Wilbur Evans and H. B. McElroy, *The Twelfth Man,* p. 29.

9. *Battalion,* Apr., 1902, p. 14.

10. Ousley, *History of the A&M College,* p. 63; *Longhorn,* 1904, p. 173. See also Dethloff, *Centennial History,* pp. 185–87.

11. *Longhorn,* 1906, p. 38.

12. Gale Oliver '05 to Dear Lamar, Jan. 4, 1960, letter in the Texas A&M Archives; *Battalion,* Oct., 1903, pp. 44–45.

13. Ousley, *History of the A&M College,* p. 63.

14. *Longhorn,* 1906, p. 38.

15. Texas A&M *Catalogue,* 1901, pp. 89–95.

16. David A. Houston, *Eight Years with Wilson's Cabinet, 1913–1920.*

17. *Battalion,* Nov., 1903, pp. 18–19, and Jan., 1904, pp. 22–23.

18. George Altgelt to Sarah and Dusty, Nov. 8, 1952, History Box 1 #0004, Texas A&M Archives. See also "Former A. & M. Boy Tells of Fight with Buffalo Bill's Show," *Bryan Daily Eagle,* July 28, 1923, p. 3.

19. Letter by Burt Hull ['04] to J. B. Hervey, June 7, 1957, Burt Hull Papers, Texas A&M Archives.

20. David L. Chapman, "The Kyle Field Chronicle: In the Beginning," *Texas Aggie,* Apr., 1996, p. 17; Chapman, "Mules and Grandstands," *Texas Aggie,* June, 1996, p. 31.

21. Chapman, "Kyle Field," *Texas Aggie,* Apr., 1996, p. 17.

22. Caesar "Dutch" Hohn, *Dutchman on the Brazos,* p. 112.

23. Burges, *A Local History,* p. 15; Texas A&M *Catalogue,* 1902, p. 84; George Altgelt to Sarah and Dusty, Nov. 8, 1952, in History Box 1, #0004, Texas A&M Archives; I. L. Allard to Rosalie, Nov. 1, 1910, Texas A&M Archives.

24. *Longhorn,* 1914, pp. 265–59; 1916, p. 246; 1908, p. 191; 1913, pp. 193–94; 1906, p. 125.

25. Perry, *The Story of Texas A and M,* p. 75–77; *Longhorn,* 1908, pp. 212–14.

26. *Longhorn,* 1903, pp. 159–60. See also Burges, *A Local History,* pp. 17, 21; and Nemo, "A Day as a Cadet," *Battalion,* Dec. 1, 1893, pp. 7–8.

27. Burges, *A Local History,* p. 18; *Longhorn,* 1903, p. 159; Perry, *The Story of Texas A and M,* p. 71.

28. Texas A&M *Catalogue,* 1907–1908, p. 15; Adams, *We Are the Aggies,* pp. 79–80.

29. Dethloff, *Centennial History,* p. 42; Burges, *A Local History,* pp. 18, 22.

30. See frontispiece of Herbert H. Sargent, *Napoleon Bonaparte's First Campaign*, 3rd ed. (Chicago: A. C. McClurg, 1894) in Texas A&M University Library. See also *Battalion*, Mar., 1904, pp. 14–16.

31. Untitled article, *Bryan Eagle*, May 19, 1904; Texas A&M *Catalogue*, 1904, p. 119, and 1906–1907, pp. 93, 120; Casey, *History of the A&M College Trouble*, p. xvi.

32. *Battalion*, Feb., 1904, pp. 22–23.

33. *Longhorn*, 1906, p. 4; B. M. Walker, "Henry Hill Harrington," *Journal of Mississippi History* 2 (July, 1940): 156–58.

34. Texas A&M Biennial Report, 1905–1906, pp. 13–14; Ousley, *History of the A&M College*, p. 67; Dethloff, *Centennial History*, p. 223.

35. Minutes of the Board of Directors, Sept. 8, 1905, Vol. 1, pp. 302–304; "President Harrington's Suggestion," *Battalion*, Oct. 3, 1906; Dethloff, *Centennial History*, pp. 194–95. See also "Student Life at the Agricultural and Mechanical College," *Houston Post*, June 10, 1906, p. 14, and "Eight Companies of Cadets at College," *Galveston Daily News*, Sept. 23, 1908, p. 7. See also "The Agricultural and Mechanical College," *Galveston Daily News*, Oct. 24, 1897, p. 4.

36. "Student Life at the Agricultural and Mechanical College," *Houston Post*, June 10, 1906, p. 14.

37. Ibid.

38. Minutes of the Board of Texas A&M, Mar. 8, 1904, and June 10, 1907.

39. *Longhorn*, 1908, pp. 7–9, 58; Casey, *History of the A&M College Trouble*, pp. xiv–xvii; "Capt. Andrew Moses," *Battalion*, May 22, 1907, p. 2.

40. Andrew Moses, Commandant, "Classification of Misdemeanors, with Demerits Charged for Them." College Station: n.d. [circa fall, 1907], Texas A&M Archives.

41. "Conquering Heroes Welcomed: Huge Bonfire and Brass Band at College," *Galveston Daily News*, Nov. 17, 1907, p. 4.

42. Ousley, *History of the A&M College*, p. 58.

43. Minutes of the Board of Texas A&M, Feb. 12, 1908, pp. 66–75.

44. Casey, *History of the A&M College Trouble*, p. 18.

45. Ibid., p. 49; "Senior Boys Report But Others Do Not," *Houston Post*, Feb. 13, 1908; A. J. "Niley" Smith '08, interview, Dec. 18, 1975.

46. Ousley, *History of the A&M College*, p. 68; Perry, *The Story of Texas A and M*, pp. 75, 78–79.

47. "Students on Strike," *Houston Post*, Feb. 11, 1908; "College Clash Involves Cadets," *Dallas Morning News*, Feb. 11, 1908, and "College Strike Is Ended by Cadets," *Dallas Morning News*, Feb. 12, 1908.

48. "Senior Boys Report But Others Do Not," *Houston Post*, Feb. 13, 1908; "A&M Students Still in Revolt," *Dallas Morning News*, Feb. 13, 1908; Ousley, *History of the A&M College*, p. 68; Casey, *History of the A&M College Trouble*, pp. 1–35.

49. Casey, *History of the A&M College Trouble*, pp. 14, 20–21, 163; Dethloff, *Centennial History*, p. 201.

50. "Harrington Upheld by the A&M Board," *Dallas Morning News*, Feb. 14, 1908; "College Deserted by Student Body," *Houston Post*, Feb. 14, 1908; Minutes of the Board of Texas A&M, Feb. 24, 1908; Adams, *We Are the Aggies*, pp. 58–65.

51. "Students Surrender to Alumni Influence," *Houston Post*, Feb. 15, 1908; Minutes of the Board of Texas A&M, June 22, 1908.

52. Furneaux to the Class of '09, Feb. 24, 1908, History Box 1 HL 0044, Texas A&M Archives.

53. Casey, *History of the A&M College Trouble*, pp. 109–12, 221. The number that actually graduated from the class of '08 ranges from forty-four to forty-nine.

54. Ibid., pp. 137–60, 200–13.

55. Ibid., pp. vi–vii, 75–76, 109–12.

56. Ibid., pp. 106–222; "Eight Companies of Cadets at College," *Galveston Daily News*, Sept. 23, 1908, p. 7.

57. Joe Utay '08, interview, May 21, 1976, Dallas.

58. Casey, *History of the A&M College Trouble*, pp. 64–108; Adams, *We Are the Aggies*, pp. 62–65.

59. Ousley, *History of the A&M College*, p. 69; "A. & M. College a Distinguished Institution," *Bryan Eagle*, June 15, 1912.

60. "Report of Committee to Visit the Agricultural and Mechanical College," *Journal of the House of Representatives* [Texas], 32nd Leg., Feb. 17, 1911, pp. 746–59.

61. "New Members of Faculty Appointed by A. & M. Director," *Bryan Eagle*, Sept. 10, 1908; "Eight Companies of Cadets at College," *Galveston Daily News*, Sept. 23, 1908, p. 7.

62. "Mothers Day at A. and M." *Eagle*, May 13, 1911. See also Hohn, *Dutchman on the Brazos*, pp. 46–116.

63. Capt. Andrew Moses [Commandant], letter to Felix B. Probandt, Feb. 14, 1909, in Texas A&M Archives.

64. "All Over at A. & M. College," *Bryan Eagle*, June 10, 1909.

65. *Alumni Quarterly*, Aug., 1918, p. 14.

66. *Longhorn*, 1904, p. 189, and 1906, pp. 154–55.

67. "A. & M. Cadet Corps Hiked out Yesterday," *Bryan Daily Eagle*, Apr. 1, 1911; *Longhorn*, 1903, p. 121; A. V. Govett '12 of Seguin, letter to John Adams, June 28, 1976.

68. *Longhorn*, 1911, pp. 93–95; "Class of '11 History," *Battalion*, reprint appearing in May 4, 1961, p. 3; "Cadets Prepare for Annual Hike," *Bryan Eagle*, Mar. 25, 1915. See also *Longhorn*, 1914, pp. 249–51; and 1918, p. 160; and Paul G. Haines, *Growing Up in the Hill Country*, p. 100.

69. "President of Board of Directors," *Bryan Eagle*, Apr. 13, 1911; Dethloff, *History of Texas A&M*, pp. 230–31; Texas A&M *Catalogue*, 1911–12, p. 128.

70. "A&M–TU Rivalry Broke Up for Four Years After 1911 Tussle," *Bryan Eagle*, in Joe Utay file, Texas A&M Archives; Hohn, *Dutchman on the Brazos*, pp. 70–82; Forsyth, *The Aggies and the Horns*, pp. 28–31; Don Lee '11, interview, Jan. 23, 1975, Bryan, Tex. Moran's six-season record (1909 to 1913) was 38–8–1, the best winning percentage at A&M.

71. "A. & M. Put the Fixin's on Varsity," *Bryan Daily Eagle*, Nov. 20, 1915, p. 3. The 1915 A&M–t.u. game was played a week prior to Thanksgiving. On Thanksgiving, November 25, 1915, A&M played Mississippi A&M at Kyle Field.

72. "Glad to Pay for It," *Bryan Daily Eagle*, Nov. 22, 1915, p. 3.

73. Texas A&M, *Report of the Committee Appointed to Investigate the Origins of the Fire that Destroyed the Main Building of the Agricultural and Mechanical College of Texas at College Station, Texas, May 27, 1912*. College Station, June 12, 1912, pp. 1–125.

74. Govett letter to Adams, June 28, 1976.

75. James M. Forsyth '12, interview, May 14, 1976, Sul Ross Reunion, College Station; H. C. Millender '12, letter to John Adams, June 1, 1976,

author's papers. Several cadets were injured fighting the mess hall fire.

76. Committee Report on the Fire at Old Main and letter to R. T. Milner, President, College Station, June 27, 1912, in the Texas A&M Archives.

77. Burges, *A Local History*, pp. 28–32; Association of Former Students, *Class of 1913: Historical Directory and Biography 1909–1913*, College Station: circa 1953, p. 14; Henry Alsmeyer, "Breakfast Late at A&M Only One Time . . . 11–11–11," *Eagle*, Nov. 18, 1962; Dethloff, *Centennial History*, p. 230; *Longhorn*, 1913, p. 193. See also Charles R. Schultz, "Old Main: A Visual Record of a Building's Life and Passing," *Texas Aggie*, Mar., 1994, p. 31.

78. "A. and M. Romped All Over Baylor," *Bryan Daily Eagle*, Nov. 29, 1912, p. 1. See also C. B. Moran, Head Coach, letter to Sawyer Nolstan, Aug. 17, 1912, in Texas A&M Archives.

79. Memorandum to Faculty Members, in Robert T. Milner Papers, Texas A&M Archives; "Serious Strike Prevails at A. & M. College," *Bryan Daily Eagle*, Feb. 3, 1913, p. 1; "Rifle Club Organized at A. and M. College," *Bryan Daily Eagle*, Nov. 20, 1912, p. 1.

80. "Students Dismissed for Hazing at A. & M.," *Galveston Daily News*, Jan. 31, 1913; "Present Status of the A. & M. College Strike," *Bryan Daily Eagle*, Feb. 4, 1913, p. 1; Proceedings of a Hearing by the Board of Directors of the A&M College of Texas, Held at Fort Worth, Texas, Feb. 24 and 25, 1913, pp. 1–57 (hereafter Hearing by the Board, Feb. 24 and 25, 1913); Victor Barraco '14, interview with Terry Anderson, June 4, 1981, in the Texas A&M Archives.

81. Barraco '14, interview.

82. Burges, *A Local History*, pp. 33–34; "Faculty Dismisses 466 A. & M. Cadets," *Galveston Daily News*, Feb. 2, 1913.

83. Hearing by the Board, Feb. 24 and 25, 1913, pp. 7–8; *Longhorn*, 1915, p. 34; "A&M Is Distinguished Again by War Department," *Houston Post*, June 14, 1913, p. 13.

84. E. E. McQuillen '20, "Remarks to the 56th Anniversary of the Class of 1913," College Station, May 5, 1969.

85. Burges, *A Local History*, p. 34; Hearing by the Board, Feb. 24 and 25, 1913, p. 15.

86. *Bryan Daily Eagle* articles: "Each A. and M. Student Costs State $740.00," Feb. 18, 1913, p. 1; "'Battle of 1913' Still Green in Bryan Mind," May 12, 1949, p. 6; and "Texas Farmers' Congress Convened Today," July 31, 1913; *Houston Post* articles: "SJR 18," July 1, 1913, p. 13; "A&M Club Raised Points of Objection," July 18, 1913, p. 9; and "Defeat of Amendments," July 21, 1913; "Amendments All Beaten 2–1," *San Antonio Express*, July 20, 1913, p. 1.

87. "Military Discipline at A. and M.," *Bryan Eagle*, Sept. 18, 1913; *Longhorn*, 1915, p. 34.

88. "A. & M. Cadets Viewed Battle," *Houston Post*, Nov. 24, 1913, p. 4.

89. "Army Sergeants Are Detailed to A. & M. College," *Houston Chronicle*, Sept. 13, 1913, p. 2; *Longhorn*, 1914, pp. 90–91; Barraco '14, interview.

90. Burges, *A Local History*, p. 24; "Annual Inspection of A. and M. Military Corps," *Bryan Eagle*, Mar. 31, 1910; "A.& M. College a Distinguished Institution," *Bryan Eagle*, June 15, 1912; U.S. War Department, General Orders No. 114, Washington, D.C.: June 21, 1910; *Longhorn*, 1914, p. 88; "To Strengthen Military Discipline at A. and M.," *Bryan Eagle*, Sept. 18, 1913; "A. & M. Military Was Inspected," *Daily Eagle*, Apr. 13, 1914; Ira L. Reeves, *Military Education in the United States*, pp. 147–48.

CHAPTER 4

1. Minutes of the Texas A&M Board of Directors, Jan. 6, 1913, Vol. 3, pp. 51–58; June 9, 10, 1913, Vol. 3, pp. 87–91; Aug. 18, 1913, Vol. 3, p. 100; Benedict, *A Source Book of the University of Texas*, pp. 431–40, 466–68, 489–95; "Scrapbook, 1913 Strike," Milner Papers.

2. *Longhorn*, 1915, pp. 96–99; 1918, pp. 60–62.

3. Dethloff, *Centennial History*, pp. 264–65; Stewart D. Hervey '17, interview, July 24, 1976, San Antonio.

4. Texas A&M, Biennial Report, 1915–16, p. 33; *Longhorn*, 1918, pp. 77–84.

5. *Longhorn*, 1918, pp. 77–84, 96.

6. Ibid.; Perry, *The Story of Texas A and M*, p. 91; *Alumni Quarterly*, June, 1916, p. 12.

7. *Longhorn*, 1914, p. 146.

8. "Campus Improvement," *Alumni Quarterly*, Oct., 1916, p. 16; Perry, *The Story of Texas A and M*, p. 91; Dethloff, *Centennial History*, pp. 265–67. Guion Hall was dedicated on May 25, 1918, cost $115,000, and was razed in 1971.

9. Dick Hughes, "November the Nineteenth: To the Cadets, the Team, the Rooters and the Coach" (poem), College Station, circa 1915, in Texas A&M Archives. In the eleven years at A&M, Coach Bible won five Southwest Conference Championships with an overall record of 72–19–9. The November 19, 1915, A&M–Texas game was the first ever in College Station. The previous twenty-one games were played in Austin or at neutral sites in Houston and San Antonio.

10. *Longhorn*, 1916, pp. 252–53, 257–60; Adams, *We Are the Aggies*, pp. 89–92; John D. Forsyth, *The Aggies and the Horns*, pp. 33–35; "Texas 'Aggies' 1917 Southwest Champions," *Alumni Quarterly*, Feb., 1918, pp. 3–4; "Nov. 20th Is the Day," *Alumni Quarterly*, Nov., 1917, p. 16; Joe Utay '08, interview, May 21, 1976, Dallas. See also the *Alcalde* articles: "'Bevo' Bedded Down," Mar., 1920, pp. 694–98; "Bevo I and Bevo II: A Brief History," Mar., 1933, pp. 139–40; and Jim Nicar, "A Rustled Steer Named Beer," July, 2000, p. 88; "Our Hats Off," *Daily Texan*, Feb. 13, 1917, and "Steer Branders May Have to Answer Criminal Charges," *Daily Texan*, Feb. 22, 1917. The use of the colors maroon and white also became more prevalent in the 1916–19 period, as the college transitioned from red to crimson to maroon (see *Longhorn*, 1916, p. 259).

11. Barbara Tuchman, *The Guns of August*, pp. 374–77.

12. William B. Bizzell, "The Service of the College to the Nation," *Alumni Quarterly*, Nov., 1917, pp. 3–6; Jacob G. Schurman, "Affirmative Discussion: Every College Should Introduce Military Training," *Everybody's*, Feb., 1915, pp. 59–67. See also Lyons and Masland, *Education and Military Leadership*, pp. 38–41.

13. W. L. Stoddard, "For a Citizen Army," *New Republic*, Sept. 4, 1915, pp. 125–27; "Why Do We Arm?" *New Republic*, Oct. 30, 1915, pp. 323–24; Walter Lippmann, "Patriotism in the Rough," *New Republic*, Oct. 9, 1915, pp. 277–79.

14. National Defense Act of 1916, *U.S. Statutes at Large* (1917), pp. 116–217.

15. Ibid.; Henry A. Wise, *Drawing Out the Man: The VMI Story,* pp. 102–103. On June 13, 1916, an underground campus newspaper called *The Court Martial* made a brief appearance on campus jabbing humor at the college, its president, and the new ROTC officers' training concept. The lead story was entitled "A. & M. to Be Training Camp for Worthless Second Lieutenants." Only one issue of this paper has been located.

16. Leonard Wood, *The Military Obligation of Citizenship,* pp. 10, 66.

17. Capt. A. Partridge to A. J. Dallas, acting secretary of war, May 9, 1815, in Lester A. Webb, *Captain Alden Partridge and the United States Military Academy 1806–1833,* pp. 207–209.

18. William A. Ellis, *Norwich University, 1819–1911,* Vol. 1, pp. 2–20; William Couper, *One Hundred Years at V.M.I.,* Vol. 1, pp. 1–230; Lyons and Masland, *Education and Military Leadership,* pp. 28–29. See also Andrew, "Soldiers, Christians, and Patriots," pp. 677–710; Dean P. Baker, "The Partridge Connection: Alden Partridge and Southern Military Education," Ph.D. dissertation, University of North Carolina, 1986, pp. 1–440; Bruce Allandice, "West Points of the Confederacy: Southern Military Schools and the Confederate Army," *Civil War History* 43, no. 4 (Dec., 1997): 310–31; and Lester A. Webb, *The Origin of Military Schools Founded in the United States in the Nineteenth Century,* Ph.D. dissertation, University of North Carolina, Chapel Hill, 1958.

19. U.S. War Department, *The ROTC Manual: Basic Course for all Arms,* pp. 30–1; Johnson Hagood, "R.O.T.C., the Key to National Defense," *Cavalry Journal,* Sept.–Oct., 1931, pp. 5–6; Lyons and Masland, *Education and Military Leadership,* p. 29.

20. Frank E. Vandiver, *Blood Brothers,* pp. 17–19; Lamar T. Beman, *Military Training Compulsory in Schools and Colleges,* pp. 61–3; Ira L. Reeves, *Military Education in the United States,* pp. 80–81; Eddy, *Colleges for Our Land and Time,* pp. 32–36, 41.

21. U.S. War Department, Headquarters of the Army, General Order No. 15, Washington, D.C., Feb. 12, 1890, pp. 1–3. See also T. B. Brown, "The Value of Military Training and Discipline in School," *School Review,* May 5, 1894.

22. U.S. War Department, General Order No. 15, pp. 1–3; Maj. Walter M. Lindsay, "Reserve Officers and Their Institutions," *U.S. Infantry Association Journal,* July 1, 1906, pp. 71–96; *Battalion,* June, 1897, pp. 56–57; Graham A. Cosmas, "Military Reform after the Spanish-American War: The Army Reorganization Fight of 1898–1899," *Military Affairs* 35 (Feb., 1971): 12–17; Wood, *Military Obligation of Citizenship,* pp. 1–40; "Fiske Says Nation Is 'Effeminized,'" *New York Times,* Feb. 11, 1917.

23. "A. & M. Changes in Military Tactics," *Bryan Eagle,* Jan. 4, 1917, p. 1; War Department, *Annual Report,* 1920, pp. 280–84; Edwin P. Parker, "The Development of the Field Artillery Reserve Officers' Training Corps," *Field Artillery Journal,* July–Aug., 1935, pp. 335–41.

24. Lyons and Masland, *Education and Military Leadership,* p. 40; Texas A&M, Biennial Report, 1916–17, pp. 16–18; War Department, *War Department Annual Reports, 1919,* Vol. 1, part 1, pp. 280–86, 304–305; "U.S. Air Service Equipment to A. & M.," *Bryan Eagle,* Dec. 2, 1920, p. 1. See also Eugene W. Schneider, "A Survey of the Administration of Air Force Reserve Officers' Training Corps Detachments," M.A. thesis, University of Texas, 1954, p. 89, and David Chapman, *Wings over Aggieland,* pp. 18–19.

25. John J. Pershing, *Final Report of Gen. John J. Pershing, Commander-in-Chief, American Expeditionary Forces* (Washington, D.C.: GPO, 1919), pp. 1–24; Frank E. Vandiver, *Black Jack: The Life and Times of John J. Pershing,* Vol. 2, pp. 703–24; *Bryan Daily Eagle,* May 26 and Apr. 6, 1917. The first American combat troops did not arrive in force in France until July 25, 1917.

26. Charles J. Crane, *The Experiences of a Colonel of Infantry,* pp. 548–59; "Col. Crane Back with Old Command," *Bryan Weekly Eagle,* Sept. 13, 1917; Charles W. Crawford '19, interview with Henry C. Dethloff, Mar. 30, 1971, Texas A&M Archives. Colonel Crane was given a "good set of quarters" on campus and a yearly salary of $600.

27. "Col. Crane to Be Commandant at A. and M. College," *Bryan Eagle*, Sept. 6, 1917, p. 1.

28. "Officer Reserve Corps," *Alumni Quarterly*, Jan., 1917, pp. 6–9.

29. E. J. Kyle, "The Agricultural Graduates' Part in the War for Democracy," *Alumni Quarterly*, Feb., 1918, pp. 10–13.

30. *Alumni Quarterly*, Oct., 1916, pp. 6, 9, and Feb., 1918, p. 4; *Battalion*, July 21, 1948; "Marion Church Pays Tribute to Late Col. E. B. Cushing," *Texas Aggie*, Mar. 15, 1924. See also Edward B. Cushing Papers, Texas A&M Archives; *Bulletin*, Feb. 1, 1918, pp. 26–27, and June 1, 1918, pp. 303–304; "E. B. Cushing, One of First Grads of A. & M. Died at Houston Today," *Eagle*, Feb. 21, 1921, p. 1; "Houston Man Had Big Part in War," *Dallas Morning News*, Oct. 31, 1920; "An Aggie to Remember," *Houston Chronicle*, Mar. 18, 1996. See also David L. Chapman, "The Cushing Restoration Project: Preserving a Past to Enhance the Future," *Texas Aggie*, Oct., 1994, p. 31; U.S. Army, *U.S. Army in the World War*, 1940, Vol. 12, pp. 174–201, and Charles G. Dawes, *A Journal of the Great War*, Vol. 1, pp. 15–17, 190–92.

31. *Alumni Quarterly*, Oct., 1916, p. 14.

32. *Alumni Quarterly*, May, 1918, p. 18, and Nov., 1918, p. 19; "Deaths: Jouine Was Houston's First French Consul," *Houston Press*, May 7, 1957; Houston A&M Mothers Club, *Yearbook: 1992–1993*, pp. 18–19; Ernest Langford, letter to E. E. McQuillen, Mar. 11, 1958, in Jouine Bio File, Texas A&M Archives.

33. "All College Captain's [*sic*] Lost to the Service," in Jack Mahan Scrapbook, SHS Corps Center, Texas A&M University; *Alumni Quarterly*, Feb., 1918, p. 24; May, 1918, p. 19; and Nov., 1918, p. 26; "Coach Bible Back on the College Campus," *Alumni Quarterly*, May, 1919, p. 14; Lou Maysel, *Here Come the Texas Longhorns*, pp. 130–32.

34. W. L. Driver, "Texas 'Aggies' 1917 Southwest Champions," *Alumni Quarterly*, Feb., 1918, pp. 3–4; *Alumni Quarterly*, May, 1918, p. 19.

35. *Longhorn*, 1918, pp. 91, 159–164; Stewart Hervey, interview, July 24, 1976, College Station; W. B. Bizzell, letter to [parents of cadets], Dec. 22, 1917, Mahan Scrapbook; *Bryan Daily Eagle*, May 4, 1917, May 18, 1917, and June 2, 1917; *San Antonio Express*, June 4, 1917; "Commencement Exercises Canceled," *Alumni Quarterly*, May, 1918, pp. 11–12.

36. Minutes of the Board of Directors, Mar. 23, 1917, Vol. 3, pp. 212–13; *Bryan Daily Eagle*, June 1, 1918; W. B. Bizzell, "The Service of the College to the Nation," *Alumni Quarterly*, Nov., 1917, pp. 3–6.

37. *Alumni Quarterly*, Feb., 1918, p. 20. According to this article, Private Crocker was "buried in a long grave with seventy other victims of the German submarine, on the Scottish coast February 13 [1918]."

38. *Bryan Daily Eagle*, May 25, 1918; *Dallas News*, July 21, 1918; Texas A&M Biennial Report, 1917–18, p. 128; Perry, *The Story of Texas A and M*, p. 94; "Signal Corps Company Training at College," *Alumni Quarterly*, Feb., 1918, pp. 16, 23; A. R. Andree, "With the School of Auto and Truck Mechanics," *Alumni Quarterly*, May, 1918, pp. 4–8; "War Work at the College," *Alumni Quarterly*, Aug., 1918, pp. 4–5, 8; "There Are Airplanes in the Pavilion," *Texas Aggie*, Apr., 1994, p. 8; Texas A&M, Meteorological Unit, *32nd Service Co. U.S. Signal Corps*, pp. 1–8. See also David L. Chapman, *Wings over Aggieland*, pp. 13–18.

39. "A. and M. Men to Organize a Cavalry Regiment," *Alumni Quarterly*, Feb., 1918, p. 12, and May, 1918, p. 15.

40. *Alumni Quarterly*, Aug., 1918, p. 13.

41. Ibid.; Joe Utay '08, interview, May 21, 1976, Dallas. Utay was the referee in the last game Knute Rockne played at Notre Dame.

42. "A. and M. Regiment Is Ready for Federalization," *Alumni Quarterly*, Nov., 1918, p. 10.

43. *Longhorn*, 1920, p. 140.

44. *Alumni Quarterly*, Nov., 1917, pp. 7, 15; "Nearly All College Men Get Commissions," *Eagle*, Jan. 31, 1918, p. 1.

45. "Fort Sheridan Training Camp," *Alumni Quarterly*, Aug., 1918, p. 10; Charles Crawford '19, interview with Henry Dethloff, Mar. 30, 1971.

46. "Reorganization of the College Program," *Bulletin*, Dec. 15, 1918, pp. 4–8; "Military Training," *Bulletin*, Aug. 15, 1918, pp. 1–15; U.S. War Department, Committee on Education and Special Training (Washington, 1920), pp. 22–30;

Eddy, *Colleges for Our Land and Time,* p. 163. A&M was officially notified on August 15, 1918, to begin SATC training, and the college was promised "additional rifles, uniforms and other equipment will be provided as far as necessary."

47. Charles Puryear, "The War and the College," *Alumni Quarterly,* Nov., 1918, pp. 3–4; *Bryan Daily Eagle,* Aug. 31, 1918; Dethloff, *Centennial History,* pp. 277–8; "1100 Enrolled in the S.A.T.C.," *Reveille,* Nov. 1, 1918; *Longhorn,* 1919, p. 93.

48. Vernor G. Woolsey, *The Gospel Truth,* pp. 50–51.

49. *Longhorn,* 1921, p. 114.

50. *Longhorn,* 1920, p. 140; *Bryan Daily Eagle,* Aug. 17, 18, 1918; "War Work at the College," *Alumni Quarterly,* Aug., 1918, pp. 4–5; Charles Puryear, "The War and the College," *Alumni Quarterly,* pp. 3–4; E. E. McQuillen '20, interview, June 2, 1975, College Station; "Commandant Is Assigned to Naval Section," *Reveille,* Nov. 1, 1918.

51. U.S. Army, *U.S. Army in the World War,* 1948, Vol. 12, pp. 144–52.

52. Ronald Spector, "'You're Not Going to Send Soldiers over There Are You!': The American Search for an Alternative to the Western Front, 1916–1917," *Military Affairs* 34 (Feb., 1972): 1–4. The Sargent plan drafted in September, 1917, and titled, "On the General Strategy of the Present War between the Allies and Central Powers," called for a second front via Turkey and the Balkans. After the war he remained on active duty as the PMS at Princeton University.

53. "Armistice Signed: War Ended at Six O'clock," *Bryan Daily Eagle,* Nov. 11, 1918. See also Lyons and Masland, *Education and Military Leadership,* p. 42.

54. *Reveille* articles: "Armistice Is Unthinkable," Nov. 5, 1918; "S.A.T.C. Units Are Unchanged," Nov. 13, 1918; and "Academic Work Given Emphasis," Nov. 14, 1918: Frank Aydelotte, "Final Report of the War Issues Course of the Students' Army Training Corps" (Washington, D.C., May, 1919), pp. 1–112.

55. "Major Miles Is Commandant," *Reveille,* Nov. 14, 1918; *Bulletin,* Dec. 15, 1918, p. 5.

56. "Students May See Texas Game," *Reveille,* Nov. 21, 1918.

57. "Demobilization of Both S.A.T.C. Sections Ordered: Process of Discharging Student-Soldiers . . . ," *Reveille,* Nov. 28, 1918.

58. Ibid.; "R.O.T.C. Basis to Be Resumed," *Reveille,* Dec. 3, 1918.

59. Charles Crawford, interview with Dethloff, Mar. 30, 1971.

60. "Every A. and M. Man a Brave Soldier," *Reveille,* May 15, 1919, p. 1; "A Defining War," *Wall Street Journal,* Nov. 17, 1998, p. A22. Colonel Ashburn commanded the unit in which a number of Aggies were killed in action: Sam Craig, Charles Hauser, Romeo Cox, and Herbert Peters—"all of them died like true soldiers."

61. "General Pershing Praises the American Farmer," *Alumni Quarterly,* Feb., 1919, p. 23; E. J. Kyle, "The Agricultural Graduates' Part in the War for Democracy," *Alumni Quarterly,* Feb., 1918, pp. 10–12; David B. Danborn, "The Agricultural Extension Service System and the First World War," *Historian,* Feb., 1979, pp. 315–31. See also John J. Pershing, *My Experiences in the World War.*

62. "Texas A. and M. Leads in Percent of Alumni in the Service," *Alumni Quarterly,* Aug., 1918, p. 10; *New York Times,* July 14, 1918; Texas A&M, *Biennial Report, 1917–1918,* pp. 22–24; "A. & M. Furnished 1233 Officers: Half Graduates Serve," *Reveille,* Feb. 26, 1919, pp. 1, 4; "Nearly All College Men Get Commissions," *Bryan Eagle,* Jan. 31, 1918, p. 1.

63. "Former Student of A. and M. Outstanding World War Hero," *Texas Aggie,* Sept. 15, 1923; Adams, *We Are the Aggies,* pp. 106–107; "Dan Edwards, Injured Veteran, Hears of Plan to Educate Him," *Bryan Daily Eagle,* July 31, 1923, p. 3; Katsy Pittman, "Aggie War Hero's Past Questioned," *Battalion,* Nov. 2, 1989, p. 6. See also Daniel Edwards Bio File, Texas A&M Archives.

64. "Our Roll of Honor," *Alumni Quarterly,* Aug., 1918, pp. 7, 21–46; N. M. McInnis, ed., *Gold Book: Agricultural and Mechanical College of Texas: A Tribute to Her Loyal Sons Who Paid the Supreme Sacrifice in the World War,* Aug., 1919, pp. 1–26; *Longhorn,* 1919, pp. 5, 155–56; Association of Former Students, *Directory of Former Students, 1876–1949,* p. xxvi; "Our Gold

Star Section," *Longhorn,* 1920, pp. 6–30. See also J. C. Nagle, "The Texas A. and M. Engineers' Part in the War," *Alumni Quarterly,* Feb., 1918, pp. 4–8; See also Marshall F. Foch, *The Memoirs of Marshall Foch,* and J. H. Johnson, *1918: The Unexpected Victory,* pp. 189–92. The number of Aggie former students included on the WWI Memorial varies from fifty to fifty-six; see Charles Puryear to Col. Ike Ashburn, Dec. 17, 1926, File HL 0050, Texas A&M Archives.

65. "Service Flag Raised at Commencement," *Alumni Quarterly,* Aug., 1918, pp. 1, 7; "A.& M. Furnished 1233 Officers," *Reveille,* Feb. 26, 1919, p. 1; "World War I Service Flag," *Texas Aggie,* Mar. 15, 1941; *Texas Aggie,* July 15, 1943; Press release by John West, Texas A&M University News Service, College Station, Oct. 13, 1976.

66. "President Bizzell Plans Fitting Memorial for Our Fallen Heroes," *Alumni Quarterly,* Nov., 1918, p. 4.

67. "The Alumni Memorial Stadium," *Alumni Quarterly,* Feb., 1920, pp. 1–3; *Alumni Quarterly,* Feb., 1921, p. 2; "A. and M. Plants [53] Live Oak Trees in Honor of Her Sons Who Died in the Great War," *Alumni Quarterly,* Feb., 1920, pp. 13–14; *Texas Aggie* articles: "Plans Accepted for Monument to Hero Dead," Oct. 31, 1923; "Memorial Dedication," Apr. 15, 1924; and David L. Chapman, "A Building Lost in Time," Mar., 1997, p. 17; "Memorial to A.& M. Student Heroes of World War Will Be Unveiled," *Dallas Morning News,* Apr. 20, 1924. See also Adams, *We Are the Aggies,* pp. 103–108, 121, and David L. Chapman, "Memorial Stadium?" *Texas Aggie,* Aug., 1996, p. 31.

68. *Longhorn,* 1919, pp. 105–106, 113–4; Johnny Johnson, "'Sully' Got First Issue of *Battalion,*" *Battalion,* Oct. 1, 1958, p. 5, and May 29, 1919.

69. "Colonel Muller Is Commandant," *Reveille,* Jan. 16, 1919; William R. Morgan '23, "Rehabilitation," manuscript, (Athens, Tex., June 15, 1975), pp. 1–30.

CHAPTER 5

1. 39 Stat. 166 (1916), 10 U.S.C. | 5 (1926); 41 Stat. 759 (1920), 10 U.S.C. || 2, 4, 5 (1926); Waldo C. Potter, "Field Artillery Units of the Reserve Officers' Training Corps," *Field Artillery Journal,* Mar., 1991, pp. 17–35.

2. Eldridge Colby, "Military Training in Land Grant Colleges," *Georgetown Law Journal* 33 (Nov., 1934): 5–7; Raymond Walters, "Field Artillery in American Colleges," *Field Artillery Journal,* Nov., 1919, pp. 543–55.

3. *Reveille,* Jan. 26, 1919, p. 1. The opposition at the University of Texas was fostered by Dean T. U. Taylor of Engineering and Dean H. Y. Benedict of the College of Arts.

4. F. C. Bolton, "Reserve Officers' Training Corps on a New Basis," *Alumni Quarterly,* Feb., 1919, pp. 7–8; "ROTC Basis to Be Resumed," *Reveille,* Dec. 3, 1918, p. 1; "Military Goes on New Basis," *Reveille,* Jan. 15, 1919, p. 1. See also "America Should Prepare for War," *Reveille,* Jan. 26, 1919, p. 1, and Colby, "Military Training in Land Grant Colleges," *Georgetown Law Journal,* Nov., 1934, pp. 1–10.

5. "History of Class of 1919," *Longhorn,* 1919, pp. 92–93; "Military Training," *Bulletin,* Aug. 15, 1918, pp. 1–15.

6. *Reveille* articles: "Signal Corps Unit Authorized," *Reveille,* Mar. 13, 1919, p. 1; "Army Equipment Arriving at College," Mar. 8, 1919, p. 1; and "ROTC Proves Financial Aid," Feb. 7, 1919, p. 1; *Longhorn,* 1919, pp. 5, 125; Charles W. Crawford, interview with Henry Dethloff, Mar. 30, 1971, College Station, pp. 11–14, Texas A&M Archives.

7. *Alumni Quarterly,* Feb., 1919, pp. 24, 26; *Bulletin,* June 1, 1919, pp. 185–88.

8. *Reveille* articles: "Field Artillery Unit Authorized," *Reveille,* Jan. 26, 1919, p. 1; "Colonel Muller Is Commandant," Jan. 16, 1919, p. 1; and "Artillery Unit Entails Big Cost," Feb. 21, 1919, p. 1; "Artillery Unit Organized at A. & M. College," *Battalion,* May 1, 1919, p. 1; Raymond Walters, "Field Artillery in American Colleges," *Field Artillery Journal,* Nov.–Dec., 1919, pp. 543–55; Paul S. Bond, *Basic Military Training,* pp. 10–12; *Longhorn,* 1919, pp. 119–24. See also Waldo C. Potter, "Field Artillery Units of the Reserve Officers' Training Corps," *Field Artillery Journal,* Jan.–Feb., 1919, pp. 17–35.

9. "Demobilization of the SATC," *Bulletin,* Dec. 15, 1918, pp. 6–8.

10. *Alumni Quarterly,* Feb., 1919, p. 35; "Demobilization of the SATC," *Bulletin,* Dec. 15, 1918, pp. 6–8; "Statue of Ross Reaches Campus," *Reveille,* Jan. 18, 1919, p. 1; "Big Social Events of the Year Come This Week," *Reveille,* Apr. 20, 1919, p. 1; *Longhorn,* 1919, pp. 105–106, 113–14, and 1920, p. 140; *Battalion,* Apr. 2, 1919; Lyons and Masland, *Education and Military Leadership,* pp. 40–43; "The Zoological Herd Increasing in Number Fast," *Battalion,* Jan. 29, 1920. See also Crawford interview with Dethloff, Mar. 30, 1971.

11. "Military Service Now Optional at A. and M. College," *Eagle,* Apr. 2, 1919; "The ROTC versus Casuals," *Battalion,* May 13, 1920, p. 4.

12. U.S. War Department, *Committee on Education and Special Training—A Review of Its Work During 1918,* pp. 22, 33, 37, 57, 143.

13. Pearson Menoher, "The Reserve Officers' Training Corps," *Cavalry Journal,* Apr., 1920, pp. 70–80; "Norwich University ROTC Unit," *Cavalry Journal,* Apr.–Oct., 1928, pp. 174–77; *The Guidon, 1979–1980* [The Citadel], pp. 29–46; Lyon and Masland, *Education and Military Leadership,* pp. 41–44; Parke R. Kolbe, *The Colleges in War Time and After,* pp. 80–82.

14. "College Once Again Designated a Distinguished Institution," *Reveille,* June 15, 1919, p. 1; "Military Training," *Bulletin,* Aug. 15, 1918, p. 11.

15. "Benefits from ROTC Many," *Reveille,* May 7, 1919, pp. 1, 4.

16. A. C. Cone, "Notes on Railroad Summer Practice, Agricultural and Mechanical College of Texas," Master's thesis, Texas A&M, June, 1917; "Artillery Horses Will Arrive in the Week," *Reveille,* June 8, 1919, p. 1; *Longhorn,* 1921, p. 200.

17. "100 Cadets Leave Thursday for Camp," *Reveille,* June 15, 1919, p. 1. The military summer camp program was halted from 1943 to 1946.

18. *Longhorn,* 1914, p. 31, and 1916, p. 26.

19. *Reveille* articles: "Huns Fail to Halt Ashburn," Oct. 29, 1918; "2 College Men Are Decorated," Nov. 21, 1918, p. 1; and "Every A. and M. Man a Brave Soldier," May 15, 1919, p. 1; "Isaac Seaborne Ashburn," *Longhorn,* 1920, p. 58, 150. See also undated article "Gen. Ike Ashburn, 72, Noted Texas Soldier, Leader, Dies," in Mahan Scrapbook, SHS Corps Center; Ike Ashburn,

"D. X. Bible: The Man Who Kept His Feet on the Ground," *Texas Parade,* Jan., 1955, pp. 8, 18; and George Wythe, *History of the 90th Division,* Ninetieth Division Association, 1920, pp. 45–70, 201.

20. *Bulletin,* Oct. 2, 14, 1919; *Texas Aggie,* Nov. 15, 1921; *Battalion,* Nov. 10, 1922.

21. *Bulletin,* Oct. 7, 14, 17, and Dec. 13, 1919.

22. *Daily Bulletin,* Oct. 7, 1919, Jan. 6, Feb. 20, and Mar. 20, 1920; "Cavalry Unit Is to Be Organized by End of Term," *Battalion,* Jan. 15 and May 13, 1920; *Longhorn,* 1920, pp. 1–2; Ernest Langford, "Here We'll Build the College," manuscript, Texas A&M Archives, pp. 1–20; "Senate Body Concludes A. & M. Hazing Inquiry," *Bryan Eagle,* Mar. 10, 1921, p. 1; "Inquiry of A. & M. Hazing Finished," *Dallas Morning News,* Mar. 10, 1921, p. 2.

23. Jerome Rektorik '28, interview, Nov. 8, 1975, Bryan, Tex.

24. "Jack Mahan Is in the Olympic Meet at Antwerp," Bryan *Eagle,* July 29, 1920; "Jack Mahan," *Battalion,* Apr. 8, 1920; *Longhorn,* 1921, p. 97.

25. "'Fish' Called to Complete the Grandstand," *Reveille,* Apr. 5, 1919, p. 1; Ashburn, "D. X. Bible," *Texas Parade,* Jan., 1955, p. 8.

26. John D. Forsyth, *The Aggies and the 'Horns: 86 Years of Bad Blood and Good Football,* pp. 33–35; Adams, *We Are the Aggies,* pp. 89–91.

27. "Bevo the Famous Longhorn Steer Mascot No More," *Daily Bulletin,* Jan. 23, 1920, p. 1; "Texas Presents A. & M. Portion of Bevo's Hide," *Battalion,* Dec. 17, 1920; Charles R. Schultz, "The Ignominious Fate of the Sacred Ox," *Texas Aggie,* Dec., 1993, p. 14.

28. "To Houston We Did Go," *Longhorn,* 1922, p. 207; "The Corps Trip" and "Corps Will Go to Dallas Game," *Battalion,* Nov. 7, 1928, pp. 1–2; "Trip to Waco Was Joyous Day for All," *Battalion,* Nov. 6, 1923, p. 1.

29. "The Waco Trip," *Longhorn,* 1920, pp. 153–54. See also "To Houston We Did Go," *Longhorn,* 1922, p. 207.

30. John Martin '29, interview, June 8, 1998. See also Dethloff, *Centennial History,* pp. 406, 513.

31. "A. and M. Will Be Here in Body for Cotton Palace Celebration Today—Expected 1700 in

Big Parade," undated article, circa 1921, seen in Sam Houston Sanders Collection, SHS Corps Center.

32. M. T. Harrington '22, "Class of 22: Then and Now," manuscript (College Station, 1985), p. 40.

33. *Longhorn,* 1920, p. 230, and 1922, p. 391.

34. "Wildcat," "Be a Wildcat Booster," and "Long Live the A&M Wildcats," *Battalion,* Dec. 17, 1920, pp. 2, 3, 6; Mart Moxley, "Cowboys Were Jinx When A&M Eked Sole Victory," *Daily Texan,* circa 1942. See also Forsyth, *The Aggies and the 'Horns,* pp. 45, 49–50.

35. "The Battle Hymn of A.& M," *Battalion,* Oct., 1920, p. 6; "Song of Genuine Aggie Origin to Be Adopted," *Texas Aggie,* Dec. 19, 1923; James Vernon "Pinky" Wilson '20, interview, Mar. 19, 1975, Burnet, Tex. Prior to enlisting as a private in the U.S. Marine Corps, Wilson had received an A&M medal for being the best drilled cadet and was one of a very few cadets to be a member of the Ross Volunteers as a sophomore.

36. "Good-bye to Texas University," *Texas Aggie,* Nov. 15, 1921.

37. Wilson '20, interview, Mar. 19, 1975; *Longhorn,* 1916, p. 246; 1908, p. 191; 1906, p. 125; 1913, pp. 193–94.

38. Sam H. Sanders '22 in Vickie G. Chelette, ed., "Class of '22: Then and Now," manuscript, College Station, 1985, p. 75; E. King Gill '24, Muster Speech, 1964, in John C. Adams, ed., *The Voices of a Proud Tradition,* pp. 84–86.

39. Wilson '20, interview, Mar. 19, 1975; *Texas Aggie,* Nov. 15, 1921; George Carmack, "'Aggie War Hymn' Is Greatest of All," *San Antonio Express-News,* Mar. 1, 1975, p. 1B, in the *Congressional Record,* Mar. 23, 1975, pp. E 1389–90; M. P. Mimms '22, "Texas Aggies Defeat Prayin' Colonels," *Collegiate World,* winter, 1922, pp. 2, 31; "School Song Copyright in Hands of Publisher," *Battalion,* Dec. 6, 1938, p. 1; "Wilson Says 'War Hymn' Can Be Used Almost as Usual," *Battalion,* Dec. 9, 1938, p. 1; "Committee to Work on School Song Problem Appointed by Bob Adams," *Battalion,* Jan. 10, 1939, p. 1. See also Sam Houston Sanders '22 Scrapbook in SHS Corps Center.

40. "Aggies Are Given Appropriate Song," *Daily Bulletin,* Oct. 1, 1925; "Mimms Recalls Background of 'Spirit Of Aggieland,'" *Battalion,* Oct. 10, 1967, p. 1; Marvin H. Mimms '26, "Birth of the Spirit of Aggieland," July 27, 1967, manuscript, Texas A&M Archives.

41. *Texas Aggie* articles: "Cadet Writes New Aggie Song Corps Pleased," Oct. 19, 1925; "'The Twelfth Man' New A & M Song Proves Popular," Jan. 31, 1942; and "Song of Genuine Aggie Origin to Be Adopted," Dec. 19, 1923; Jack H. Littlejohn, *I'd Rather Be a Texas Aggie,* sheet music (College Station, 1940); *Battalion* articles: Curtis Vinson, "There Shall Be No Regrets," Oct. 1, 1930, p. 1; "Yell Songs," Sept. 19, 1934, p. 21; and "Committee to Work On School Song Problem Appointed by Bob Adams," Jan. 10, 1939, p. 1.

42. *Longhorn,* 1921, pp. 198, 204; *Longhorn,* 1922, pp. 170–71, 178; "Silver Cup for Best Drilled Cavalry Unit," *Battalion,* Nov. 20, 1920; "U.S. Air Service Equipment to A&M," *Eagle,* Dec. 2, 1920, p. 1. The U.S. Air Force was created in 1947.

43. "Freshman ['26] Class History," *Longhorn,* 1923; "Texas A. and M. College Needs No Building Program," *Texas Aggie,* Apr. 15, 1923, p. 1; Jim Peter, "Tent City," *Battalion,* Mar. 8, 1976. See also "Hazing at A&M College," *Sterling City News-Record,* Nov. 9, 1923, in Texas A&M Archives.

44. "History of the Class of 1927," *Longhorn,* 1924, pp. 150, 217; *Longhorn,* 1923, p. 134; Peters, "Tent City," *Battalion,* Mar. 8, 1976; "My Shack," *Battalion,* Mar. 13, 1929, p. 6; "Colonel C. C. Todd Commandant A. and M. after Jan. Second," *Bryan Eagle,* Jan. 3, 1924, p. 1; "Charles Todd '97 Follows Ashburn [in] Important Post," *Texas Aggie,* Jan. 15, 1924, p. 1; "Association of Former Students," *Longhorn,* 1924, p. 217, and 1925, p. 296; John Martin '29, interview, June 8, 1998.

45. "Report Shows Large Growth of College," *Battalion,* Feb. 12, 1924, p. 1; "A. & M. Cadet Corps Largest in Country," *Battalion,* Feb. 19, 1924, p. 5; *Longhorn,* 1925, pp. 161–69; U.S. Army, Center for Military History, *Correspondence Relating to the War with Spain,* p. 981. The 1925 *Longhorn* was dedicated to Col. C. C. Todd.

46. Lowell Jones, "Cadet Corps Not the Same," *Battalion,* Oct. 5, 1966, p. 1.

47. College Station, *Bulletin,* June 1, 1923, pp. 63–64.

48. "Air Service Summer Camp," *Longhorn,* 1923, p. 179; "Texas A. and M.," *Longhorn,* 1925, pp. 20–21. See also Chapman, *Wings over Aggieland,* pp. 18–20.

49. A. C. Taylor '24, letter to John Adams, Apr. 9, 1992.

50. *Battalion* articles: "Junior Banquet Was Great Success," Apr. 8, 1920; "RV Week Was Complete Success," Apr. 23, 1924, p. 1; and "Junior Class Has First Banquet of the Year," Mar. 6, 1929, pp. 1, 7.

51. "College Radio Plans Work," *Battalion,* Oct. 7, 1920, p. 1; Charles R. Schultz, "First Aggie Football Broadcast," *Texas Aggie,* Nov., 1993, p. 37.

52. Dethloff, *Centennial History of Texas A&M,* Vol. 1, pp. 305–306.

53. "Commander of Eighth Corps Area Inspects A. and M. Cadets," *Texas Aggie,* Mar. 1, 1923, p. 1.

54. *Battalion* articles: "Governor Neff Addressed the Cadet Corps," Apr. 22, 1924, p. 1; "Aggie Band Wins Praise in Ft. Worth," Oct. 16, 1923, pp. 1, 2; and "President Talks to Corps for Last Time This Year: Hazing Pledge Read," May 20, 1921, p. 1; "The Aggieland Six," *Longhorn,* 1923, p. 408; "History of the Class of 1927," *Longhorn,* 1924, p. 150; "History of the Freshman Class [1928]," *Longhorn,* 1925, p. 128.

55. A. C. Taylor '24, letter to John Adams, Mar. 31, 1992. See also Haynes W. Dugan, *The History of the Great Class of 1934,* pp. 52–53.

56. Joseph G. Rollins, Jr., *Aggie! Y'all Caught That Dam' Ol' Rat Yet?* pp. 19–22. See also Dugan, *Great Class of '34,* p. 7.

57. Rollins, *Aggie! Y'all Caught That Dam' Ol' Rat Yet?* pp. 19–22. A brief look at the role of a "fish corporal or sergeant" can be seen in the movie *We've Never Been Licked.*

58. "Fish Don'ts," Louis Hartung Scrapbook, Texas A&M Archives.

59. Article titled "A Parting Message," source and date unknown [circa June–July, 1925] in Bizzell Papers, Texas A&M Archives. See also Heidi A. Knippa, "Salvation of a University: The Admission of Women to Texas A&M," M.A.

thesis, University of Texas at Austin, 1995, pp. 26–28.

60. "The Commandant of Cadets," *Longhorn,* 1927, p. 23; *Dallas News,* June 3, 1925; "Col. C. C. Todd, Former A and M Commandant, Dies," *Battalion,* Apr. 3, 1935, p. 1.

61. Texas A&M, "Program of the Semi-Centennial Celebration and Inauguration of Thomas Otto Walton as President of the College, Oct. 15, 16, 17, 1926," 1926, pp. 1–5, in Texas A&M Archives.

62. *Longhorn,* 1927, p. 99, and 1928, n.p.; *Battalion* articles: "The Booster's Club," Mar. 3, 1923, p. 3; "Increase of 400 in Enrollment," Sept. 26, 1928, p. 1; "Fish Learning Aggie Customs," Oct. 3, 1928, p. 1; "Home Again," Sept. 19, 1928, p. 4; and "Campus Lumber Supply Suffers as First Year Students Garner Material for Annual Bonfire," Nov. 23, 1932, p. 1.

63. Minutes of the Board of Directors, Dec. 3, 1928, Vol. 4, p. 113; Feb. 5, 1929, Vol. 4, p. 115; Nov. 26, 1930, Vol. 4, p. 159.

64. "Seniors to Invite Inspection by Legislatures," *Battalion,* Dec. 10, 1930, p. 1; "The Invitation to the Legislature," *Battalion,* Dec. 10, 1930, p. 2.

65. "The Commandant," *Longhorn,* 1928, p. 19; "Professors of Military Science and Tactics," *Longhorn,* 1929, p. 149; *Battalion* articles: "A More Comfortable Uniform," Oct. 10, 1928, p. 4; C. J. Nelson, "Suggestions," Mar. 27, 1929, p. 2; "Uniform Changes," Sept. 18, 1929, p. 3; and "Dormitories Assigned to Military Units," p. 7.

66. *Battalion* articles: "Ain't It Awful," Jan. 30, 1929, p. 4; "Junior Class Has Meeting," Feb. 13, 1929, pp. 1, 10; "Old Style Ring Wins Class Vote," Mar. 5, 1930; "The Ring Again," Dec. 17, 1930, p. 5; "Regarding the Senior Ring," Feb. 4, 1931, p. 4; "Modified Ring Is Chosen by Juniors," Feb. 11, 1931; and "New Standardized Senior Ring Now Being Exhibited," Feb. 28, 1934, p. 6; Adams, *We Are the Aggies,* pp. 74–77. See also Jerry Cooper, "100 Years of Aggie Rings," *Texas Aggie,* Oct., 1969, pp. 6–10.

67. "Our Senior Ring," *Battalion,* Jan. 30, 1929, p. 4. See also the Aggie Ring collection at the Clayton Williams Building offices of the Association of Former Students, and at the Sam Houston Sanders Corps of Cadets Center.

68. "4,272 Graduates of A & M Since Founding," *Battalion,* Dec. 3, 1930, p. 3.

69. "Fourteen Project Houses to Be Built for Accommodations of Next Semester's Enrollment Increase," *Battalion,* Apr. 21, 1937, p. 1; Dethloff, *Centennial History,* pp. 426–30.

70. *Battalion* articles: "New Coastal Artillery Unit Assigned Equipment and Basic Course Students," May 20, 1931, p. 1; "Five-Cent Fare on Bus Line," Nov. 21, 1934, p. 1; "Band Uniforms," Oct. 21, 1931, p. 2; and "First Corps Trip Will Be Made to Centenary Game," Oct. 21, 1931, p. 1.

71. *Battalion* articles: "Prexy's Moon," Mar. 4, 1931, p. 4; "Meet the Colonel," Jan. 17, 1934, p. 10; and "Reserve Officers Form Battle Plan in Sham Skirmish," Nov. 9, 1932, p. 1.

72. See C. C. Todd, Legal Brief on Women at Texas A&M Papers, circa 1933, Texas A&M Archives.

73. *Battalion* articles: T. O. Walton, "Texas A. & M. College Offers—Training in Mind and Body—for Life's Battle Fronts," Apr. 12, 1933, pp. 7, 31; "Changes Made in Uniforms for Next Year," May 10, 1933, p. 1; "230 Juniors Will Attend Summer Camp," May 30, 1933, p. 1; "A&M Likely to Remain a Boy's School Even Though State Courts Affirm Girl's Right to Entrance," Nov. 29, 1933, p. 3; and "College Radicals Start Bonfire Three Weeks Early; Fire Quenched after Student Body Is Aroused," Nov. 22, 1933, p. 1.

74. "Traditional 'Elephant Walk' on Eve of Thanksgiving Game Marks the Passing of the Senior Class," *Battalion,* Nov. 6, 1933, p. 6.

75. Dethloff, *Centennial History,* pp. 432–36.

76. Dugan, *Great Class of '34,* pp. 3, 173n.

77. "Systematic Military Training Is Objective in Seven Branches of Reserve Officer Courses Here," *Battalion,* May 1, 1935, p. 5; "Chemical Warfare Unit Here Is Second University Branch to Be Organized under Army Control," *Battalion,* Oct. 23, 1935, p. 2.

78. *Battalion* articles: "Things Military," Sept. 19, 1934, p. 12; "Enrollment Is Nearly Three Thousand," Sept. 26, 1934; "Proposed Entrance to A&M College," Jan. 24, 1934; and "Discipline," Sept. 26, 1934, p. 2.

79. "Individuals and Companies Get Trophy Awards," *Battalion,* May 29, 1935, p. 1.

80. *Battalion* articles: "Total Registration of 3400 Student Body Twenty-one Percent Larger Than That of Last Session," Sept. 25, 1935 p. 1; "We Need Room!" p. 2; "4,075 Enrollment Sets New Record," Sept. 30, 1936; and "Colonel Anderson Sets Second Term Estimate at 4000," Feb. 10, 1937, p. 3.

81. Dethloff, *Centennial History,* pp. 436–37.

82. Ormond R. Simpson '36, "Reflections of a Cadet in the Corps—1932–1936," manuscript, College Station, 1997. See also Paul McKay, "Ormond Simpson Fights Ainning Battles," *Bryan Eagle,* May 1, 1983, p. 1D, and M. W. Malley, "Military Head Says Corps 'Prepared,'" *Battalion,* Dec. 13, 1979, p. 2.

83. W. O. Thompson, "Military Training at Educational Institutions," *Infantry Journal,* May, 1929, pp. 496–502.

84. "Faculty-Student Conference Had Several Interesting Features," *Battalion,* Jan. 7, 1931, p. 3.

85. "ROTC Is Upheld by Supreme Court of United States," *Battalion,* Dec. 6, 1934, p. 1. See also Colby, "Military Training in Land Grant Colleges," *Georgetown Law Journal,* Nov., 1934, pp. 1–36.

86. Maj. Gen. Ray Murray '35, interview, Oct. 16, 1999, College Station; USMC news release, "Somewhere in the Southwest Pacific," PR Section, Atlanta, Ga., circa 1943, author's papers. See also Maj. Gen. Wood B. Kyle '36, "Muster Speech—1968," in John C. Adams, ed., *Voices of a Proud Tradition,* pp. 101–107.

87. John W. Killigrew, "The Impact of the Great Depression on the Army, 1929–1936," Ph.D. dissertation, Indiana University, 1960, chapters 12 and 13. In March, 1933, President Roosevelt pushed Congress for authorization of funding for the CCC for the purpose of reforestation and unemployment relief.

88. Gen. Bernard A. Schriever '31, telephone interview, Aug. 9, 1999. See also Dugan, *Great Class of '34,* pp. 36–37. Reflecting on the mid-1930s and the CCC Dugan wrote, "This experience [CCC] obviously stood them in good stead in the years ahead when an army had to be made of civilians."

89. U.S. Air Force, *Of Flight and Bold Men* (Washington, D.C.: GPO, 1968), n.p. General

Schriever was one of the earliest proponents of manned space flight. See also "ICBM—Step Toward Space Conquest," Feb. 19, 1957, speech, Space Flight Symposium, San Diego, Calif., author's collection.

90. *Battalion* articles: "Aggies Oppose Disarmament and Prohibition But Uphold ROTC," Apr. 6, 1932, p. 2; "A&M Cadets to Be Offered One Year Regular Training," Mar. 31, 1937, p. 1; and "A Reason for Military Training," Mar. 10, 1937, p. 2; "Aviation Hall of Fame Honors Schriever," *Texas Aggie*, Sept., 1980, p. 21.

91. "A & M Has Largest Ag. Enrollment in World," *Battalion*, June 8, 1936, p. 1.

92. "War in Europe Is Inevitable Says Editor of *Nation*," *Battalion*, Nov. 3, 1933, p. 3; "War!" *Battalion*, Feb. 3, 1932, p. 2.

93. "Roosevelt Appears Today," *Battalion*, May 11, 1937, pp. 1–2; "Congratulations Corps," *Battalion*, May 19, 1937, p. 2; "President Roosevelt Ends His Gulf Fishing Cruise" and "Cadet Corps Is Reviewed as 1st Move," *Bryan Daily Eagle*, May 11, 1937, p. 1; "Housing Problem," *Battalion*, Sept. 30, 1936, p. 2; "We Need a Union Building," *Battalion*, Mar. 11, 1936, p. 2; Col. Victor M. Wallace '38, interview, Sept. 19, 1998, Austin.

CHAPTER 6

1. Text of President Roosevelt's May 11, 1937, address in President's Personal File, 1053, Franklin D. Roosevelt Papers, Hyde Park, N.Y.; "FDR Tells A&M Students of Opportunities to Serve," *Houston Post*, May 12, 1937, p. 1; *Longhorn*, 1938, pp. 2–3, 38–39.

2. *Battalion* articles: L. E. Thompson, "Great Increase in Air Fighting Would Probably Be Main Feature of Another Big World War," Sept. 30, 1939, p. 1; "Coastal Artillery Will Probably Be More Important in Air Defense," Nov. 8, 1938, p. 1; and "A. & M. Holds Important Place in Training Infantry Officers," Oct. 2, 1938, p. 1. Also see "Madman in Europe," *Battalion*, Nov. 9, 1939, p. 4, and "Will There Be a War?" *Battalion*, July 21, 1939, as reprinted from the *Daily Texan* newspaper at t.u.

3. "A Reason for Military Training," *Battalion*, Mar. 10, 1937, p. 2; "College Has an Enviable Record in War," *Battalion*, Sept. 17, 1940, p. 11; Dethloff, *Centennial History*, pp. 450–51. Antimilitary pacifists had a significant impact in preventing the addition of a ROTC field artillery branch at the University of Texas in 1936.

4. "ROTC at UT," *Fort Worth Star-Telegram*, in the *Battalion*, Feb. 3, 1939, p. 2. *The Daily Lariat* at Baylor stated that ROTC was a waste of time: "The idea of placing one student in ranks above another is bad business. ROTC leads to bad spirit. Military training is all right in high school, but college is not the place for it. There are enough freaks in Baylor already without ROTC" (seen in *Battalion*, Nov. 24, 1937, p. 2).

5. *Battalion* articles: "War in Europe Is Inevitable Says Editor of *Nation*," Nov. 8, 1933, p. 3; "World War Soon to Come, Says Vanderbilt," Feb. 17, 1939, p. 1; and "Another World War Soon to Come Predicts Max Brauer, German Refugee, during Lecture Here Monday Night," Mar. 7, 1939, p. 1.

6. *Longhorn*, 1908, p. 166, and 1943, p. 69; "Know These Men," *Battalion*, Sept. 22, 1937, pp. 12, 20. In a May, 1976, interview Joe Utay '08 said that he felt he had a major role in securing Colonel Moore as the commandant. According to Utay, then a member of the board of directors of the college, he was in Washington and dropped by to visit with the army chief of staff about the need for a new commandant at A&M. Utay was informed there was a "good man for the job, a colonel named Moore." Utay asked the general, "What's his first name . . . George? Well he's my classmate and we called him Maude!"

7. "Four Officers Are Added to Military Staff for Session: Colonel Moore Takes Role of Commandant and P.M.S. and T.," *Battalion*, Sept. 29, 1937; "Enrollment Should Exceed 5200—4,933 Here Last Term," *Battalion*, Jan. 26, 1938, p. 1; *Longhorn*, 1938, 1939, and 1940, n.p.; "Seven ROTC Branches Are Taught in College with Country's Largest Units, Four Thousand Students Take Military," *Battalion*, May 4, 1938, p. 4.

8. James E. Miller, "A Survey of the Cooperative Housing Plan at the Agricultural and Me-

chanical College of Texas," Master's thesis, Texas A&M College, 1936.

9. Minutes of the Board of Directors, June 4, 1937, Vol. 5, pp. 98–99; Nov. 24, 1937, Vol. 5, pp. 123–26; June 14, 1938, Vol. 5, pp. 147–54; Nov. 5, 1938, Vol. 5, p. 164; "Plans for 12 New Dorms at A&M Discussed," *Battalion,* Mar. 23 and May 4, 1938, p. 1; Langford, "Here We'll Build the College," pp. 163–66; *Bryan Daily Eagle,* Sept. 4, 1941; Chapman, *Wings over Aggieland,* pp. 27–28. Note: Nine Corps dorms originally had 111 rooms each; one, 103; one, 106; and one, 107.

10. Perry Luth '42, interview, San Antonio, Oct. 13, 1998; *Battalion* articles: "Dormitories Will Be Built for A&M, Says Ashburn," Sept. 20, 1938; "Crowded A&M Campus Extended 23 Miles to Navasota," Sept. 27, 1938, p. 4; and "Additional Dormitories on Campus May Relieve Crowded Conditions," Sept. 29, 1937, p. 1.

11. James B. "Dick" Hervey '42, interview, San Antonio, Oct. 13, 1998; "Ike S. Ashburn Heads Publicity Department Here," *Battalion,* Sept. 23, 1937, p. 4; *Longhorn,* 1939, p. 20. Dick Hervey, in addition to being the 1942 senior class president, was executive director of the Association of Former Students, 1947–63.

12. J. Wayne Stark, "Aggie Scrapbook," *Battalion,* Oct. 6, 1937, p. 2.

13. "Aggies Ready for Turkey Day Invasion," *Battalion,* Nov. 24, 1937, p. 1; "Aggie Team Beats Texas by 7–0 Score," *Battalion,* Dec. 1, 1937, p. 1; Forsyth, *The Aggies and the 'Horns,* p. 68; Maysel, *Here Come the Texas Longhorns,* pp. 130–34.

14. "4,000 Aggie Cadets, Largest Number to Attend Corps Trip, Parade before Horned Frog Tussle," *Battalion,* Oct. 20, 1937, p. 1; "5,187 Enter College for 62nd Session," *Battalion,* Sept. 16, 1938, p. 1; "Day Student Dilemma," *Battalion,* Oct. 6, 1937, p. 2; "Organization of the Student Body, A. and M. College of Texas," *Battalion,* Sept. 3, 1938, p. 4; "Why Come to A&M?" *Battalion,* May 4, 1938; "Aggies Select Waco as Corps Trip Location," *Waco Times Herald,* Oct. 4, 1938, p. 1. The Waco Corps trip was the first time since 1923 that an official Corps trip was made to Waco.

15. "Large Crowd to Witness Bonfire," *Battalion,*
Nov. 22, 1938, p. 1; "Seniors before Game [Elephant Walk]" *Battalion,* Nov. 29, 1938, p. 3.

16. "Largest Senior Class in History to Be Graduated" and "Official Notice: Memorandum 32, 34," *Battalion,* June 2, 1939, pp. 1, 4.

17. "A&M Assured NYA Aid during Coming Year," *Battalion,* July 14, 1939, pp. 1, 4; "College NYA Funds Increased," *Battalion,* Aug. 4, 1939, p. 1.

18. "T.T.'s and Swastikas Have Rival in the 'Clickety Clack Club,'" *Battalion,* Apr. 28, 1939, p. 1; "Upstreaming Is Not Aggie Way," *Battalion,* May 9, 1939; "'Thank You, It Has Been a Pleasure to Ride with You,' Say Aggie Hitch-Hiking Cards Issued by 'Y' Cabinet," *Battalion,* Nov. 23, 1939, p. 1; "Hitch-Hiking and Ethics," *Battalion,* Mar. 14, 1940, p. 1; "Fish-Eating Craze Spreads—Soph Gobbles 17 Live Frogs to Win $1 Bet," *Battalion,* May 19, 1939, p. 2.

19. "Radical Changes Now Being Planned in Infantry Drill, Likely to Be Adopted at A&M Next Year," *Battalion,* Feb. 28, 1939, p. 1; George Fuermann, "Corps Organization Changes Made—Day Students Outfits Are Broken Up," *Battalion,* July 21, 1939, p. 1; *Bulletin of the A&M College 1938–39 Session,* June 1, 1939, pp. 37–39.

20. "War! Bombs Blast Poland," *Houston Chronicle,* Sept. 1, 1939, p. 1; Henry Herge, *Wartime College Training Programs of the Armed Services,* p. 3.

21. *Longhorn,* 1943, p. 69; George Fuermann, "Anne Moore, Passenger in the Ill-Fated Athenia, Arrives Home after Thrilling Tour," *Battalion,* Oct. 7, 1939, p. 1; "Anne Moore Recounts Dramatic Rescue and Aftermath of Athenia Torpedoing," *Battalion,* Oct. 12, 1939, p. 6; James H. Parker, Jr., '41, telephone interview, Oct. 17, 1998.

22. "Official Notices: Exams for Appointments as Flying Cadets," *Battalion,* Apr. 16, 1940, p. 4; "Texas Aggies and the World War—No. 2," *Battalion,* June 14, 1940, p. 2; "School Prepared before Bombing of Pearl Harbor," *Bryan Eagle,* Nov. 11, 1942, in the World War II File, Texas A&M Archives.

23. A&M College, *Facilities for National Defense,* College Station, July, 1940, pp. 1–43, with an introduction from the 1940 valedictory address by Cadet Colonel of the Corps Durward B. "Woody" Varner '40.

24. *Battalion* articles: "Colonel Moore Notified of Transfer," Apr. 30, 1940, p. 1; "Leaving . . . A Real Aggie," May 2, 1940, p. 2; "Marine Corps Commissions Now Offered," May 30, 1940, p. 1; "'Fish Sergeant,' Book about A&M, to Be Published Soon," May 7, 1940, p. 4; and "Roosevelt Offered All A&M Facilities for National Defense Use," June 30, 1940, p. 1; *Longhorn,* 1940, n.p.

25. *Battalion* articles: "And Still the War," June 20, 1940, p. 2; "Too Many ROTC Officers," June 27, 1940, p. 1; "A&M Loses," July 4, 1940, p. 2; and "A&M Can Boast of Present Set-up and Past Record for Training Army Officers," Aug. 8, 1940, p. 1.

26. *U.S. Statutes at Large,* 79th Cong., 3rd sess., chap. 720, Sept. 16, 1940, pp. 885–97. See also Lewis B. Hersey [director of Selective Service, 1940–41], *Selective Service in Peacetime,* pp. 3–7.

27. *Battalion* articles: "LtCol James A. Watson Succeeds Moore as Commandant and PMS&T," Sept. 17, 1940, p. 1; "Record Enrollment Looms Next Year," Aug. 22, 1940, Sept. 21 and 24, 1940, p. 1; and "Fish Class Responsible for the Big Increase," Sept. 21, 1940, p. 1; William A. Becker '41, telephone interview, Oct. 5, 1998; *Longhorn,* 1941, pp. 1–10, 24, 269–73, 374. See also Frederick L. Allen, *The Big Change: America Transforms Itself, 1900–1950,* pp. 158–76.

28. *Bulletin of the A&M College, 1940–41,* Apr. 1, 1941, pp. 44–45; "Military Department Is Now Comfortably Situated in Ross Hall," *Battalion,* Oct. 3, 1940, p. 1.

29. The Militia Act of 1792 is sometimes referred to as a peacetime conscription law; however, no citizens were called up for duty. The Selective Service Act of 1917 was a wartime measure.

30. "1,455 Aggies Register for Selective Service," *Battalion,* Oct. 19, 1940, p. 1; "$18,000,000 Plant Trains U.S. Officers," *Battalion,* Jan. 7, 1941, p. 1; "Juniors, Seniors Leave en Masse for Corps Trip," *Battalion,* Oct. 5, 1940, p. 1; *Longhorn,* 1941, pp. 24, 375–76. This was the third Corps trip to San Antonio; the first was in 1902 for a game versus Texas (0–0 tie) and the second witnessed a 26–13 win over Michigan State in 1934.

31. George Fuermann, "National Defense Coopera-

tion," *Battalion,* Jan. 7, 1941, p. 1; "Cotton Bombay Slacks Added as Drill Uniform by Senior Class Decision," *Battalion,* Sept. 17, 1940, p. 2; *Bulletin of the A&M College,* Apr. 1, 1941, pp. 44–45; William A. Becker '41, interview, Oct. 5, 1998; *Longhorn,* 1941, p. 24 (see special section on each branch of training at A&M).

32. "Aggies Prepared," *Battalion,* July 30, 1941, p. 2; R. W. Steen, "As the World Turns," *Battalion,* Aug. 6, 1941, p. 2.

33. "After the War," *Battalion,* June 1, 1941, p. 1. See also "Student Conscription," *Battalion,* June 25, 1941, and "Col. Ike Ashburn Resigns from College," *Battalion,* June 11, 1941, p. 2. See also Col. Ike Ashburn, "Educating for Fullness of Living," *Proceedings of the 1941 Annual Meeting,* Association of Texas Colleges, Dallas, July 15, 1941, pp. 56–60.

34. "ROTC Contract Exempt Aggies from Army Draft," *Battalion,* Aug. 22, 1940, p. 1; "War Time Log of A&M," *Battalion,* Mar. 7, 1942.

35. *Battalion* articles: "Military Organizations for Next Year Nearly Full," Aug. 13, 1941; "'42 Dormitory Assignments," Aug. 6, 1941, p. 2; and "Bricklayers in Demand for New Dormitories," Sept. 30, 1941, p. 1; James H. Parker, Jr., '41, interview, Dallas, Oct. 17, 1998. For background on Corps area dorm names see Dethloff, *Centennial History,* pp. 445–46.

36. *Longhorn,* 1942, pp. 363–75.

37. "Aggie Slanguage Rules at Camp," *Battalion,* Aug. 13, 1941, p. 1.

38. John A. Salmond, *The Civilian Conservation Corps, 1933–1942: A New Deal Case Study,* pp. 159, 172–73, 197–98. See also *Washington Star,* Dec. 30, 1939; *Washington Post,* Feb. 16, 1940; and *New York Times,* Aug. 17, 1941.

39. "Tom Gillis Cadet Colonel for '42 College Session," *Battalion,* Sept. 9, 1941, p. 1; "Hervey Vows to Lead Class of '42 in Aggressive Program," *Battalion,* Sept. 20, 1941, p. 1; "LtCol James A. Watson Relieved as Commandant," *Battalion,* Aug. 13, 1941, p. 1; "Senior Meet in Guion Set on Thursday," *Battalion,* Sept. 16, 1941, p. 1; "Special Train Carries Aggie Band and Students to San Antonio Saturday," *Battalion,* Sept. 30, 1941, p. 1; "The Twelfth Man in Action," *Battalion,* Nov. 5, 1941, p. 2; *Annual*

Report of the A&M College of Texas for 1940–41, Nov. 15, 1941, pp. 11–31; *Longhorn,* 1942, pp. 41–42, 182, 192–93, 286.

40. "Like One Student All Aggies Call Themselves the 12th Man," *Houston Post,* Nov. 11, 1942, p. 21.

41. *Battalion* articles: "Colonel Welty Delayed, Will Not Arrive Until Middle of Nov. to Take Post," Sept. 12, 1941, p. 1; "New Commandant Already Admires Aggie Spirit and Plans to Make A&M Best ROTC Unit in Nation," Nov. 6, 1941, p. 1; "'The Twelfth Man' New Song of Corps presented Tuesday," Sept. 18 and 20, 1941, p. 1; "Turkey Day Clash Dedicated to Former Aggies Now in Service of Uncle Sam," Sept. 25, 1941, p. 1; and Pete Tumlinson, "Fish Blotto," Oct. 3, 1941, p. 13; *Longhorn,* 1942, pp. 172, 285.

42. "The US and World War II," *Battalion,* Oct. 7, 1941, p. 2.

43. "No Chance of Getting Out of That Chemistry Lab Because of Shortage of Materials Here," *Battalion,* Oct. 23, 1941, p. 2; "Special Defense Courses Planned for A. & M. Summer Session," *Battalion,* June 11, 1941, p. 1; James Robert Latimer, Jr., '44, interview, Dallas, May 22, 1976.

44. *Battalion* articles: "Spare the Spoons!" Sept. 9, 1941, and "SOS—Save Our Spoons," Sept. 11, 1941, p. 2; "Coordination Group Appointed to Solve Mess Hall Problems," Nov. 20, 1941, p. 1; "Mess Hall Cooperation," Nov. 22, 1941, p. 2; and "Defense Taxes Raise Prices of Senior Ring," p. 1. See also "War Time Log of A&M," *Battalion,* Mar. 7, 1942, p. 4.

45. "3000 Ex-Aggies Are Now on Active Duty with US Army," *Battalion,* Oct. 14, 1941, p. 1; "School Prepared before Bombing of Pearl Harbor," *Bryan Eagle,* Nov. 11, 1942.

46. President T. O. Walton, letter to the parents of Fred W. Dollar '44, College Station, n.d. [Jan. 30, 1942], in the Fred W. Dollar Papers, Texas A&M University Archives.

47. *Battalion* articles, Dec. 9, 1941, p. 1: "Remain in School Profs Tell Students," "'If Called to Duty Full Credit Will Be Given'—Welty," and "Faculty Cooperates with Selective Service Boards."

48. *Battalion* articles: "Commanders Pledge Corps Loyalty to Nation in Crisis," Dec. 9, 1941, p. 4; "Ole Army Gets the Spirit," Dec. 11, 1941, p. 1;

"Cadet Corps and War," Dec. 11, 1941, p. 2; and "Aggie Officers," Dec. 13, 1941, p. 2.

49. *Battalion* articles: "Three Semester Year," Jan. 8, 1942, p. 1, and "12 Month School Year: System Will Continue for Duration of War Emergency," Jan. 13, 1942; U.S. Office of Education and Federal Security Agency, "Recent Action in the Land-Grant Colleges and State Universities," *Education for Victory,* Mar. 3, 1942, p. 10. By mid-February, 1942, a survey indicated that forty-four colleges and universities had modified their academic calendar for the war effort, with thirteen adopting the annual "three-semester plan."

50. "School Prepared before Bombing of Pearl Harbor," *Bryan Eagle,* Nov. 11, 1942, pp. 1, 10.

51. *Annual Report of the Agricultural and Mechanical College of Texas for the Year 1941–1942,* Nov. 15, 1942, p. 34; *Battalion* articles: "No Finals, May 26 Graduation, Three Semester Year," Jan. 8, 1942, p. 1, and "Board Approves No Finals, 12 Month School Year," Jan. 18, 1942.

52. *Battalion* articles: "Settle Down, Ole Army," Jan. 12, 1942, p. 2; "Sweeping Changes to Effect Every Student Monday," Jan. 22, 1942, p. 1; James Holekamp, "They Capitalize on the War," Feb. 3, 1942, p. 1; "From President Walton, A Message," Jan. 31, 1942; "Military Plan Junked by Commanders; Many Radical Suggestions Predominate," Jan. 31, 1942; "Monday *Reveille* Started New and Unknown Life for Aggies," Jan. 27, 1942, p. 1. See also A&M College *Bulletin,* Jan. 1, 1942.

53. *Battalion* articles: "Hello Navy—We're Glad You're Here," Mar. 31, 1942; "Marines Land, Enjoy Mess Hall; Puzzled by Uniforms, Customs," Apr. 28, 1942: "Navy Blue Blend with Army and Marine Corps Khaki on Campus as College Trains Seamen," June 2, 1942; "Cooperation with Navy," June 11, 1942; "War Time Log of A&M," Mar. 7, 1942, p. 4. See also U.S. Congress, "Detailing Personnel of the Army of the United States to Educational Institutions, Etc.," 77th Cong., 2nd sess., Report No. 1704, Jan. 28, 1942; and Felix R. McKnight, "A. & M. College Gets Its First Navy Students," *Dallas Morning News,* Apr. 1, 1942.

54. *Battalion* articles: "Quartermaster Corps Training Will Be Inaugurated," Feb. 19, 1942, p.1;

"Ordnance Unit Started Here in March, 1942," Mar. 3, 1942, p. 1; and "A&M Only School with All Branches of ROTC Courses," Feb. 28, 1942, p. 1; "Our Country Needs Trained Men," *Bulletin of the Agricultural and Mechanical College,* Feb. 1, 1942, n.p.

55. "Training Defense Experts," *Dallas Morning News,* Mar. 14, 1942, p. 11. See also "War Theater Students Learn the Ropes at A.& M.," *Houston Post,* Nov. 11, 1942, n.p.

56. Col. Tom Dooley '35, telephone interview, Oct. 21, 1991; "Flash—Corregidor," *Texas Aggie,* Apr. 22, 1942; "The Last Stand," *Time,* Jan. 12, 1942; John C. Adams, ed. *Voices of a Proud Tradition;* Muster speech by Tom Dooley '35, College Station, Apr. 21, 1978; *Houston Press,* Apr. 22, 1942; "Historic Meeting Held by Texas Aggies on Corregidor," *Bryan Eagle,* Nov. 11, 1942, n.p.; John A. Adams, Jr., *Softly Call the Muster,* pp. 17–33. See also Jonathan M. Wainwright, *General Wainwright's Story,* ed. Robert Considine (Garden City: Doubleday, 1946), pp. 1–306, and E. M. Flanagan, *Corregidor: The Rock Force Assault* (Novato, Calif.: Presidio Press, 1988), pp. 1–77. Ironically, General Moore, after surviving captivity, was reassigned to the "Rock" for a fourth tour of duty in 1947–48.

57. General MacArthur quote in John A. Adams, Jr., *Softly Call the Muster,* pp. 29–30.

58. "Raiders Used 20 Cent Bombsight at Tokio [*sic*] to Protect Norden Sight," n.d., n.p., clipping seen in World War II Documents, Texas A&M Archives; Ted W. Lawson, *Thirty Seconds over Tokyo,* pp. 44–52. See also Carroll V. Glines, *The Doolittle Raid.*

59. Sidney L. James, "Torpedo Squadron 8," *Life,* Aug. 31, 1942, pp. 70–80; "Downed Flier Gives Account of Sea Battle," unknown paper, Westchester, Conn., June 9, 1942, in the Texas A&M Archives; George H. Gay, *Sole Survivor: A Personal Story about the Battle of Midway.*

60. "Aggie Heroes Get Ovation at College Station," July 11, 1942, clipping seen in World War II files, Texas A&M Archives; Melinda Rice, "MacArthur: A&M Wrote History with Blood," *Bryan Eagle,* June 6, 1994. See also World War II clipping file with pictures of Gay and Hilger in Kyle Field, Texas A&M Archives.

61. Perry Luth '42, interview, San Antonio, Oct. 13, 1998.

62. *Battalion* articles: "Corps Faces Shortage of Collar Brass," July 30, 1942, p. 1; "What the Well Dressed Aggie Will Not Wear This Summer—Cuffs, Sports Shirts, Rubber Shoes," June 6, 1942, p. 2; and "Leftover Sugar at Breakfast Is Used in Lunch Time Tea," June 6, 1942, p. 2.

63. Minutes of the Board of Directors, Mar. 13, 1943, Vol. 6, p. 208; "Corps Trip Canceled Because of Transportation," *Battalion,* Oct. 22, 1942, p. 1; Col. Maurice D. Welty, "1364 A.& M. Students to Receive Army Reserve Commissions Upon Graduation," *Houston Post,* Nov. 11, 1941. See also "Former Students of A. & M. Decorated for Deeds of Courage in World War II," *Houston Post,* Nov. 11, 1942.

64. "Hollywood Will Bring Aggieland to Screen," *Battalion,* July 12, 1942; "Filming Will Supersede Classes If Weather Permits," *Battalion,* Dec. 5, 1942. See also Lisa Rush, "A. & M. at War: How World War II Affected the Campus of the Agricultural and Mechanical College of Texas," manuscript, College Station, circa 1993, pp. 18–20, in Texas A&M Archives.

65. *Battalion* articles: Knox quoted in "Hollywood Will Bring Aggieland to Screen," July 12, 1942; "Synopsis of 'We've Never Been Licked' Reveals Unusual Story," Dec. 3, 1942, p. 1; "Boston Herald Razzes Universal—Aggie Film as an Injustice to Texas A&M College," Dec. 3, 1942, p. 1; "NYT Reporter Criticizes WNBL," Aug. 14, 1943; and "Comments on We've Never Been Licked Continue," Aug. 14, 1943; William A. McKenzie '44, telephone interview, Aug. 10, 1999. Note: The original title for the movie was "American Youth Has Never Been Licked"; it was changed for the final release.

66. "Enlistment in Reserves Is Required for Contracts," *Battalion,* May 2, 1942, p. 1; "Corps of Cadets Organized into Seven Regiments," *Bryan Eagle,* Nov. 11, 1942; A&M College of Texas, *Annual Report of the Agricultural and Mechanical College of Texas for the Fiscal Year 1941–1942,* College Station, Nov. 15, 1942, p. 7; "ASTP Reservists Program Is Announced by War Department for Boys of 17–18," *Battalion,* June 29, 1943, p. 1; Dethloff, *Centennial History,* pp. 457–59.

67. Minutes of the Board of Directors, Mar. 13, 1943, Vol. 6, p. 3; "1,306 ROTC Students at Texas A. and M. Will Be Activated," Mar. 18, 1943, clipping seen in World War II Papers, Texas A&M Archives; *Battalion* articles: "Ex-Students Offer College Plan for War," Nov. 5, 1942, p. 1; "Enlisted Reserve Corps Are Subject to Call," Sept. 12, 1942; "Statement by Walton, Welty Squelches Rumors," Dec. 3, 1942, p. 1; and "Juniors Called to Active Duty Here," Dec. 31, 1942. See also State of Texas, "A. & M. College—Military Record," *General and Special Laws of the State of Texas,* 78th Leg., reg. sess., Austin, 1944, p. 1063. See also *Congressional Record,* 77th Cong., 2nd sess., vol. 88, part 10, pp. A3784–85, and Dethloff, *Centennial History,* pp. 456–58.

68. T. O. Walton to the Parents of A. & M. Students, with attachment "Status of A. & M. Students under Army-Navy Plans," College Station, Jan. 16, 1943, and T. O. Walton to the Parents of Students in the A. &. M College of Texas, College Station, Apr. 5, 1943, in the T. O. Walton Papers, Texas A&M Archives.

69. Texas A&M, *Annual Report,* 1942–43, pp. 7–8; "Contract Men to Be Sent to Induction Centers for Activation When Specialization Training Begins," *Battalion,* Mar. 6, 1943; "Contract Men Receive Orders," *Battalion,* Mar. 18, 1943; Mike Haikin, "Texas A & M Transforms Self into Military Camp; 5000 Enrollees in Service," *Houston Post,* June 13, 1943; Minutes of the Board of Directors, May 21, 1943, Vol. 6, p. 3; *Longhorn,* 1946, pp. 49–50.

70. George "Ray" Alderman '47, J. Duff Pitcock '47, and Robert W. "Bob" Coffins '47, interview, College Station, Sept. 11, 1998.

71. Duke Hobbs '47, letter to author, Dec. 13, 1999. See also Forsyth, *Aggies and the 'Horns,* p. 81.

72. "Summer Enrollment Reaches 1,655," *Battalion,* June 3, 1943, p. 1; Texas A&M, *Annual Report,* 1942–43, Mar. 1, 1945, p. 5; "Total Enrollment for the Semester Goes over 2,000," *Battalion,* Sept. 30, 1943, p. 1; Texas A&M, *Bulletin,* May 1, 1943, pp. 40–43.

73. Fernando F. Zuniga '47, interview, Aug. 10, 1999, Laredo, Tex.

74. "Band Membership Available to All Branches," *Battalion,* Jan. 19, 1943, p. 1; "The Texas Aggie Band . . . ," *Battalion,* Oct. 14, 1943, p. 1; John Forsyth, *Aggies and the 'Horns,* p. 81; *Longhorn,* 1946, pp. 49–50.

75. Minutes of the Board of Directors, Aug. 7–8, 1943, Vol. 6, pp. 228–29, and Oct. 9, 1943, p. 8; Dethloff, *Centennial History,* pp. 476–83; *Houston Post,* Aug. 10, 19, 1943; *Dallas Morning News,* Aug. 10, 13, 1943; *San Antonio Express,* Aug. 10, 1943.

76. *Dallas Morning News,* Aug. 13, 14, 19, 1943, and Apr. 28, 30, 1944; *Houston Post,* Apr. 30, 1944. See also Dethloff, *Centennial History,* pp. 476–83.

77. Dethloff, *Centennial History,* p. 481.

78. "When Country Called, A. & M. Was Ready," *Houston Press,* Apr. 15, 1946.

79. *Battalion* articles: "Deferment Cancellation Affects Majority at A&M," Apr. 11, 1944, p. 1; "Total Registration Reaches 1665 Mark," June 15, 1944, p. 1; "Enrollment Climbs to 2073; 779 New," Oct. 6, 1944; "Cadet Program Curtailed to Meet Quotas for AAF," Apr. 4, 1944; and "All on Campus Can Benefit from Improved Hitch Hiking Manners," May 9, 1944; Texas A&M *Annual Report,* 1943–1944, Mar. 1, 1945, pp. 7, 12.

80. "Protest by 90 Per Cent of A.M. Cadets Reported," *Bryan Eagle,* May 12, 1944; "Cadet Corps Rises in Protest to Executive Order," *Battalion,* May 13, 1944, p. 1; "When the Yanks March in . . . ," *Battalion,* Sept. 12, 1944; Minutes of the Board of Directors, Oct. 13, 1944, p. 5.

81. E. E. McQuillen to Chairman, Apr. 21 Aggie Muster, n.d., 1944, Muster File, Association of Former Students, Clayton Williams Center, College Station. See also Adams, *Softly Call the Muster,* pp. 35–37.

82. Wick Fowler, "Aggie-Exes Rouse Jerries with War Hymn While Ack-acks Chunk Steel over Naples," *Dallas Morning News,* circa Apr. 21–22, 1944, in the World War II Papers, Texas A&M Archives.

83. Jim Carroll, "An Aggie May Die But He Never Leaves the Ranks," *Houston Press,* Apr. 16, 1946; "29 Generals from A. & M. in World War II," *Houston Press,* Apr. 15, 1946; "Joe Routt Will Become Aggie Legend," *Houston Post,* Dec. 12, 1944; "Beyond the Call," *San Antonio Express,*

Nov. 5, 1950; Stephen E. Ambrose, "Pointe du Hoc," *The Victors,* pp. 143–57; U.S. Department of the Army, *European Theater of Operations: Cross-Channel Attack,* pp. 322–24; Evans and McElroy, *The Twelfth Man,* pp. 133–36. See also detailed presentation on Aggies in World War II in Dethloff, *Centennial History,* pp. 450–75, 643–48; *Texas Aggie* articles: "Six Aggies Among 'Lost Battalion' POWs Who Built Death Railway," Nov., 1996, p. 21; "Reunion Brings Walker Memories of World War II," Aug., 1992, p. 32; and "Hiram Broiles Helped Chart Course to Allies' Victory," Nov., 1992, pp. 8–9; "First Statue on Quad Honors Hollingsworth '40," *First Call,* fall, 1999, p. 5; Cornelius Ryan, *The Last Battle,* pp. 306–309; Texas A&M Foundation, "Beaumont's Heroes," *Spirit,* fall, 1999, pp. 6–9.

84. U.S. Senate, *Medal of Honor Recipients 1863–1973,* 93rd Cong., 1st sess., no. 15, Oct. 22, 1973, pp. 516, 557, 575, 584, 593, 609, 717; "These Are the Brave," *Houston Press,* Apr. 19, 1946; Richard Newcomb, *Iwo Jima,* pp. 222–23; Dethloff, *Centennial History,* pp. 463–70; David Nunnelee, "Hero Says He Just Did His Job," *Bryan Eagle,* Nov. 10, 1984, p. 2; "Honor Medal Winner Eli Whiteley Dies at Age 72," *Texas Aggie,* Mar., 1987.

85. SHAFE Headquarters, Confidential, "Unconditional Surrender," May 7, 1945, in the Gen. Roderick R. Allen '15, CG Twelfth Army Division, Bio File, Texas A&M Archives.

86. Wick Fowler, Austin, Tex., to Earl Rudder, College Station, Jan. 27, 1967, World War II Papers, Texas A&M Archives.

87. "Aggie Heros Freed from Japs," *Battalion,* Sept. 6, 1945, p. 1; Dethloff, *Centennial History,* p. 475.

88. Visit of President Franklin Roosevelt, Kyle Field, Texas A&M, May 11, 1937 (film), in Texas A&M Archives.

89. Speech by General of the Army Dwight D. Eisenhower, Victory Homecoming and Muster, Apr. 21, 1946, A&M College of Texas, College Station, and "Victory Homecoming: Texas A&M College Program, Apr. 19–21, 1946, in Muster Papers, Texas A&M Archives; "Eisenhower to Become an Aggie, as A. and M. Muster 'Ends War,'" *Houston Press,* Apr. 21, 1946; Gen.

William Becker '41, telephone interview, Oct. 5, 1998. See also Adams, *Softly Call the Muster,* pp. 45–50.

CHAPTER 7

1. Dethloff, *Centennial History,* pp. 490–93; John E. Orr '49, telephone interview, Nov. 4, 1998. It was estimated that 85 percent of the returning veterans to Texas A&M had at some time been a member of the A&M Corps of Cadets.

2. Minutes of the Board of Directors, May 13, 1944, p. 3, and May 25, 1944, p. 5.

3. Texas A&M, "A Report of the Sub-Committee on Student Life," Aug. 31, 1944, pp. 1–5, and "A Report of the Sub-Committee on Growth of the College," Aug. 1, 1944, pp. 4–7, both in *The A. and M. College of Texas Committee on Postwar Planning and Policy,* College Station, Aug., 1944; Dethloff, *Centennial History,* p. 478.

4. Minutes of the Board of Directors, Sept. 7, 1945, p. 3; "Newly Created Dean of Men Is Charged with Administration of Directors' Student Life Policy," *Battalion,* Sept. 27, 1945, p. 7.

5. Minutes of the Board of Directors, May 11, 1944, p. 3, Sept. 7, 1945, p. 3, and Sept. 27, 1946, Vol. 8, p. 98, and Nov. 17, 1946, Vol. 8, p. 118; Texas A&M, "Sub-committee on Student Life," *Postwar Planning and Policy,* Aug. 31, 1944, pp. 1–6.

6. *Longhorn,* 1946, pp. 220, 290. Enrollment on Oct. 4, 1945, was 2,572 students.

7. "Varner, '40, Named Extension Director," *Texas Aggie,* May 15, 1952, p. 2; Minutes of the Board of Directors, Sept. 7, 1945, p. 3; Texas A&M, "Sub-committee on Student Life," *Postwar Planning and Policy,* Aug. 31, 1944, p. 6. During World War II Varner rose from second lieutenant to lieutenant colonel in fifty-four weeks and in the early 1970s became president of the University of Nebraska.

8. "Col. R. J. Dunn, Author of the 'Spirit of Aggieland,' Retires Feb. 1," *Battalion,* Jan. 10, 1946, p. 1; "Precision Marching and Military Airs Spread Fame of A&M Band, College," *Texas Aggie,* Oct. 28, 1950.

9. Texas A&M, "President's Report to the Board of Directors, May 23, 1945," in the Texas A&M

Archives; Texas A&M, *Bulletin,* Feb. 1, 1946, pp. 6–11; *Battalion* articles: "Rusty Now in Training for Appearance as Aggie Mascot," July 3, 1946, p. 1; "Record Enrollment: 9200 Students—Vets 3 to One to Corps," Sept. 12, 1946; "Promotions Out; Ed Brandt Cadet Colonel; Self [Allen Self '47] Exec."; and "F [I] Company Is First Winner of Moore Flag," May 13, 1946, p. 1; Col. G. S. Meloy to Assistant to the President E. L. Angell, College Station, Dec. 30, 1946, Ross Volunteer Papers, Texas A&M Archives; *Longhorn,* 1947, pp. 10, 181–82, 235, 389; *Longhorn,* 1946, p. 177; Robert B. MacCullum '47, telephone interview, July 24, 2000.

10. "Advanced ROTC Training to Be Resumed in February," *Battalion,* Dec. 13, 1945, p. 1. See also President Earl Rudder, memo draft to Steven Ailes, Under Secretary of the Army, College Station, circa 1962, pp. 1–4, James Earl Rudder Papers, Texas A&M Archives.

11. "Requirement for Advanced Military Contracts Listed by Col. Meloy," *Battalion,* Aug. 1, 1946, p. 1.

12. Texas A&M, "President's Report to the Board, May 23, 1945," Texas A&M Archives, pp. 7–8; Texas A&M, *Bulletin,* Feb. 1, 1946, pp. 6–11; F. C. Bolton to Bureau of Naval Personnel, Washington, D.C., June 21, 1949, and W. A. Leny, Captain USN, to President F. C. Bolton, July 7, 1949, President's Papers.

13. Texas A&M, "President's Report to the Board, May 23, 1945," Texas A&M Archives.

14. Minutes of the Board of Directors, June 26–27, 1947, pp. 1–2; Perry, *The Story of Texas A and M,* pp. 101–102; Gen. William Becker '41, telephone interview, Oct. 5, 1998. The old Bryan AFB has been known as the "Riverside Campus" since 1998.

15. Minutes of the Board of Directors, Sept. 27, 1946, p. 98, Nov. 17, 1946, p. 118, and Jan. 11, 1947, pp. 1–2; *Battalion,* Jan. 21, 1947; *Houston Press,* Jan. 29, 1947; *Bryan Eagle,* Jan. 29, 1947.

16. Robert "Sack" Spoede '48, telephone interview, Oct. 29, 1998, and Newton V. Cole '48, telephone interview, Dec. 31, 1998.

17. *Longhorn,* 1947, p. 310; *Dallas Morning News,* Jan. 30, 1947; *Waco Times-Herald,* Jan. 29, 1947; Minutes of the Board, Mar. 7, 28–29, 1947, pp.

1–3; Gibb Gilchrist, "Gibb Gilchrist: An Autobiography," manuscript, 1991, p. 74, Texas A&M Archives; John E. Orr '49, telephone interview, Nov. 4, 1998. See also Perry, *The Story of Texas A and M,* pp. 102–103.

18. *Dallas Morning News,* Jan. 30, 1947.

19. *Waco Times-Herald,* Feb. 1, 1947; *Bryan Eagle,* Feb. 1, 1947; *Dallas Morning News,* Jan. 31, 1947.

20. Minutes of the Board, Feb. 21 and Mar. 28–29, 1947; *Houston Chronicle,* Feb. 1 and 2, 1947; *Dallas Morning News,* Jan. 31, 1947, and Feb. 1 and 2, 1947; *San Antonio Light,* Jan. 30, 1947; *Houston Post,* Jan. 31, 1947; Newton V. Cole '48, telephone interview, Dec. 31, 1998, and Thomas R. Parsons '49, telephone interview, Feb. 5, 1999.

21. Dethloff, *Centennial History,* p. 495.

22. Minutes of the Board, Feb. 21 and Mar. 28–29, 1947; Dethloff, *Centennial History,* p. 495; *Fort Worth Star-Telegram,* Feb. 9, 11, 1947; *Bryan Eagle,* Feb. 9, 1947.

23. *Houston Chronicle,* Mar. 25, 1947; *Dallas Morning News,* Mar. 25, 1947.

24. Gilchrist, *Autobiography,* p. 74.

25. *Dallas Morning News,* Mar. 27, 1947; *Bryan Eagle,* Mar. 26, 28, 1947; *Longhorn,* 1947, p. 311.

26. *Waco Times-Herald,* Apr. 8, 1947; *Houston Chronicle,* Apr. 10, 11, 15, and May 14, 1947; *Bryan Eagle,* Apr. 8, 10, 22, and May 15, 1947.

27. "'No Basis for Charges,' Committee Reports," *Bryan Eagle,* June 7, 1947, p. 1; *Dallas Morning News,* May 12 and June 7, 1947; *Houston Chronicle,* May 12, 1947.

28. Robert "Sack" Spoede '48, telephone interview, Oct. 29, 1998.

29. T. R. Fehrenbach, *This Kind of War: A Study in Unpreparedness,* pp. 134–39; "Meloy Awarded DSC for Heroism," *Battalion,* Sept. 26, 1950, p. 1; "Maj. Gen. Meloy to Lead 4th Army," *Battalion,* Sept. 11, 1958, p. 1. See also Clay Blair, *The Forgotten War: America in Korea, 1950–1953,* pp. 127–31; Dean R. Heaton, *Four Stars: The Superstars of United States Military History,* p. 95.

30. Fehrenbach, *This Kind of War,* pp. 509–10.

31. Barbara W. Tuchman, *Stilwell and the American Experience in China—1911–45,* pp. 248, 328, 393; "Col. Boatner, New Commandant at College Station, Veteran of Much Action in C-B-I

Theatre," *Texas A&M System News,* Aug., 1948, p. 1; *Longhorn,* 1949, pp. 7–9, 23, 268; Donald G. Kasper '49, telephone interview, Nov. 4, 1998, and John E. Orr '49, telephone interview, Nov. 4, 1998. See also Leslie Anders, *The Ledo Road,* pp. 37, 125.

32. Dean Reed '52, telephone interview, Nov. 15, 1998; Richard Ingels '52, telephone interview, Dec. 13, 1998. See also, W. Pat Kerr, "As Good As It Gets," manuscript, College Station, 1998, pp. 1–11.

33. Col. Haydon L. Boatner, "Gentlemen of the Class of '52" and "Manual for Cadet Officers and Non-Commissioned Officers, A&M College 1948–49," in Corps of Cadets Information, 1948–49, President's Office Papers, Texas A&M Archives (hereafter cited as POP).

34. Maj. Gen. A. D. Bruce ['16], HQ Fourth Army, Fort Sam Houston, to Dr. F. C. Bolton, president, Aug. 1, 1949, and Maj. Gen. A. D. Bruce to Dr. Marion T. Harrington, president, Oct. 15, 1949, both in POP.

35. "Appointments and Assignments of Cadet Officers," Office of the Commandant, General Order No. 2, Aug. 20, 1948, Texas A&M Archives; H. L. Boatner, Colonel Infantry, to All Instructors, Adm. Memo No. 4, Aug. 30, 1948, College Station, Corps of Cadets Papers, Texas A&M Archives; Perry, *The Story of Texas A and M,* pp. 135–37. Total campus enrollment was 7,238, including veterans and fifth-year seniors.

36. Headquarters 4519 ASU ROTC, Texas A&M, "Distinguished Military Students," Adm. Memo No. 13, Sept. 22, 1948, and "To All Instructors," Adm. Memo No. 61, Mar. 16, 1949, both in Texas A&M Archives. To be eligible for DMS a cadet needed a grade point ratio (GPR) of 1.5 (on 3.0 scale) or better and be in the upper 30 percent of his class in military science.

37. Congressman Olin E. Teague to Dr. F. C. Bolton, College Station, June 25, 1948, and "Implementation of Selective Service Act of 1948 in Relation to ROTC," in Selective Service Act 1948, POP; "Advanced Contract Quotas, 1948–1949," Headquarters 4519 ASU ROTC, A&M College of Texas, Admin. Memo No. 8, Sept. 10, 1948, POP. In addition to the U.S. Air Force, the U.S. Army offered branch training in infantry, coast artillery corps, field artillery, armored cavalry, engineer corps, signal corps, chemical corps, quartermaster corps, ordnance, transportation corps, and Army Security Agency (military police). See also T. O. Walton to Col. O. E. Teague, M.C., May 6, 1946; Travis B. Bryan to Colonel Teague, Jan. 18, 1947, and President F. C. Bolton to Hon. Olin E. Teague, Nov. 15, 1948, all in Olin E. "Tiger" Teague Papers, Texas A&M Archives.

38. Travis Bryan to Col. Olin E. Teague, M.C., Mar. 21, 1949, Teague Papers.

39. "A&M Not Opposed to National Guard at Bryan Field," *Bryan News,* Mar. 18, 1949, p. 1; Travis Bryan to Olin Teague, M.C., Jan. 8, 1949, Teague Papers; The White House, Executive Order, "Organization of the Reserve Units of the Armed Forces," Washington, D.C., Oct. 15, 1948. See also Edward T. Miller, M.C., "The Role of the Reserve Officer," *Congressional Record,* May 13, 1948, app., pp. A3148–49.

40. Travis Bryan to Harold E. Talbott, Secretary of the Air Force, Apr. 8, 1954, and M. T. Harrington, chancellor, to Harold Talbott, Apr. 8, 1954, both in Teague Papers; Col. John A. Adams, interview, Norcross, Ga., Nov. 22, 1998. See also U.S. House of Representatives, Committee on Armed Services, "Full Committee Hearings on H.R. 5337, to Provide for the Establishment of a United States Air Force Academy," Jan. 13, 1954, 83rd Cong., 2nd sess., pp. 2969–3068.

41. See "Letters to the Editor," *Battalion,* Apr. 6, 7, 11, 1949, concerning reprimand for mix-up in Corps review ceremony; H. L. Boatner to Corps of Cadets, military review, Nov. 17, 1949, POP; Dean Reed ['52], "Former PMS&T Now in Hot Seat," *Battalion,* May 27, 1952, p. 2; Dean Reed '52, interview, Loudoun Heights, Va., Nov. 15, 1998.

42. Col. Haydon Boatner, "The Instructional Policy of the School of Military Science"; "Correlation of Discipline with Counseling and Guidance"; "Technique of Instruction"; "Christmas Greetings," 1948; "Cadet Officer and Noncommissioned Officer Orientation Course 1950–51"; and "Wearing of the Uniform," Nov. 14, 1950, POP; John E. Orr '49, interview, Fort Worth, Tex., Nov. 4, 1998.

43. "Report of Discussion on 'Aggie Social Customs and Courtesies,'" Jan. 11, 1949, pp. 1–3, POP; "Top Generals Visit A. & M. to See Cadet Corps Review and Attend Military Ball," *Texas A. & M. System News,* Mar., 1949, p. 5; Perry, *The Story of Texas A and M,* p. 141; Don Kasper '49, interview, Shiner, Tex., Nov. 6, 1998. See also YMCA, *The Freshman Handbook 1949–50,* pp. 1–52.

44. Col. H. L. Boatner, Western Union Telegram to Stanley J. Baker, Sept. 25, 1950, POP; Col. H. L. Boatner to President M. T. Harrington, Apr. 24, 1951, POP; Col. Donald L. Burton, interview, Temple, Tex., Oct. 25, 1998.

45. Col. H. L. Boatner, PMS&T, to President M. T. Harrington, Feb. 19 and 20, 1951, POP.

46. Boatner, "The Instructional Policy of the School of Military Science," 1948, Texas A&M Archives; "United States Army Salutes A. & M. Student Body, Cadet Corps," *Texas A. & M. System News,* May, 1949, p. 7; See also "Leadership," *Manual for Cadet Officers and Non-Commissioned Officers,* Texas A&M, 1948–49, pp. 3–4, POP.

47. Joe Fenton '58, "In the Beginning: Fish Days and Nights," manuscript, SHS Corps Center.

48. Col. H. L. Boatner to M. T. Harrington, Feb. 12, 1951, POP. See also William H. Harris, Adjutant General, HQ Fourth Army, to M. T. Harrington, Sept. 25, 1950; Boatner, "Wearing of the Uniform," College Station, Nov. 14, 1950; Harrington to Boatner, Jan. 4, 1951, POP; Boatner to the Athletic Battalion, Mar. 21, 1951, POP; Maj. Gen. Tom Darling '54, interview, Sept. 8, 1998, College Station. See also *Aggieland,* 1952, p. 32.

49. "Former Commandant of A&M Sends Letter from Korea Post," *Battalion,* Oct. 9, 1951, p. 1; Fehrenbach, *This Kind of War,* p. 515.

50. "New 'Stern' Policy Is Instituted," *Houston Chronicle,* May 19, 1952; "'Bull' Boatner to Stomp in Texas," *Houston Chronicle,* Sept. 2, 1952; "Soft Measures Out, Says Gen. Boatner," *Dallas Morning News,* May 16, 1952; Fehrenbach, *This Kind of War,* pp. 509–10, 582–84; Arned L. Hinshaw, *Heartbreak Ridge: Korea 1951,* pp. 63–64, 96, 103–108, 125–27; U.S. Department of the Army, *Truce Tent and Fighting Front,* pp. 233–67; "'Boat' Saw Koje without Any Prob-

lems," *Battalion,* Oct. 29, 1958, p. 1. After Korea Boatner was promoted to major general in command of the Third Infantry Division at Fort Benning.

51. Fehrenbach, *This Kind of War,* pp. 3–6; *Texas Aggie* articles: "We Shall Never Forget Their Deeds," Nov., 1953, p. 2; "Lt. Helm Killed in Korean War," Nov., 1953, p. 8; and "Posthumous Hero Is Awarded DSC," Dec., 1953, p. 1; David McCullough, *Truman,* p. 790. Robert H. Moore '50 was credited as being the ninth air force ace in Korea on April 1, 1952.

52. "A&M Could Field 40,000–Man Army," *Dallas Morning News,* Aug. 29, 1951; David Halberstam, *The Reckoning,* pp. 224–35, and *The Fifties,* pp. 116–87.

53. Halberstam, *The Fifties,* p. 28; McCullough, *Truman,* pp. 474, 607. See also *New York Times,* Sept. 7, 1945, and Samuel P. Huntington, *The Soldier and the State,* pp. 282–88, 354–73. See also U.S. House of Representatives, Committee on Armed Services, *Reserve Components,* "Statement in Behalf of the Association of Land-Grant Colleges and Universities," 82nd Cong., 1st sess., Jan. 15, 1951, pp. 182–84.

54. "College Celebrates Anniversary Date," *Texas A&M System News,* Nov., 1950, p. 2; U.S. House of Representatives, Committee on Armed Services, "Universal Military Training," H.R. 1752, 82nd Cong., 1st sess., Jan. 23—Mar. 8, 1951, pp. 533–35, 641–59. See also Dethloff, *Centennial History,* pp. 528–40; James B. Kelly '52, interview, Houston, Feb. 6, 1991.

55. "General Bradley Credits A&M as Outstanding Military School," *Texas A&M System News,* July, 1950, p. 3; "450 A&M Cadets Get Commissions," *Texas A&M System News,* July, 1950, p. 4; "M. T. Harrington Inaugurated as President of A&M College," *Texas A&M System News,* Dec., 1950, p. 1; "A&M's 75th Anniversary Day Draws Estimated 14,000 Crowd," *Battalion,* Oct. 4, 1950, p. 1.

56. *Texas A&M System News* articles: "New Memorial Student Center Holds 3-Day Informal Opening," Oct., 1950, p. 1; "Dreams Come True," Oct., 1948, p. 2; and "A&M's War Dead Honored in San Jacinto Day Ceremony," May, 1951, p. 1; "New A and M Center Cost $1,700,000," *Houston Post,* Sept. 20, 1950; Dani

Presswood, "MSC Hot Spot of Controversy Since 1950 Birth," *Battalion,* Jan. 11, 1966, p. 1; "Teague Eulogizes Classmate Rudder," *Battalion,* Nov. 27, 1973, p. 1.

57. "Davis Resumes Asst. Comm. Duties," *Battalion,* Dec. 12, 1946; "Col. Joe E. Davis to Be Commandant at A.& M. College," *Texas A&M System News,* Aug., 1951, p. 2.

58. Boatner to Harrington, Feb. 12, 1951, POP.

59. Boatner to Harrington, Jan. 15, 1951, POP; Col. H. P. Bonnewitz, Air Adjutant General, to Professor of Air Science and Tactics, AFROTC, Jan. 23, 1951, POP; Harrington to Lt. Gen. LeRoy Lutes, CG, Fourth Army, Mar. 29, 1951, POP; *Aggieland,* 1953, p. 474.

60. Harrington to Lt. Gen. LeRoy Lutes, Mar. 29, 1951, POP; Lutes to Harrington, Apr. 9, 1951, POP; Harrington to Davis, May 7, 1951, POP; Harrington to the Adjutant General of the Army, Washington, D.C., May 15, 1951, POP. No announcement of the Davis appointment as commandant was made until a new Army PMS&T could be identified.

61. "Major Davis Leaves for Staff, Command School," *Battalion,* June 13, 1944; "Davis Resumes Asst. Comm. Duties," *Battalion,* Dec. 12, 1946; "Col. Joe E. Davis to Be Commandant at A. & M. College," *Texas A&M System News,* Aug., 1951, p. 2; "Commandants at A&M," *Texas A&M System News,* Sept., 1955, p. 8.

62. "Teague Battles for A-M Support," *Bryan Eagle,* Jan. 1, 1954; House Committee on Armed Services, "Universal Military Training," 82nd Cong., 1st sess., H.R. 1752, Jan. 23–Mar. 8, 1951; Hon. Olin E. Teague, "Universal Military Training," *Congressional Record,* June 7, 1951, in Teague Papers; M. T. Harrington, chancellor, to Harold E. Talbott, Secretary of the U.S. Air Force, Apr. 8, 1954, POP; *Aggieland,* 1953, pp. 6–7. See also H.R. 1168 and 1775, cited as the "Reserve Officers' Training Corps Act of 1951," 82nd Cong., 1st sess., Jan. 9 and 19, 1951.

63. *Battalion* articles: Joel Austin, "Commanders Want Action on Discipline Problems," Oct. 28, 1952, p. 1; Joe Hipp, "Hazing Approved in Student Poll," Oct. 29, 1952 p. 1; and "Specialized ROTC Continues—Myers," Nov. 2, 1952, p. 1.

64. David Morgan to John A. Hannah, assistant secretary of defense, Washington, D.C., Apr. 19, 1954, POP; Gen. Mark Clark, president, The Citadel, to David Morgan, Apr. 27, 1954, POP; Olin E. Teague to Dr. David H. Morgan, July 12, 1954, POP; U.S. House Committee on Armed Services, Subcommittee No. 3, ROTC Programs in Military Colleges, Hearing, 83rd Cong., 2nd sess., Jan. 18, 1954, pp. 3075–96. Morgan was in active contact with Norwich, The Citadel, VMI, VPI, North Georgia College, Clemson, New Mexico Military Institute, and Pennsylvania Military College.

65. "Most Aggies Will Serve Army 2 Years," *Bryan Eagle,* Apr. 4, 1956; David H. Morgan, "Meeting of Corps of Cadets with president of A & M College," College Station, Mar. 8, 1956; "What Is the College Attitude toward Hazing?" *Battalion,* May 14, 1954, p. 2. Note: Bennie Zinn, assistant dean of men, reported that between Sept., 1950, and May, 1954, 130 students had been suspended (of this total 31 were "indefinitely suspended for hazing").

66. "A Proposed Program for the Military Colleges," n.d., seen in Box 19–14, POP.

67. Departments of the Army, "Enrollment of Noncitizen of the United States in ROTC," Memo 145–10–44, Washington, D.C., Aug. 10, 1948.

68. "Resolution: Regulation Concerning Enrollment of Non-Citizens in the ROTC," Minutes of the Board, July 8, 1949; F. C Bolton, president, to Gibb Gilchrist, Chancellor, June 28, 1949; Gibb Gilchrist to Gen. Omar N. Bradley, chief of staff, U.S. Army, Sept. 11, 1948; Omar N. Bradley to Gibb Gilchrist, Aug. 15, 1949, POP.

69. "Teague Battles for A-M Support," *Bryan Eagle,* Feb. 2, 1954.

70. John D. Kraus, Jr., "The Civilian Military College in the Twentieth Century: Factors Influencing Their Survival," M.A. thesis, University of Iowa, July, 1978, pp. 68–79.

71. Thomas K. Finletter, secretary of the air force, to Gibb Gilchrist, Mar. 27, 1952; Gilchrist to Finletter, Mar. 15, 1952; Col. E. W. Napier to Harrington, Feb. 13, 1952; Harrington to Gilchrist, Feb. 13, 1952, all in POP.

72. "Dr. Morgan Made a Good Impression in

Capital," Mar. 11, 1956; Morgan, *Meeting of the Corps of Cadets,* Mar. 8, 1956, pp. 7–8; Maj. Gen. Hugh M. Milton to Harrington, Jan. 7, 1952; Gilchrist to Finletter, Mar. 15, 1952; Memo of Agreement signed by Frank Pace, Jr., secretary of the army, and Thomas K. Finletter, secretary of the air force, Jan. 6, 1953, all in POP.

73. H. Lee White, secretary of the air force, to Dr. David H. Morgan, president, Feb. 19, 1954, POP.

74. Morgan to White, Mar. 22, 1954, POP; House Committee on Armed Services, "Subcommittee Hearings Pertaining to ROTC Problems in Military Colleges," 83rd Cong., 2nd sess., Jan. 18, 1954, pp. 3075–97.

75. Morgan to White, Mar. 22, 1954, POP.

76. Jim Mousner, "A&M's Cadet Corps Assumes New Look under Reorganization Plan," *Houston Post,* Nov. 18, 1956.

77. Ibid.; David Morgan, "Meeting of Corps of Cadets with President of A&M College," College Station, Mar. 8, 1956, pp. 5–7.

78. David H. Morgan, "Recommendations for Improving Status of Military Colleges," June, 1954; Melvin S. Brooks and John R. Bertrand, "Student Attitudes Toward Aspects of the A&M College of Texas," spring, 1954, pp. 1–21; Col. Joe Davis to Morgan, "Reorganization of the School of Military Science," Apr. 12, 1954, pp. 1–5; "Report of the Special Academic Council Committee on the School of Military Science," June 15, 1954, pp. 1–7, all in POP; *Aggieland,* 1954, pp. 14, 16; Dr. David H. Morgan to Corps Commander Fred H. Mitchell, Dec. 8, 1953, Morgan Papers, POP; Fred Mitchell '54, telephone interview, Oct. 30, 1999.

79. Morgan to Fred Mitchell '54, Dec. 8, 1953, POP.

80. Mousner, "A&M's Cadet Corps Assumes New Plan," *Houston Post,* Nov. 18, 1956; Morgan, "Meeting of Corps of Cadets," p. 8. See also Dethloff, *Centennial History,* pp. 549–51. Morgan's address was printed in booklet format and circulated statewide. Subsequent A&M administrators often refer to this document on matters involving the Corps.

81. "'Mr. A. and M.' Is Biggest Booster College Ever Had," *Houston Chronicle,* Nov. 25, 1952; "Little Man with Big Voice Is A&M's Official Greeter," *Texas A&M System News,* Mar., 1951, p. 2; "Cadet Slouch—An Aggie Who Never Left Despite Changes," *Bryan Eagle,* Aug. 28, 1975; John West, "Slouch Survives A&M Progress," *Texas Aggie,* Sept., 1979; Ray Bowen '58, interview, College Station, Sept. 7, 1998; Joe W. Tindel '58, telephone interview, Dec. 5, 1998.

82. Joe West '54 quoted in "Tech Game Anniversary of First Yell Practice," *Battalion,* Oct. 25, 1996, p. 2.

83. Mickey Herskowitz, "Aggies Double up on Best of Times," *Houston Post,* Mar. 5, 1995.

84. "A Special Place in History: 10 Days of Sacrifice at Junction Made Bear's Boys Famous," *Bryan Eagle,* July 3, 1994; Clifford Broyles, "Aggie Agony: Bryant's 1954 Experiment in Junction," *San Antonio Express-News,* July 31, 1994, pp. C1, 15. See also Steve Pate, *John David Crow: Heart of a Champion,* pp. 1–100. The first live television broadcast of the annual A&M–Texas game was in November, 1956: A&M 34, t.u. 21.

85. "Army, Air Force Prepare for 12th Man Bowl Game," *Bryan Eagle,* Dec. 5, 1954. The Twelfth Man Bowl game first originated in the 1940s, died during the war, and was revived in 1952. Proceeds from ticket sales were put in the student aid fund and the Twelfth Man Scholarship program. In the early 1970s, the rivalry was renamed the Elephant Bowl.

86. Fred Myers, "Texas A. & M. Band Puts on Dazzling Show," *Marching Bands,* Oct., 1955, pp. 14–15; "Texas [A&M] Also Has Country's Largest Marching Band," *The Baton,* fall, 1955, pp. 16–20.

87. Richard G. McPherson, interview, College Station, Sept. 6, 1999.

88. Jon Kinslow, "Ex-Student Killed in Saturday Crash," *Battalion,* Nov. 23, 1954, p. 1; Hugh Cunningham, "Plane Crash Victim Ex-Ag," *Bryan Eagle,* Nov. 22, 1954, p. 1.

89. Joe Tindel '58, "Man to Man," *Battalion,* n.d., seen in Texas A&M Archives. See also Tom Allen, "The Day Jack Pardee Saved a 'Fish,'" *Washington Post,* n.d., n.p., author's papers.

90. Office of the President to Chancellor Harrington (For Submission to the Board of Directors), "Report to the Board of Directors on the Tonkawa Tribe (TT's)," College Station, July 3, 1954, pp. 1–6, in Texas A&M Archives; "Secret

Group Offers Full Expose," *Battalion,* May 18, 1954, p. 1.

91. *Battalion* articles: "TTs Disbanded, Confess to Authorities," Apr. 4, 1952, p. 1; "Kala Kinasis Gave Parties, Ran Corps," May 13, 1953, p. 2; "Secret Group Offers Full Expose," May 18, 1954, p. 1; "A Look at the TTs Record," May 20, 1954, p. 2; and "Texas A&M Has Rich History of Secret Societies," Apr. 25, 1996, p. 2; David Morgan to Fred Hickman, Robert O. Murray, and Lt. Col. Robert L. Melcher, Apr. 9, 1954, POP; Office of Campus Security memo to President Morgan, "Secret Organizations Investigation," June 16, 1954, POP; Richard A. Ingels '52, telephone interview, Dec. 13, 1998; Richard Word Roberson '58, telephone interview, Dec. 22, 1998; *The Mole,* various issues, circa 1957–58, in documents of SHS Corps Center.

92. "Cadets Take Oath," *Bryan Eagle,* Sept. 22, 1954. G. Rollie White Coliseum was completed in 1954 at a cost of $670,000.

93. "Cadets Officers Receive Commissions in Public Ceremony at White Coliseum," *Texas Aggie,* Oct., 1954.

94. "Anti-Hazing Oath Sworn by Cadets Officers," *Bryan Eagle,* Oct. 23, 1955.

95. "Leadership Trained by Responsibility," *Battalion* clipping, circa 1954, in Corps of Cadets Papers, Texas A&M Archives.

96. Dethloff, *Centennial History,* pp. 551–2.

97. Leon Hale, "Main Issue Is Faculty, Board Split," *Houston Post,* Feb. 9, 1958, p. 1, and "Problem of Military Is Ironic One," *Houston Post,* Feb. 10, 1958, p. 1; Dethloff, *Centennial History,* pp. 551–59; *Aggieland,* 1957, p. 14; "Association Council Opposes Coeducation," *Texas Aggie,* Apr., 1958, p. 1; "Association Council Takes Strong Stand Against Co-Eds," *Texas Aggie,* May, 1958, p. 1.

98. *Battalion* articles: "What's Next?" July 3, 1958, p. 2; "Col. Elder Takes Job as New PMS&T," July 17, 1958; and Johnny Johnson, "Drill Set for Saturday Morning: Academic Council Votes Changes," p. 1.

99. *Battalion* articles: Johnny Johnson, "Cadet Corps to Drill Saturdays; No Classes," and "Changes at A&M," Nov. 11, 1958, pp. 1, 2; "Corps Officers Given Orientation," Sept. 4,

1958, p. 1; "Typical Aggie Day Anything But Dull," Aug. 21, 1958, p. 6; "Good Rating Vital," Oct. 10, 1958, p. 2; and Jack Teague, "Bill Heye's Grades High," Oct. 9, 1958, p. 2; *Aggieland,* 1957, p. 14, and 1959, p. 6.

100. *Battalion* articles: "'Fish' Drop Outs," Oct. 14, 1958, p. 1; "Doherty Pledges Aid in Anti-Coed Battle," Sept. 18, 1958, p. 1; and "'New' Friends," editorial, Sept. 16, 1958, p. 2; "What Goes on in A&M's Famous Corps of Cadets," *Texas Aggie,* Nov., 1959, p. 3; *Houston Post* articles: Leon Hale, "One Trouble Is Shortage of Students," Feb. 11, 1958, p. 1; "Tradition Is Top Hurdle for Co-Eds," Feb. 12, 1958; and "'Spirit' Feared Imperiled by Any Change," Feb. 13, 1958, p. 1.

101. Hale, "Problem of Military Is Ironic One," *Houston Post,* Feb. 10, 1958, p. 1. See also Brooks and Bertrand, "Student Attitudes Toward Aspects of the A. & M. College of Texas," College Station, June 10, 1954, pp. 1–21.

102. Hale, "Problem of Military Is Ironic One," *Houston Post,* Feb. 10, 1958, p. 1; Hale, "Spirit Feared Imperiled by Any Change," *Houston Post,* Feb. 13, 1958, p. 1; "What Goes on in A&M's Famous Corps of Cadets," *Texas Aggie,* Nov., 1959, p. 3.

CHAPTER 8

1. "The Cold War: From Containment to Commonwealth," *New York Times,* Feb. 2, 1992, p. 8; David M. Young, "ROTC: Required or Elective?" *Military Review,* Feb., 1962, pp. 21–32; Charles H. Blumenfeld, "The Case for ROTC," *Military Review,* pp. 1–13.

2. "Association Council Opposes Coeducation," *Texas Aggie,* May, 1958, p. 1; "Association Council Takes Strong Stand Against Co-Eds," *Texas Aggie,* May, 1951, p. 1.

3. Fred Meusa, "Corps to Stress Academic Work," *Battalion,* Sept. 1, 1958, p. 1; "No Excuse Now," *Battalion,* Sept. 18, 1958, p. 2; Bob Saile, "Co-Ed A&M Supported in Report," *Houston Post,* Mar. 19, 1962, p. 1; Dethloff, *Centennial History,* pp. 550–54.

4. *Longhorn,* 1931, pp. 89, 280, and 1932, pp. 39, 234, 263; "James Earl Rudder," *Houston*

Chronicle, Mar. 25, 1970, p. 4. *Sports Illustrated* named Rudder to the 1956 Silver Anniversary All-American football team.

5. "A&M's President Upholds Aggie Traditions of Soldier, Statesman, Knightly Gentleman," *Battalion,* Mar. 25, 1960, p. 5; "Army Sophs Study Rudder's Rangers" *Battalion,* Dec. 7, 1962, p. 3; Stephen E. Ambrose, *D-Day—June 6, 1944: The Climactic Battle of World War II,* pp. 398–426; Omar N. Bradley, *A Soldier's Story,* pp. 1–145. See also Rudder Biographical Papers, Texas A&M Archives.

6. Rudder Biographical Papers, Texas A&M Archives; Dethloff, *Centennial History,* pp. 555–57; *Aggieland,* 1959, p. 14; "Aggie of Aggies," *Amarillo Globe-Times,* Mar. 30, 1970, p. 18; "James Earl Rudder," *Houston Chronicle,* Mar. 25, 1970, p. 4. See also display of General Rudder's military decorations and awards at first floor of Rudder Tower on the Texas A&M campus.

7. Ray Bowen '58, interview, College Station, Feb. 26, 1999.

8. Earl Rudder, Speech on Reorganization and Retention, May 20, 1958, Rudder Papers, Texas A&M Archives; Minutes of the Academic Council Meetings, Apr. 29, 1958, Texas A&M Archives; "Muster 1965," *Battalion,* Apr. 21, 1976, p. 1. See also "No Excuse Now," *Battalion,* Sept. 18, 1958, p. 2.

9. Walter H. Bradford, "Aggie Fashion: A Survey of the Texas A&M Cadet Uniform Through the Years," manuscript, College Station, 1975, pp. 9–11, in Texas A&M Archives.

10. Memo, C. M. Taylor to the Commandant, "Suggestions for Improving Conditions in the Corps and in the College," College Station, circa 1959–60, William B. Heye Papers, Texas A&M Archives.

11. "Juniors Get Warning of Challenge in '59," *Battalion,* May 22, 1958, p. 1. See also Jon L. Hagler, cadet colonel, to Members of the Academic Council, "Scholastic Report of the Corps of Cadets," Apr. 8, 1958, Academic Council Papers, Texas A&M Archives.

12. "New Corps Insignia," *Texas Aggie,* Dec., 1960, p. 1, and notes seen in the Heye Papers.

13. Earl Rudder, Talk to Seniors, Juniors, and Sophomores, Sept. 15, 1958, Rudder Papers; Johnny Johnson, "Rudder Appeals to Class of '61 to Ease Tension," *Battalion,* Feb. 25, 1959, p. 1.

14. Don Cloud '59, telephone interview, Dec. 15, 1998.

15. "Girls at the A. and M.," *Houston Daily Post,* June 7, 1899, p. 10; Tommy DeFrank, "Early Coeducation Tries Failed to Materialize," *Battalion,* Feb., 1966, p. 1. This is the first article of an eight-part front-page series, "History of Coeducation," prepared by DeFrank in February, 1966.

16. Ernest Langford, "It's History—Women at A&M," manuscript in Texas A&M Archives, pp. 2–3; Stan Redding, "Texas A&M's First Coed Graduates," *Houston Chronicle Magazine,* Mar. 2, 1975, pp. 5–9; Tommy DeFrank, "All-Male Status Broken as Hutson Twins Enroll," *Battalion,* Feb. 16, 1966, p. 1. While the Hutson twins did not receive degrees, they did participate in the graduation ceremony for the class of 1903 and received certificates of completion in civil engineering as well as letters of recommendation to any prospective employer.

17. Dethloff, *Centennial History,* pp. 556–70; *Aggieland,* 1965, p. 131; DeFrank, "Women Lose 1933 Lawsuit," *Battalion,* Feb. 17, 1966, p. 1.

18. Todd Brief, in Texas A&M Archives; Polly Westbrook, "A History of Coeducation at Texas A&M University," manuscript, Texas A&M Archives, n.d., pp. 1–42; Dugan, *Great Class of '34,* pp. 162–64; DeFrank, "Women Lose 1933 Lawsuit," *Battalion,* Feb. 17, 1966.

19. Polly Westbrook, "A History of Coeducation at Texas A&M University," pp. 13–16; Dethloff, *Centennial History,* pp. 550–52; DeFrank, "Fight Shifts to Legislature," *Battalion,* Feb. 22, 1966, p. 1.

20. Bob Saile, "Co-Ed A&M Supported in Report," *Houston Post,* Mar. 19, 1962, p. 1; James Holley, "A&M Review General Says Corps Still Vital," *Houston Post,* Mar. 13, 1960, p. 2; "No Coeds for Aggieland," *Battalion,* Apr. 7, 1959; DeFrank, "Appellate Courts Reverse 1958 Barron Suit Victory," *Battalion,* Feb. 18, 1966, p. 1; *Aggieland,* 1959, p. 8.

21. *Inauguration of James Earl Rudder as President of the Agricultural and Mechanical College of Texas,* Mar. 26, 1960; Texas A&M, *Report of the*

Century Council, 1962, pp. 1–46; Texas A&M, *Faculty-Staff-Student Study of Aspirations*, 1962, pp. 3–4, all in Texas A&M Archives.

22. Texas A&M, *Study of Aspirations*, pp. 23–27; Saile, "Co-Ed A&M Supported in Report," *Houston Post*, Mar. 19, 1962, p. 1.

23. Texas A&M, *Report of the Century Council*, 1962, pp. 45–46; Gerry Brown, "Thousands Expected at A&M Convocation," *Battalion*, Nov. 15, 1962, p. 1. The Century Council report was prematurely "leaked" to the press on March 18, 1962, much to the irritation of Rudder and the A&M Board of Directors (Texas A&M Office of Public Information press release, "Former Student Once 'Suspended' for Proposing Coed A&M," Apr. 10, 1982).

24. "Wow! We Get More Dorms," *Battalion*, Sept. 13, 1962, p. 8; *Aggieland*, 1959, pp. 9–10.

25. Robert Sherrill, "Texas Athletic & Military," *Time*, Sept. 28, 1962, p. 79.

26. Allan Payne, "Aggies Gigged . . . Literally— Time's Article—How It Happened," and "Time's Story Disappointed Most Students," *Battalion*, Sept. 28, 1962, p. 1; Nicholas C. Chriss, "They're Proud: Texas Aggies Beat Jokers to the Punch," *Los Angeles Times*, May 7, 1977, p. 1. See also Dan Louis, "Best Hits 'Unethical' Doctors in Corps Resignations," *Battalion*, Dec. 7, 1962, p. 1; "Aggie Jokes—Again and Again," *Battalion*, Oct. 29, 1965, p. 4.

27. Thomas R. Hargrove, *A Dragon Lives Forever*, pp. 186–88. See also Frank W. Cox III '65, *I Bleed Maroon*, pp. 1–11. See also Thomas G. Hargrove '66, "We Happy Few Brothers or We Are the Aggies," *Texas Aggie*, May, 1993, pp. 2–7, and Hargrove, "The Class of '66?" *Texas Aggie*, Nov., 1993, pp. 12–15.

28. *Battalion* articles: Gerry Brown, "Thousands Expected at A&M Convocation," Nov. 15, 1962, p. 1; "The Corps vs. Civilians: Where Are We Headed?" Dec. 6, 1962, p. 2; and "Rudder, Seniors Eye Corps Problems," Dec. 19, 1962, p. 6. See also President Earl Rudder, letter to Steven Ailes, Under Secretary of the Army, circa 1962, Rudder Papers, Texas A&M Archives.

29. *Houston Post*, Apr. 28 and 30, 1963; "Sterling C. Evans, President A&M Board of Directors, to Association of Former Students, April 1963,"

Texas Aggie, Apr., 1963, p. 1; "Staying Awake Nights," *Texas Aggie*, Jan., 1993, pp. 6–7; Paul Dresser '64, telephone interview, Dec. 27, 1998.

30. Paul Sherrill, "The Proudest Squares," *Sports Illustrated*, Nov. 18, 1962, p. 102; "Limited Co-education Legality Questioned," *Battalion*, Sept. 16, 1965, p. 1.

31. David L. Morgan, "Beaumont Aggies Erupt into Fiery Co-ed Debate," *Battalion*, Nov. 6, 1963, p. 1; Paul Dresser '64, interview, Dec. 27, 1998. See also Dethloff, *Centennial History*, pp. 568–70.

32. Dugan, *Great Class of '34*, p. 105. See also Texas A&M Office of Public Information Press release, "Former Student Once 'Suspended' for Proposing Coed A&M," Apr. 10, 1982.

33. "Name Change Was Effective 23rd," *Battalion*, Aug. 8 and 23, Sept. 12, 1963, p. 1; "AMC Shoulder Patch Going-Going-Gone," *Texas Aggie*, Dec., 1967, p. 7; Polly Westbrook, "A History of Coeducation at Texas A&M University," pp. 1–21; *Houston Post*, Nov. 23, 1967. Note: Seniors in the class of 1968 were the last to wear the maroon-and-white "AMC" shoulder patch. All others were changed to the "AMU" patch.

34. *Battalion* articles: Steve Wilkes, "Cyclotron Construction Now Well Underway," Sept. 16, 1965, p. 6; "$1.9 Million Space Center Is Being Built," Sept. 16, 1965; and "Air Base Shakeup Revealed," Oct. 2, 1965, p. 1; Sam Kinch, "Texas A&M to Build Giant Atom Smasher," *Fort Worth Star-Telegram*, Feb. 13, 1964, p. 2; Henry C. Dethloff, *Suddenly, Tomorrow Came . . . A History of the Johnson Space Center*, p. 259.

35. *Battalion* articles: "Change," Sept. 12, 1963, p. 1; Dani Presswood, "Liberal Arts College Finds Deserved Status," Oct. 13, 1965, p. 2; "A&M Produces Many Leaders," Aug. 23, 1962, p. 2; and "Story of 'Old Army' Fish Recalls Past," Sept. 10, 1964, p. 2; *Aggieland*, 1965, p. 60; Dethloff, *Centennial History*, pp. 574–75.

36. "Corps Reorganization Plan Gets Fall Semester Test," *Battalion*, Sept. 12, 1963, p. 1; Charles Hornstein '53, interview Feb. 27, 1999, College Station. Three of the former tactical officers were retained as "civilian Corps advisors": Reese, Charles Hornstein '53, and Bill Presnel.

37. "New MS Head Will Also Be Commandant,"

Battalion, Sept. 12, 1963, p. 1; "Baker Selected to Head Corps," *Bryan Eagle,* Aug. 1, 1963, p. 1; Blair, *The Forgotten War,* pp. 417, 690–93.

38. Dethloff, *Centennial History,* pp. 575–76; interviews with Paul Dresser '54, Dec. 28, 1998, Bob Boldt '68, June 15, 1999, and Eddie Joe Davis '67, Jan. 19, 1999.

39. "Esten Opens 14th Season," *Battalion,* Sept. 16, 1965, p. 1.

40. "Defiant Bonfire," *Bryan Daily Eagle,* Nov. 24, 1960, p. 1; *Battalion* articles: "Show Fish How It's Done," Oct. 9, 1962, p. 4; "Pinky Opens New Season," Jan. 4, 1963, p. 1; and Gerry Brown, "Aggies Invade Dallas Area for Wild, Woolly Weekend," Nov. 8, 1962, p. 1.

41. Jim Earle, "Cadet Slouch," *Battalion,* Oct. 10, 1958, p. 2; "Fish Gets Ready for Ponies," *Battalion,* Nov. 2, 1965, p. 1; Bob Boldt, interview, June 15, 1999.

42. *Battalion* articles: "Owl Home—Steer Home," Nov. 13, 1963, p. 1; "Rendezvous with Texas' Bevo Ends as Aggies Surrender UT Mascot," and "Steer Returned in Good Shape," Nov. 14, 1963, p. 1; and "No Heavy Charges Expected for Student Mascotnappers," Nov. 15, 1963, p. 1; Cox, *I Bleed Maroon,* pp. 91–92; "The Excitement of Football . . . ," *Aggieland,* 1964, pp. 28–29; Frank Muller '65, interview, Houston, Dec. 9, 1998; Charles Hornstein '53, interview, College Station, Feb. 27, 1999.

43. *Battalion* articles: Lani Presswood, "Ranger—An A&M Campus Legend," Nov. 5, 1965, p. 3; "Reveille Replacement Sought," Oct. 22, 1965, p. 1; "Ranger for Mascot," Oct. 26, 1965, p. 2; Glenn Dromgoole, "Ranger Dies After Short Illness," Dec. 10, 1965, p. 1.

44. *Battalion* articles: "Reveille Replacement Sought," Oct. 22, 1965, p. 1; "Ranger for Mascot," Oct. 26, 1965, p. 2; Presswood, "Ranger—An A&M Campus Legend," Nov. 5, 1965, p. 3; Dromgoole, "Ranger Dies After Short Illness," Dec. 10, 1965, p. 1; John Fuller, "Reveille Put to Rest," Sept. 22, 1966, p. 1; and "Reveille III Makes Memorial Stadium," Nov. 22, 1966, p. 1.

45. *Battalion* articles: "Bonfire Work Is Organized for Fast Two-Day Effort," Sept. 27, 1962, p. 1; "Volunteer Work Begins in Bonfire Cutting Area," Nov. 13, 1962; "Bonfire Burns Tonight,"

Nov. 20, 1962, p. 1; and Tommy DeFrank, "What It Takes to Build World's Largest Bonfire," Nov. 24, 1965, p. 2; Myron Cope, "The Proudest Squares," *Sports Illustrated,* Nov. 18, 1968, p. 104.

46. *Battalion* articles: "It's Time to Build the All Aggie Bonfire," Nov. 20, 1963; "Bonfire Center Pole Goes Up," and "Traditional Fire Well Organized," Nov. 23, 1963, p. 1.

47. "Bonfire Construction Represents Aggie Men," *Battalion,* Nov. 22, 1963, p. 2. See also John Tyson, "History of the Class of 1967," *Aggieland,* 1967, p. 131.

48. "Bulletin—Connally Critical; Johnson Not Hurt," *Battalion,* Nov. 22, 1963, p. 1; "Tradition Stacks Up with Bonfire," *Battalion,* Nov. 26, 1997, p. 2; "John F. Kennedy," *Aggieland,* 1964, p. 5; Glenn Dromgoole, telephone interview, Dec. 29, 1998.

49. A&M College of Texas, *Articles of the Cadet Corps,* Sept., 1961, pp. 1–79.

50. Larry Jerden, "New Corps 'Standard' Revises Cadet Policy," *Battalion,* Sept. 12, 1964, p. 1; Tommy DeFrank, "Fish Don't Approve of New Privileges Survey Indicates," *Battalion,* Nov. 11, 1964, p. 1; Corps of Cadets, *The Standard,* 1968, pp. 1–114.

51. *Standard,* 1968, p. 79.

52. Allan Dees '64, letter to the editor, *Battalion,* Nov. 29, 1962, p. 2; Cox, *I Bleed Maroon,* pp. 5, 63; "History of the Class of 1967," *Aggieland,* 1967, p. 131; "History of the Class of '69," *Aggieland,* 1969, n.p.

53. Rudder quoted in Tommy DeFrank, "Rudder Assures Prominent Corps," *Battalion,* May 18, 1965, p. 1.

54. Texas A&M *Catalogue,* 1965–66, p. 63.

55. Young, "ROTC: Required or Elective?" *Military Review,* Feb., 1962, pp. 21–32; Theodore Wyckoff, "Required ROTC: The New Look," *Military Review,* Nov., 1964, pp. 24–26.

56. Glenn Dromgoole, "Cadet Corps Goes Non-Compulsory as Administrators Voice Approval," *Battalion,* Apr. 27, 1965, p. 1; "'65 Campus News," *Battalion,* Jan. 4, 1966, p. 1.

57. *Battalion* articles: Dromgoole, "Cadet Corps Goes Non-Compulsory as Administrators Voice Approval," Apr. 27, 1965, p. 1; "Full Coeducation

Voluntary Military Narrowly Approved," Feb. 26, 1965, p. 1; and DeFrank, "Rudder Assures Prominent Corps," May 18, 1965, p. 1; Frank Muller, interview, Dec. 9, 1998; "Some Cadets Cry, 'Maggie Go Home!'" *Texas Magazine, Houston Post,* May 9, 1965, p. 7. The April, 1965, board meeting once again discussed the "limited" enrollment of women, but any action on "full" coeducation was "deferred" to a future date.

58. *Battalion* articles: "Enrollment Set New Mark, Eclipses Old 1946 Record," Sept. 21, 1965, p. 1; "State Increases Funds for 1965–66 A&M Budget," Sept. 16, 1965, p. 1; "Rudder Assumes New Duties as President of A&M System," Sept. 2, 1965, p. 1; Editorial, "Growing Pains," Sept. 24, 1965, p. 2; "Enrollment Tops Old Record as 9,385 Students Register," Sept. 28, 1965, p. 1; "Limited Coeducation Is Discriminatory, Waggoner Carr Says," Oct. 15, 1965, p. 1; "Heldenfels's Coed Policy Letter Releases by President Rudder," Dec. 7, 1965, p. 1; and Michael White, "Commandant Says No Girls Allowed in Cadet Corps," Nov. 4, 1965, p. 1; *Aggieland,* 1966, p. 99.

59. New Students Files, 1963–65, Rudder Papers; Dethloff, *Centennial History,* pp. 567–74; Angus Martin, Cushing Library display, "Brothers and Sisters in Arms: African Americans in the Corps of Cadets," fall, 2000.

60. *Battalion* articles: Tommy DeFrank, "Corps Freshman Hike Foreseen: Commanders Get Increased Power," Sept. 16, 1965, pp. 1, 12; "'65 Campus News," Jan. 4, 1966, p. 1; and "Enrollment Sets New Mark, Eclipses Old 1946 Record," Sept. 21, 1966, p. 1.

61. "A&M Gets Largest Percentage Increase," *Texas Aggie,* Dec., 1967, p. 8; Ralph "Fil" Filburn '66, telephone interview, Jan. 3, 1999; Robert Beene '67, telephone interview, Jan. 3, 1999; Ken Nicolas, assistant to the commandant, 1963–76, telephone interview, Jan. 3, 1999; *Aggieland,* 1966, pp. 14, 99. See also *Battalion* articles: "Commandant's Reception," Sept. 16, 1965, p. 4; "Baker Says Draft Concern Evident in Advanced ROTC," Sept. 23, 1966, p. 2; and "A&M Sees Growth," Jan. 10, 1967, p. 3.

62. "Telegram Endorses Stand in Viet Nam: 2,148

Sign Message to LBJ," *Battalion,* Oct. 22, 1965, p. 1; Tommy DeFrank, "Anti-Viet Nam Protests Border on Treason," *Battalion,* Oct. 29, 1965, p. 3.

63. Eddie Joe Davis, interviews, College Station, July 24, 1998, and Jan. 19, 1999; "Corps-Civilian Melee Criticized by Dean," *Battalion,* May 10, 1966, p. 1. Eddie Joe Davis '67 is not kin to Col. Joe Davis. See also Tommy DeFrank, "Non-Regs: True Aggies?" *Battalion,* Oct. 13, 1965, p. 2.

64. "Col. Baker's Spirit Remains with Corps," *Bryan Daily Eagle,* May 28, 1967; *Battalion* articles: "Col. Baker Receives Award During Corps Final Review," June 1, 1967, p. 1; "McCoy Picked to Fill Corps Commandant Post," Aug. 3, 1967, p. 1; and "Mrs. Cook," Sept. 30, 1966, p. 1; Bill Lee, "The A&M Story: Aggie Traditions Deeply Rooted in Cadet Corps," *Houston Chronicle,* Oct. 1, 1968, p. 1; Ken Nicolas, telephone interview, Jan. 3, 1999, and Malon Southerland, interview, College Station, Jan. 17, 1999.

65. Lee, "The Story of A&M," *Houston Chronicle,* Oct. 1, 2, 3, 1968, p. 1; *Aggieland,* 1969, p. 4. See also Blumenfeld, "The Case for ROTC," *Military Review,* May, 1963, pp. 3–13.

66. Steve Forman, "31 Windows in 4 Corps Dorms Broken in Early Morning Blast," and "4 Admit 'Bombing,' Receive Suspension," *Battalion,* Oct. 8 and 9, 1969, p. 1. See also Edward B. Glick, *Soldiers, Scholars, and Society: The Social Impact of the American Military,* pp. 81–95.

67. David W. Levy, *The Debate over Vietnam,* pp. 107–108, 148–51, 156–58; Alexander Kendrick, *The Wound Within: America in the Vietnam Years,* pp. 151–52, 205–206, 273–74; Ronald Frazer, ed., *1968: A Student Generation in Revolt,* pp. 100–22, 285–96; telephone interviews: Jerry Geistweidt '70, Jan. 3, 1999; Hector Gutierrez '69, Dec. 31, 1998; and J. Malon Southerland, Jan. 8, 1998.

68. "Rudder Sees Cadet Corps as First Target of 'Kooks,'" *Battalion,* Sept. 10, 1969, p. 1; "Rudder: 'Hell of a Fight' for Any Troublemakers," *Houston Chronicle,* Apr. 1, 1969, p. 2; "Feud over A&M Campus Newspaper Continues Unabated," *Dallas Morning News,* Apr. 2, 1967, p. 26A. See also Earl Rudder, "What Price Disarmament?" *Ordnance,* Nov., 1965, pp. 288–90.

69. "Rudder Sees Cadet Corps as First Target of 'Kooks,'" *Battalion*, Sept. 10, 1969, p. 1; "Selective Service Classifications Listed," *Battalion*, Sept. 29, 1965, p. 1; Jerry Geistweidt '70, interview, Jan. 3, 1999. See also 1965–67 Correspondence File, Rudder Papers.

70. "Baker Says Draft Concern Evident in Advanced ROTC," *Battalion*, Sept. 23, 1966, p. 2.

71. E. L. Reilly Company, *Texas A&M Today: As Seen by High School Seniors, Guidance Counselors, Former Students and the General Public* [hereafter cited as Reilly Report], 1967, pp. C-14-15, 13, 63; James A. Donovan, *Militarism, USA*, p. 29.

72. Dan Kennerly, "History," *Aggieland*, 1968, p. 116.

73. "Aggie Record Tops at Summer Camps," *Battalion*, Sept. 15, 1966, p. 5; "Aggie Hero Talks to ROTC Educators," *Battalion*, Sept. 21, 1967, p. 3.

74. Michael Lee Lanning, telephone interview, Jan. 26, 1999; Eddie Joe Davis, interview, College Station, Jan. 19, 1999. See also Lanning, *The Only War We Had*, pp. 1–4.

75. Larry C. Kennemer '66, telephone interview, Jan. 6, 1999. See also John J. Tolson, *Airmobility 1961–1971*, pp. 136–50, and *Aggieland*, 1966, p. 14.

76. *Battalion* articles: Malcolm W. Browne, "Viet Nam War in Doubt," Oct. 17, 1962, p. 2; Peter Arnett, "Red Surge in Viet Nam Worries U.S. Authorities," Oct. 3, 1962, p. 2; and "U.S. 'Copter: Experimental Ace Now in Viet Nam," Oct., 11, 1962, p. 2. See also "South Viet Nam: Their Own Battle," *Time*, Sept. 21, 1962, p. 28.

77. The UH1A had the official name "Iroquois."

78. "Johnson Urges Study of SCONA Topics," *Battalion*, Dec. 14, 1962, p. 1; Lyndon B. Johnson, "Remarks by the Vice President," Bryan [College Station], Tex., Dec. 13, 1962, Statements of LBJ, LBJ Library, Austin, Tex.

79. Statement by LBJ, Oct. 14, 1964, Public Papers of the President, Lyndon B. Johnson, 1963–1964, pp. 1336–37; Stanley Karnow, *Vietnam*, pp. 357–86; Robert S. McNamara, *In Retrospect*, pp. 127–43; U.S. Department of the Army, *Report of the Department of the Army Board to Review Army Officer Schools*, Vol. 2, Feb., 1966, pp. 150–53. See also Louis Galambos, *America at Middle Age*, pp. 120–24, and Daniel B. Brewster, "A Senator Looks Back," *Vietnam*, Aug., 1999, pp. 22–28.

80. Ted Lowe '58, interview, Jan. 20, 1999; George Kelly, "Footnotes on Revolutionary War," *Military Review*, Sept., 1962, pp. 31–39; "Sixty-one A&M Students Killed in Viet Nam," *Bryan Pictorial Press*, June 8, 1969, in the Vietnam War Papers, Texas A&M Archives.

81. "How Goes the War?" *Newsweek*, Jan. 1, 1968, pp. 17–26 and cover; Hargrove, *A Dragon Lives Forever*, pp. 180–200; Col. Ted Lowe '58, telephone interview, Jan. 20, 1999.

82. U.S. Army, *Report of the Department of the Army Board of Review*, Vol. 1, Feb., 1966, p. 14. This report on the sources of army officers between 1961 and 1965 provided the following data: ROTC, 70.1 percent; medical officers, 14 percent; OCS, 8.2 percent; direct appointment, 4.3 percent; and the USMA, 3.5 percent.

83. James Fallows, "Vietnam: Low Class Conclusions," *Atlantic*, Apr., 1993, pp. 38–44.

84. "What Kind of Draft?" *Washington Post*, Dec. 23, 1966, p. A12. According to Baskir and Strauss, "During the Spanish-American War, 13 percent of all troops were trained for non-combat roles; during World War II, 61 percent were non combat; and at the 1968 height of the Vietnam war, 88 percent of all servicemen were assigned to noncombat occupational specialties" (Lawrence M. Baskir and William A. Strauss, *Chance and Circumstance: The Draft, The War, and the Vietnam Generation*, p. 52).

85. Robert L. Acklen, Jr. "Service Record," DA-24, Jan. 5, 1975, Papers of the Association of Former Students.

86. James Fallows, "Vietnam: Low Class Conclusions," *Atlantic*, Apr., 1993, pp. 38–44; Hargrove, *A Dragon Lives Forever*, pp. 184–85; Baskir and Strauss, *Chance and Circumstance*, pp. 3–61; Comments by Hon. Olin E. Teague, "Memorial Garden Dedicated at Texas A & M University," *Congressional Record*, Nov. 19, 1969, pp. E9813–4; *Aggieland*, 1970, p. 44.

87. "Objective of the Corps of Cadets," *Aggieland*, 1962, p. 37.

88. Kenneth Tomlinson, "ROTC under Attack," *Reader's Digest*, Nov., 1969, pp. 231–36; Dromgoole, "'65 Campus News," *Battalion*, Jan. 4, 1966, p. 1.

89. The Reilly Report, pp. 1–63.

90. David Reed '70, Commander First Brigade, "The State of the Corps—An Analysis," College Station, Jan., 1970, in Texas A&M Archives. See also Bill Lee, "The A&M Story: Tradition Means Much in Spirit of Aggieland," *Houston Chronicle*, Sept. 29, 1968, p. 1, and Don W. Bonifay '69, "History of the Class of '69," *Aggieland*, 1969, p. 278.

91. A. Phillips Brooks, "Tradition Is the Eye of A&M Storm," *Austin American-Statesman*, June 24, 1996; Kendrick, *The Wound Within*, p. 342.

CHAPTER 9

1. *Battalion* articles: Tommy Thompson, "Luedecke Appointed," Apr. 1, 1970, p. 1; "A&M to Mark 94th Year," Aug. 5, 1970, p. 1, and "Luedecke Greets Seniors of 1974," Aug. 5, 1970, p. 1; *Aggieland*, 1970, pp. 12–13, 17, 69; Jim Carroll, "Texas A&M in World War II," *Houston Press*, Apr. 15, 1946, p. 20.

2. "Inauguration," *Texas Aggie*, May, 1971, pp. 2–6. Williams was discharged from the Marine Corps at the rank of major.

3. "The Inauguration of Jack Kenny Williams as Seventeenth [Eighteenth] President of Texas A&M University," Apr. 16, 1971, College Station, bio documents seen in Texas A&M Archives; "Inauguration," *Texas Aggie*, May, 1971, pp. 2–8; Dethloff, *Centennial History*, pp. 580–81. See also John Adams, "A&M Beginnings Part of Event-Filled Year," *Battalion*, Oct. 5, 1976, p. 5.

4. Lynn Ashby, "In Aggieland," *Houston Post*, Nov. 2, 1972, p. D-1; *Aggieland*, 1971, pp. 52–53, 368–69, 408; Dave Mayes, "He Must Find A&M Familiar Place," *Bryan Eagle*, Nov. 12, 1970, p. 1; "Putting 'Life' in the Campus," *Texas Aggie*, Feb., 1973, pp. 8–11; "Shapes of Things," *The Review*, winter, 1972, pp. 9–10; "The Faculty behind the University," *TAMU Today*, Sept., 1973, p. 3; "'TAMU—2001' Subject of Williams Lecture Tonite," *Battalion*, Nov. 7, 1972, p. 1; Adams, *We Are the Aggies*, pp. 193–95. By the fall of 1973 the number of "faculty with title" had doubled to 1,339 members during the previous seven-year period.

5. Col. Tom Parsons '49, telephone interview, Feb. 5, 1999; "They're Proud: Texas Aggies Beat Jokes to the Punch," *Los Angeles Times*, May 7, 1977, p. 1; "Teague Eulogizes Classmate Rudder," *Battalion*, Nov. 27, 1973, p. 1; *Aggieland*, 1971, p. 408; "Inauguration," *Texas Aggie*, May, 1971, pp. 6.

6. Headquarters Corps of Cadets, "Modification of Corps Policy for Spring Semester, 1970," Feb. 2, 1970, Papers of the Corps of Cadets, Texas A&M Archives. See also David Reed, "The State of the Corps—An Analysis," College Station, Jan., 1970, and Dr. Rod O'Connor, "Studying without Wheel-Spinning," College Station, n.d. [circa 1971], pp. 1–7. Tuition was $50 per semester for Texas residents and $150 for nonresidents. Room, board, and laundry was $328 per semester; an air-conditioned room was an extra $45.

7. *Aggieland*, 1970, p. 398.

8. Col. Tom Parsons '49, telephone interview, Feb. 4, 1999; "New Corps Commandant Named," *Battalion*, June 23, 1971, p. 1; Sally Hamilton, "Cadet Commandant Recalls Prediction of Corps Doom," *Battalion*, Aug. 7, 1974, p. 2.

9. Debi Blackmon and Bruce Lawrie, "The Camp—Home of the Fighting Texas Aggies," *The Review*, winter, 1972, pp. 13–15.

10. Stephen L. Baker '72, "Working Out Civilian, Corps Polarization," *Battalion*, Feb. 23, 1972, p. 2.

11. Dethloff, *Centennial History*, pp. 581–83; "Corpsmen Go Recruiting during Christmas," *Battalion*, Feb. 8, 1973, p. 2; William G. Bell, *Department of the Army Historical Summary: FY 1970* (Washington, D.C., 1973), pp. 42–59; Robert K. Griffith, Jr., *The U.S. Army's Transition to the All-Volunteer Force* (Washington, D.C., 1997), p. 256; Drew Middleton, "Armed Forces' Problem: Finding Good Volunteers," *New York Times*, Apr. 17, 1974, p. 21.

12. Corps of Cadets, " . . . Something Extra," circa 1973, pp. 1–10, author's collection.

13. Dethloff, *Centennial History*, pp. 585–610; Ken Stroebel, "Texas A&M Is Biggest, Oldest University in State," *Bryan Eagle*, Aug. 26, 1976, p. 9D; *Aggieland*, 1972, pp. 18–19; See also Baker, "Civilian, Corps Polarization," *Battalion*, Feb. 12, 1972, p. 2.

14. Blackmon and Lawrie, "The Camp," *The Review*, winter, 1972, p. 14.

15. Association of Former Students, *Leadership Manual*, College Station, n.d. (circa 1970), pp. 1–55; Adams, *We Are the Aggies*, pp. 190–94; "Inauguration," *Texas Aggie*, May, 1971, p. 6. President Williams noted of the alumni, "Texas A&M is blessed with a legion of former students—tremendous Aggies, whose fierce loyalty to this university is no less material than it is vocal . . . this being a formal way of saying that Aggies put their dollars where their mouths are."

16. Texas A&M, "Organizational Chart: Vice President for Student Services," College Station, Jan. 1, 1975. The formal name of the Texas A&M Board of Directors was changed to the Board of Regents in 1975.

17. *Battalion* articles: Clifford Broyles, "Aggies Nudge LSU in Final Seconds," Sept. 22, 1970, p. 1; "Final Score in Elephant Bowl: Zip-to-Zip," Mar. 2, 1973, p. 1; "Army Beats Air Force for Fourth Straight Year," Mar. 4, 1975, p. 6; Michelle Scudder, "Cadets Raise $11,500 in March to the Brazos," Apr. 3, 1978, p. 1; and "Elephant Bowl," Apr. 16, 1984; "Elephant Bowl," *Texas Aggie*, Feb., 1972, p. 8; "Elephant Bowl," *The Quadrangle*, Mar. 10, 1976, p. 2.

18. *Battalion* articles: "Yell Practice Leads Fall Football Fever," Aug. 25, 1976, p. 5C; Sanford Russo, "Plays, Concerts, Music Crowd Fall Season," p. 1B; "SCONA," Jan. 23, 1973; and "Students! Sound Off!" Mar. 8, 1973, p. 8; "Gen. Simpson Wonders What 'Sully' Might Think about Changing Campus," *Bryan Eagle*, Aug. 30, 1974, p. 5C; Combat Ball, Lone Star Company, "Let's Keep Aggie Traditions: Combat Ball '75," College Station, n.d.; Paul Dresser '64, interview, Oct. 16, 1999, SHS Corps Center, College Station, and confirmed with Bill Heye '60 on Oct. 28, 1999.

19. Adams, *We Are the Aggies*, p. 90–91; "Stolen 'Beevo' [*sic*] Missing in Action," *Battalion*, Nov. 7, 1972, p. 2; "White Stuff Brought Rare Fun to Aggieland," *Battalion*, Jan. 16, 1973, p. 1. See also "A Crushing Wave of Wood," *Newsweek*, Nov. 29, 1999, pp. 44–45.

20. Adams, *Softly Call the Muster*, pp. xviii–xix.

21. *Battalion* articles: "Heart Attack Fells 36th President," Jan. 23, 1973, p. 1; "U.S. Officially Ends Decade of Intervention in Vietnam," Mar. 29, 1973, p. 6; and "Vietnam Cease-Fire Could Be Truly 'At Hand,'" Jan. 23, 1973, p. 4; "Bulletin," *Texas Aggie*, Feb., 1973, p. 1; "Ex-POW [Capt. James Ray] Receives Four Medals," *Bryan Eagle*, Jan. 14, 1975, p. 1; "Aggies Come Home: Return to Life," *Texas Aggie*, July, 1973, pp. 2–9. See also Vietnam War Papers; Maclear, *The Ten Thousand Day War—Vietnam: 1945–1975*, p. 310; and Bruce Palmer, *The 25-Year War: America's Military Role in Vietnam*, pp. 128–43.

22. "Cavalry Show Colors," *Battalion*, Mar. 4, 1975; PMC, "Parsons' Mounted Cavalry '76: Scheduled Appearances—Spring 1976," College Station, n.d.; "PMC Carries on Proud Military Tradition," *Battalion*, Nov. 12, 1997, p. 1; Texas A&M University Relations, press release, "Cavalry Building Dedicated to 'Buddy' Seewald '42," College Station, Oct. 7, 1996. See also Parsons Mounted Cav Papers, Texas A&M Archives; Col. Tom Parsons, letter to author, July 19, 1999.

23. Kathy Brueggen, "A&M Going to Cotton Bowl Says Williams at G. Rollie," *Battalion*, Sept. 3, 1974, p. 1.

24. Ibid.

25. Paul McGrath, "Ag Defense Dominates Texas Game," *Battalion*, Dec. 2, 1975, p. 1; "Bellard Enjoys Aura of a King," *Battalion*, Dec. 2, 1975.

26. A. F. "Tony" Pelletier '75, letter to author, Feb. 18, 1999; *Battalion* articles: "Monday Afternoon Shower [quadding]," Sept. 9, 1975, p. 1; "Getting Even," Sept. 9, 1977; "Quadding," Nov. 21, 1979; and "Quad Players," Oct. 3, 1983, p. 11; Al Reinert, "Being an Aggie Is No Joke," *Texas Monthly*, Jan., 1981, p. 80.

27. Headquarters, Corps of Cadets, Corps Bulletin No. 16, "Hauling Off," Oct. 20, 1975; Memorandum from Joseph M. Chandler, Jr., Cadet Corps Commander to All Cadets, "Policy on Quadding," Sept. 25, 1975; Col. James R. Woodall '50, interview, College Station, Sept. 7, 1998. According to the *Texas A&M University Student Rules 1999–2000*, activities such as "any form of quadding" are deemed "hazing."

28. H. L. Bechington, Deputy Director of Personnel

to Deputy Chief of Staff (Manpower) [O. R. Simpson], Sept. 10, 1971, interoffice memo, Washington, D.C.; Dr. Jack K. Williams to Lt. Gen. Ormond R. Simpson, USMC, Washington, D.C., Oct. 22, Nov. 12 and 17, 1971, NROTC File, Corps of Cadets Papers, Texas A&M Archives; Gen. Ormond R. Simpson '36, interview, College Station, Oct. 3, 1990; Ashby, "In Aggieland," *Houston Post,* Nov. 2, 1972. The Marine PLC program has produced officers for the USMC since the early 1960s, yet at no time was it a formal part of NROTC.

29. Col. Tom Parsons '49, telephone interview, Feb. 5, 1999; Jack K. Williams to Chief of Naval Operations, "NROTC University/College Survey and Application Data Pertaining to Institution," Nov. 17, 1971; E. R. Zumwalt, Chief of Naval Operations to Olin E. Teague, M.C., Jan. 21, 1972, Washington, D.C., NROTC File, Texas A&M Archives; Lt. Gen. Ormond R. Simpson, Deputy Chief of Staff (manpower), letter to Dr. Jack K. Williams, Aug. 22, 1972, in the papers of Col. Tom Parsons. The Texas Maritime Academy, a branch of Texas A&M in Galveston, offered commissions in both the navy and the coast guard.

30. Memo for record, "Establishment of an NROTC Unit at Texas A&M University," Feb. 17, 1972, Washington, D.C., NROTC File, Texas A&M Archives; E. D. Foxworth, PNS, The Citadel, to Lt. Gen. O. R. Simpson, Feb. 21, 1972, Washington, D.C., NROTC File, Texas A&M Archives; Col. Tom Parsons, "Memo: Establishment of NROTC," Apr. 20, 1972, and Col. C. E. Hogan to Rear Adm. J. L. Abbot, Jr., Oct. 15, 1973, Texas A&M Archives; "Marine Corps–oriented ROTC Initiates Operations This Fall," *Battalion,* Feb. 15, 1972, p. 1; Ken Stroebel, "Marines Invade Aggie Corps," *Bryan Pictorial Press,* Dec. 14, 1972, p. 2; *Aggieland,* 1972, pp. 64–65; *Aggieland,* 1973, pp. 319, 331; Texas A&M, "Commissioning Exercises," Dec. 14, 1974, p. 4. Note: General Simpson did not feel the Texas A&M PNS should be an A&M former student. Not until 1984 did an Aggie, Col. Richard G. McPherson '62, become PNS. Also, Gun. Sgt. Gaudencio Viloria '83 returned to A&M as a civilian in the early 1980s to complete his degree.

31. Col. C. E. Hogan to Rear Adm. J. L. Abbott, Oct. 15, 1973, and Lt. Gen. O. R. Simpson to Vice Adm. James E. Wilson, Feb. 11, 1976, in NROTC File, Texas A&M Archives; Col. Tom Parsons, interview, Feb. 5, 1999; "73–74 Corps Heads Named," *Battalion,* Mar. 2, 1973, p. 1; *Aggieland,* 1973, pp. 271, 298–99, 302.

32. Jack K. Williams to Brig. Gen. Phillip Kaplin, Military Personnel Management, Department of the Army, Mar. 22, 1976, Corps of Cadets Papers, Texas A&M Archives. See also P. Kaplin to Williams, Mar. 12, 1976; O. R. Simpson to Dr. J. K. Williams, Mar. 25, 1976; and Rear Adm. James B. Wilson to Dr. Williams, Mar. 2, 1976. Note: General Simpson was the primary correspondent and contact with Washington and the Pentagon.

33. Rod Speer, "Corps-persons at Texas A&M: Female Cadets on the March in Aggieland," Scene Magazine, *Dallas Morning News,* Mar. 21, 1976, pp. 6–9; Fritz Lanham, "Changing the Rules," Saturday Magazine, *Bryan Eagle,* Apr. 23, 1983, pp. 3–7; *Aggieland,* 1972, p. 315, and 1973, pp. 138–42, 145, 169: Debi Blackmon and Gary Aven, "Profile Mrs. Schreiber, Dean of Women," *The Review,* Nov., 1971, pp. 5–6.

34. Colonel Parsons, interview, Feb. 5, 1999; Department of the Army (hereafter DoA), memorandum, "Assignment of Women's Army Corps Officers for Duty with ROTC," Oct. 17, 1972; DoA, "Female Enrollment in ROTC," Nov. 22, 1972; DoA, "Expanded Enrollment of Women in ROTC," May 18, 1973; DoA, "Request for Waivers of Female Enrollment Policy," Dec. 14, 1973; Col. Thomas R. Parsons, letter to Brig. Gen. Robert Arter, commander, U.S. Army Third ROTC Region, Fort Riley, Kans., Dec. 18, 1973, in Papers of the Professor of Military Science, Texas A&M University, "Historical File," Military Science Building, College Station.

35. Title IX of the Education Amendments of 1972, P. L. 92–318, as amended (20 U.S.C. 1681 et seg.)

36. Gil Sewell, "The New Sex Rules," *Newsweek,* Dec. 3, 1979, p. 84; Lynn Ashby, "WACs Melted," *Houston Post,* Oct. 9, 1975; Rickey Gray '75, telephone interview, Feb. 17, 1999. See also United States, GAO, *Intercollegiate Athletics:*

Status of Efforts to Promote Gender Equity, HEHS-97-10, Nov. 25, 1996.

37. Mark Mattox, "Justice Dept. Counters VMI Appeal," *Lexington News-Gazette,* Nov. 18, 1992, p. 1; "Virginia Military Institute to Establish Courses at Women's College," *New York Times,* Sept. 26, 1993, p. 10; "Military College Is Ordered to Admit Women," *New York Times,* Sept. 26, 1993, p. B16; Suzanne Fields, "Brother Rat, Sister Snake and the VMI War," *Washington Times,* Oct. 7, 1993, p. A19; Ronald Smothers, "All-male Citadel Ordered to Enroll Women at Once," *San Diego Union-Tribune,* July 23, 1994, p. A1; "Keep The Citadel for Men Only?" *Houston Chronicle,* Aug. 20, 1994, p. 31A; Amy Bernstein, "Shannon Faulkner Should Have Come Here," *U.S. News & World Report,* Aug. 22, 1994, p. 16; James L. Kilpatrick, "Years before Citadel Suit Will Be Settled," *San Antonio Express-News,* Aug. 25, 1994, p. 2; Peter Finn, "Women Reach Rat [VMI] Finish Line," *Washington Post,* Mar. 17, 1998, pp. 1, 8. See also U.S. GAO, *Military Academy: Gender and Racial Disparities,* NSIAD 94–95, Mar. 17, 1994, and *Air Force Academy: Gender and Racial Disparities,* NSIAD 93–244, Sept. 24, 1993.

38. Corps of Cadets, "The Minerva Plan: A Plan to Integrate Women into the Corps of Cadets, Phase I–III," College Station, 1973–74, in papers at SHS Corps Center; "Corps Comment," fall, 1974, pp. 1–3; Mary A. Woodham, "W-1: Minerva's Fairest and Finest," *Texas Aggie,* May, 1976, pp. 2–7; Terry W. Rathert '75, telephone interview, Feb. 15, 1999.

39. Corps of Cadets, "The Minerva Plan, Phase I," Nov. 30, 1973, p. 2. See also Parsons, letter to Arter, Dec. 18, 1973.

40. Corps of Cadets, "Minerva Plan, Phase I and II," 1973–74; Lisa Messer, "Women Cadets Celebrate 20th Year in Corps," *Battalion,* Nov. 18, 1994, p. 9; Kathy Brueggen, "Corps Women Move Slowly into Military Participation," *Battalion,* Sept. 17, 1974, p. 1; W-1, "Standard Operating Procedure and Basic Information, 1974–75," fall, 1974, pp. 1–15, in the SHS Corps Center; Speer, "Corps-persons at Texas A&M," Scene Magazine, *Dallas Morning News,* Mar. 21, 1976, p. 8; *Aggieland,* 1975, pp. 290–91, 331;

527; Rickey Gray '75, telephone interview, Feb. 17, 1999. Note: In sharp contrast the service academies employed a cadre of female active duty military personnel to phase women into each academy.

41. Ron Kleinsasser, "Corps Women: A&M ROTC Struggle with Change When Females Joined," *Bryan Eagle,* Dec. 26, 1979, pp. 1, 8; *Aggieland,* 1974, p. 483; 1975, p. 73; and 1977, p. 744.

42. Mary A. Woodham, "Minerva's Fairest and Finest!" *Texas Aggie,* May, 1976, pp. 3–5. See also W-1 Papers, SHS Corps Center.

43. W-1, "W-1 fish privileges," and "Company W-1: Organization and Activities," both in W-1 File, SHS Corps Center; Memo: Class of '76 and Class of '77, Company W-1 to Cadet Rickey A. Gray, Cadet Colonel of the Corps, "Allotment of Class Privileges and Responsibilities," Dec. 11, 1974, SHS Corps Center; *Aggieland,* 1975, pp. 4, 73; "Women's Corps Apparel Arrives," *Texas Aggie,* Feb., 1975, p. 13; Carol Jones, "Heresy: Women in Aggie Corps?" *Dallas Times Herald,* Apr. 14, 1976, p. 8E.

44. Kleinsasser, "Corps Women: A&M ROTC Struggled with Change When Females Joined," *Bryan Eagle,* Dec. 26, 1979, pp. 1, 8; Roxie R. Pranglin '78, telephone interview, Feb. 15, 1999; "An Army First at Texas A&M," *Houston Chronicle,* May 12, 1977, p. 1.

Pranglin was the second woman to command W-1 in 1977–78. By comparison, of the 119 female cadets that entered West Point in the fall of 1976, 62 graduated in the class of 1980. The first female commissioned in a formal ceremony at Texas A&M was Lt. Colette K. Boyd '75 in May, 1975. Boyd was a member of a special army program and not a part of the ROTC program at A&M.

45. *Battalion* articles: Don Middleton, "Corps Changes Whip-out Policy," May 1, 1975, p. 1; Amy Rowlett, "Corps Adjusting to Rituals," Sept. 3, 1975, p. 2; and Mike Kimmey, "Corps Direction Changes," Apr., 1975, in 1974–75 Commandant's Office Scrapbook, SHS Corps Center.

46. "Top ROTC Grad Honored," *Army ROTC Newsletter,* Sept., 1976, p. 19.

47. Texas A&M, "Statement of Expenses: 1975–

1976," College Station, fall, 1975; See also Glenn Dromgoole, "A&M Revisited: It's Really Changed in 10 Years," *Fort Worth Star-Telegram,* Oct. 23, 1975, p. 5A.

48. "Housing Limited '75 Admissions," *Battalion,* Feb. 5, 1976, p. 1; "Office Space Scarce, Barracks Being Used," *Battalion,* Feb. 5, 1976, p. 1; "Aggieland Spirit Remains Despite Recent Change," *San Antonio Express-News,* Nov. 6, 1975. In April, 1975, the old water tower with "Welcome to Aggieland" was razed.

49. Hank Wahrmund, "Centennial Opens at A&M," *Bryan Eagle,* Feb. 3, 1976, p. 1; Don Middleton and John Adams, "Semi-Centennial Fair Celebrated Fifty Years," *Battalion,* Feb. 5, 1976, p. 2; "Feb. 2 Proclamation Day Begins Events," *Fortnightly,* College Station, Jan. 23, 1976, p. 1; "A&M Centennial Celebration Starts Feb. 2," *Texas Aggie,* Feb., 1976, pp. 4–5; "Academic Variety," *Dallas Times Herald,* Sept. 3, 1976, p. F-5.

50. Lisa Junod, "Convocation Ends 100 Years," *Battalion,* Oct. 5, 1976, p. 1; "100 Education Years Celebrated," *Bryan Eagle,* July 4, 1976, p. 7A; "Centennial Plans About Set," *Fortnightly,* Dec. 5, 1975, p. 1; "Centennial Muster Held Around the World," *Texas Aggie,* July, 1976, pp. 4–6; John B. Connally, "Draft of Centennial Remarks at Texas A&M," College Station, Oct. 4, 1976, in author's papers. W. Clyde Freeman, executive vice president, acted on behalf of President Williams.

51. Colonel Parsons quoted in "Cadets Follow Traditions," *Dallas Times Herald,* Sept. 3, 1976, p. F-7.

52. Col. James R. Woodall '50, interview, Sam Houston Sanders Corps Center, College Station, Sept. 7, 1998; Ken Stroebel, "'Fish Days' Recalled by Head of Corps," *Battalion,* Aug. 7, 1977, p. 1.

53. Memo of Designation of Distinguished Military Student, Lt. Col. William F. Lewis to PMS&T, A&M College of Texas, Oct. 17, 1949, in POP; "Commandant of Cadets Assigned," *Bryan Eagle,* Apr. 27, 1977, p. 3; Colonel Woodall, interview, Sept. 7, 1998.

54. "Regents Made Wise Choice," *Battalion,* Aug. 3, 1977, p. 2; Jane Smith, "New President Has

Aggie Background," *Bryan Eagle,* July 30, 1977, p. 2; Ken Herman, "Miller Foresees New Corps Popularity," *Eagle,* Oct. 12, 1977, p. 9A.

55. Mike Gentry '77, letter to author, Re: Fish Orientation Week, June 3, 1999.

56. Ibid.; Colonel Woodall, interview, Sept. 7, 1998; Ken Stroebel, "'Fish Days' Recalled by Head of Corps," *Bryan Eagle,* Aug. 7, 1977, p. 3; *Aggieland,* 1977, p. 744. See also Dan Quinn '81, *The Spirit Within: A True Story of Spirit and Camaraderie of Life at Texas A&M University,* pp. 1–133.

57. Diane Blake, "Committee Formed to Review, Advise on Problems Women Find in the Corps," *Battalion,* Feb. 1, 1979, p. 1.

58. Ibid.; Roy Kleinsasser, "Aggie Women Cadets Upset Traditionalists," *Battalion,* Mar. 21, 1979, p. 1; Scot K. Meyer, "Corps Traditions Face Modern Challenge," *Battalion,* Sept. 1, 1980, p. 10.

59. Blake, "Committee Formed," *Battalion,* Feb. 1, 1979.

60. Ibid.

61. *Battalion* articles: Jack Anderson, "Corps Puerility, Not Virility Rides at Texas A&M," Mar. 21, 1979, p. 10; Diane Blake, "Columnist Blasts Corps," p. 1; Blake, "A&M Official Raps Column on Corps," Mar. 23, 1979, p. 1; Blake, "BCLU Probing Sex Discrimination Involving A&M, Women in Corps," Apr. 5, 1979, p. 1; and Roy Kleinsasser, "Corps Leaders Say Situation Improving," Mar. 22, 1979. See also Sandra Englert, "Women or Men First?" *Battalion,* Mar. 27, 1979, p. 2.

62. *Battalion* articles: Blake, "BCLU Probing Sex Discrimination," Apr. 5, 1979; "A&M Official Raps Column on Corps," Mar. 23, 1979; and Blake, "Corps Discrimination Suit Slowed by Red Tape; Results Expected Soon," Apr. 26, 1979, p. 1; Clarence Page, "Title IX Has Changed Us as Much as It Has Changed Sports," *Dallas Morning News,* July 18, 1999, p. 5J.

63. Minutes of the Texas A&M Board of Regents, May 22, 1979, pp. 25–26; Kim Tyson and Diane Blake, "Women Charge Discrimination: Corps Suit Seeks Injunction," *Battalion,* May 16, 1979, p. 1; Becky Swanson, "Federal Judge Dismisses Portion of Zentgraf Suit," *Battalion,* June 24, 1980, p. 1; Richard Vara, "Turmoil at Texas

A&M: Aggie Corps Split by Charges of Discrimination," *Houston Post,* May 13, 1979, pp. 1, 3.

64. Col. James Woodall, telephone interview, May 5, 1999; Roy Kleinsasser, "Justice Department Seeks to Join Suit Against A&M," *Bryan Eagle,* Nov. 20, 1979, p. 1; Rhonda Walters, "Justice Dumps Woodall, Enters Suit Due to Title IX," *Battalion,* Nov. 21, 1979, p. 1.

65. Swanson, "Federal Judge Dismisses Portion of Zentgraf Suit," *Battalion,* June 24, 1980, p. 1.

66. "Handshake Snub Shocks Aggie Grad," *Dallas Morning News,* May 18, 1980.

67. Dillard Stone, "Hubert-Miller Conflict Was Key to Firing," "Former Students Blast Regents," and "Miller a Victim of System Politics," all in *Battalion,* July 15, 1980, p. 1; Jarvis E. Miller '50, interview, College Station, Jan. 23, 2000.

68. Col. Donald Burton, Memo, "Sexual Harassment/Discrimination," to All Members of the Cadet Corps, May 21, 1985, Corps of Cadets File, Texas A&M Archives. Judge Ross Sterling died before the conclusion of the final decree.

69. U.S. District Court for the Southern District of Texas, "Consent Decree: Melanie Zentgraf v. Texas A&M University, et al.," No. H-79-943, Houston, Jan. 24, 1985, pp. 1–13; John J. Koldus, Vice President of Student Services, to Judge Ross N. Sterling, Oct. 27, 1986, letter seen in the Zentgraf Case File, Office of the Commandant, Texas A&M.

70. Cyndy Davis, "Final Review Last for Woodall," *Battalion,* May 5, 1982, p. 1; Dan Chiszar, "Draft Registration Blows Uncertainty over ROTC," *Idaho Free Press,* Feb. 18, 1980, in Col. James B. Woodall Papers, Texas A&M Archives; Office of the Commandant, "A Report to the President on the Corps of Cadets," College Station, Nov. 27, 1979, pp. 1–11. See also Will Van Overbeek, *Aggies: Life in the Corps of Cadets at Texas A&M.*

71. Fritz Lanham, "New Commandant Sets Priorities of Academics, Larger Enrollment," *Bryan Eagle,* Sept. 1, 1982, p. 2; Memo, "Policy Changes in the Corps of Cadets," Donald L. Burton, commandant, to Dr. John J. Koldus, vice president for Student Services, College Station, Jan. 24, 1984; *Aggieland,* 1956, pp. 173, 179.

72. United States Military Academy, Memo for the Assistant Dean for Student Administration, "Study of Academic Attrition," West Point, N.Y., June 6, 1986, pp. 1–17, in William Mobley Papers, Texas A&M Archives.

73. Col. Donald Burton, interview, Temple, Tex., Oct. 25, 1998; Kelley Smith, "Corps Represents More Than Tradition," *Battalion,* Mar. 28, 1983, and "First Year Tough for Cadets," *Battalion,* Mar. 31, 1983.

74. Burton, memo to Koldus, Jan. 24, 1984.

75. Ibid.

76. Minutes of the Corps Development Council (CDC) Steering Committee Meeting, Aggieland Inn, College Station, Oct. 7, 1983; Texas A&M, *Texas A&M University Annual Report, 1982,* College Station, Nov., 1982, p. 16; "Most Students Come from Major Metropolitan Areas," *Fortnightly,* Nov. 7, 1980, p. 7.

77. Don Burton, interview, Oct. 25, 1998; Minutes of the CDC, Rudder Tower, College Station, Mar. 31, 1984; "Houston Lawyer to Head A&M Panel," *Houston Chronicle,* Apr. 4, 1984, n.p.; *Pass in Review* 1 (Oct., 1986): 1–12.

78. "Motivational Exercise Fatal," *New York Times,* Aug. 31, 1984, p. B9; "Grand Jury Indicts Four," *New York Times,* Sept. 29, 1984, p. 9; Daniel Puckett, "Cadet's Death Sparks Recollection of Corps," *Bryan Eagle,* Sept. 9, 1984, p. 1; "Cadet's Case to Start after Mistrial," *Bryan Eagle,* Jan. 29, 1985, p. 1; Sheila Taylor, "The Aggie Hazing Rite Off the Roll of Honor," *Dallas Morning News,* Sept. 13, 1984, p. 4C.

79. Roy Bragg, "Tradition of Hazing Is Still Entrenched in A&M Corps," *Houston Chronicle,* Mar. 8, 1985, p. 25.

80. Charles H. Rollins III, '85, "A Report to the Board of Regents on Hazing," College Station, Dec., 1984.

81. Ibid.

82. Fred Bonavita, "Bill Would Increase Penalties for Hazing," clipping from *Houston Post,* n.d., p. 12A.

83. Leonard S. Goldberg, "ROTC and the College: A Complex Relationship," *NASPA Journal,* fall, 1985, pp. 15–18; Daniel Puckett, "Exercises Lead to Suspension of A&M Cadet," *Bryan Eagle,* Mar. 1, 1985, p. 1; Col. Donald L. Burton, memo

to Dr. Frank E. Vandiver, "Annual Army ROTC Report," College Station, July 1, 1985.

84. Presentation to the Faculty Senate, Frank E. Vandiver, "State of the University," College Station, Sept. 10, 1984; Virginia Kirk, "New Cadets Undergo Physicals," *Bryan Eagle*, Aug. 26, 1985, p. 1; Brian Pearson, "Freshman Enrollment in Corps Falls," *Battalion*, Aug. 28, 1985, p. 1; Corps of Cadets, "The Fish Buddy," College Station, 1985.

85. "Fish Reflects on Freshman Orientation Week," *The Saber*, Dec., 1985, p. 1; Texas A&M Public Information Office, "Three Female Band Members Begin a New Tradition," Sept. 2, 1985; Col. Donald Burton, memo, "Administrative Regulations for Women in the Aggie Band," Aug. 8, 1985, Company W-1 Papers; Col. Richard G. McPherson '62, interview, College Station, Sept. 6, 1999. The enrollment was down due in large part to an increase in tuition for nonresident students.

86. Edward C. "Pete" Aldridge, Jr. '60, Office of the Under Secretary of the Air Force to Dr. Frank Vandiver, Sept. 9, 1985; Frank Vandiver, former A&M president, interview, College Station, Oct. 10, 1999.

87. Office of the Commandant, memo, "ROTC/D&C Enrollment," Dec. 3, 1985; Vandiver interview, Oct. 10, 1999.

88. Minutes of the Board of Regents, Jan. 24, 1984, p. 7; Sept. 25, 1985, p. 7; Jan. 18, 1986, p. 12; May 27, 1986, pp. 11–12; Nov. 24, 1986, pp. 9–10.

89. Draft: "Mission Statement: Corps of Cadets," College Station, Nov., 1985.

CHAPTER 10

1. Frank E. Vandiver, "And Now, Mr. President Bring Back the Draft," *Christian Science Monitor*, Mar. 30, 1982, and reprinted in the *Houston Post* on Apr. 1, 1982; Corps of Cadets, "The Corps of Cadets and Vision 2020," Office of the Commandant, Nov. 16, 1998, pp. 1–2. Vandiver noted that the "ready reserve," the citizen-soldiers to whom the nation historically has turned in times of trouble, was 370,000 men short in strength.

2. Lance Morrow, "The Long Shadow of Vietnam," *Time*, Feb., 24, 1992, pp. 18–21; Joe Hyde, "Is There Over Regulation?" *The Saber*, Feb., 1986, p. 2; Tony Best, "Cadet Attitudes: A New Perspective," *The Saber*, May, 1986, p. 3.

3. President Frank Vandiver, address to the Corps of Cadets, Rudder Auditorium, audio tape, Dec. 5, 1984, Texas A&M Archives.

4. Mary Jo Rummel, "Officials, Cadet to Meet, Discuss Saber Incident," *Battalion*, Nov. 2, 1981, p. 1; John Makeig, "Frank Vandiver—A Firefighter for Troubled Campuses," Houston *Chronicle*, Dec. 6, 1981, p. 1; Col. James Woodall, interview, College Station, May 5, 1999.

5. Jade Boyd, "Senate Hears Reports on Bonfire, Corps," *Battalion*, May 15, 1990, p. 1.

6. Frank E. Vandiver, "State of the University," manuscript, College Station, May 5, 1988; Jade Boyd, "Senate Hears Reports on Bonfire, Corps," *Battalion*, May 15, 1990, p. 1.

7. Vandiver, interview, Oct. 10, 1999; Brad Owens, "Vandiver Says Commandant Should Be A&M Employee," *Bryan Eagle*, Mar. 2. 1986, p. 2; Mona Palmer, "Vandiver Recommends Making Corps's Leader A&M Employee," *Battalion*, Apr. 11, 1986, p. 2.

8. Buck Henderson, interview, College Station, July 22, 1999.

9. Genevieve Blute, "Corps Leader Says Search for Commandant Nearly Over," *Battalion*, Nov. 13, 1986, p. 1; "Southerland Is Interim Corps Commander," *Battalion*, Aug. 26, 1986, p. 1; Vandiver, interview, Oct. 10, 1999, and J. Malon Southerland, interview, Laredo, Tex., Jan. 8, 1998.

10. Connie O'Conner '88, "Old Army Duncan," *The Saber*, circa fall, 1987, p. 4. The reference to "look-on-the-tray" was a major departure from the old-style "hot corner" in which cadets requested food they desired by name and then it was passed. Seniors were served first, then juniors, followed by the sophomores and fish.

11. J. Malon Southerland, interview, Jan. 8, 1998; "Messages to Members of the Corps Development Council," *First Call*, fall, 1986, pp. 2–6; Jerry Cooper, "John Koldus to Retire in Aug.," *Texas Aggie*, Aug., 1993, pp. 42–45; Corps of

Cadets, "Cadre Leadership Training," 1987, pp. 1–16; Garland W. Wilkinson, "The Role of the Corps at A&M," *Battalion,* Feb. 11, 1987, p. 12; Minutes of the Board of Regents, Jan. 28, May 27, and Nov. 24, 1986.

12. Corps of Cadets, "Visitation Schedule for Major General Thomas G. Darling," Sept. 25–26, 1986, College Station; "A New Commandant Comes to the Corps," *The Saber,* Oct., 1987, p. 1; Maj. Gen. Tom Darling, interview, College Station, Sept. 8, 1998.

13. Thomas Darling, "This Semester in Review," *The Saber,* Dec., 1987, p. 1; Derrick Grubbs, "At the Heart of Texas A&M," *Texas A&M Today,* summer, 1988, pp. 1–3; "General Darling Says 1987–88 Best Year Yet," *First Call,* Sept., 1988, p. 1; Memo, "Renovation of the Commandant's Office," Jan. 11, 1988, POP.

14. "Spend the Night with the Corps," *First Call,* Sept., 1988, p. 7; Office of the Commandant, "Accomplishments for 1988," Commandant's File, POP; Maj. Gen. Thomas Darling, "Commandant's Report: The Year in Review," *First Call,* summer, 1989, p. 2; Memo, Dr. John J. Koldus to Doug DeCluitt, Board of Regents, Feb. 14, 1989, POP. In 1988 the Corps Sul Ross Scholarship provided an annual stipend of $1,000 for tuition and books.

15. Minutes of the Board of Regents, Sept. 21 and Nov. 16, 1987, and Jan. 25, 1988; Ormond R. Simpson to Dr. William H. Mobley, Mar. 13, 1989, POP. See also the following letters, Royce E. Wisenbaker to Dr. William H. Mobley, Nov. 1, 1989; John J. Koldus to Roger Haverson, Dec. 8, 1989; and Memo, "Budget Situation," William H. Mobley to Students, Faculty, Staff, Former Students and Friends of Texas A&M University, Apr. 3, 1991, POP; Interviews with Vandiver, Oct. 10, 1999, and Gen. Tom Darling, Oct. 11, 1999. See also "The Corps of Cadets and Vision 2020," p. 1.

16. Minutes of the Board of Regents, Jan. 24, 1984, Jan. 25, 1988; Karen Kroesche and Lee Schexnaider, "Vandiver's Resignation Provokes Mixed Reactions at A&M," *Battalion,* Jan. 13, 1988, p. 1; Steve Vinson, "A&M President for 7 Years Plans to Leave Office by Sept. 1," *Bryan Eagle,* Jan. 8, 1988, p. 1; David Elliot, "A&M President

Keeps Focus on 21st Century," *Bryan Eagle,* Aug. 6, 1989, p. 1.

17. Robert E. Wagner, major general commanding, letter to Dr. William H. Mobley, Mar. 15, 1990, POP; Carrie Wiedenfeld, "Exceptional Aggies," *Hullabaloo! Magazine,* summer, 1991, p. 15. The ROTC Washington celebration in May, 1991, was to commemorate the 1921 ROTC Act. As part of these ceremonies President Bush commissioned ten graduating cadets at the White House, among them Conrado Alvarado III '91.

18. Mobley, letter to Wagner, Apr. 19, 1990, POP; Gen. Tom Darling, interview, College Station, Apr. 24, 1999. See also Joe West, "Cutbacks May Hurt ROTC Recruitment," *Air Force Times,* Apr. 8, 1991, p. 10.

19. Keith Bradsher, "R.O.T.C. Doesn't Want You," *New York Times,* Jan. 6, 1991, pp. 36–37; Col. John C. Parrish, Third ROTC Region, letter to Mobley, Aug. 10, 1992, POP; Sam H. Sanders, M.D. '22, letter to Robert L. Walker, July 12, 1989, POP; "Center Planned for Spence Park," *First Call,* winter, 1990, pp. 3–4. Texas A&M in March, 1990, became one of five universities (the others were The Citadel, Norwich, Prairie View A&M, and VMI) to offer a direct commissioning program in the U.S. Coast Guard, in the Mobley Papers.

20. Fernandez quoted in Texas A&M University Relations press release, "Corps Center Dedication," College Station, Mar. 8, 1991.

21. "The Cold War: From Containment to Commonwealth," *New York Times,* Feb. 2, 1992, p. 8; William E. Odom, *The Collapse of the Soviet Military,* pp. 388–404. Gorbachev resigned on December 25, 1991, and the USSR formally ceased to exist.

22. Gretchen Kruger, "Aggie Leads Air Blitz in Persian Gulf," *A&M Magazine,* Feb., 1991, p. 12; Jerry C. Cooper, "List of Aggies in Middle East Grows Larger," *Texas Aggie,* Mar., 1991, pp. 2–9.

23. Lane Stephenson, "Tradition of Military Service Continues," *A&M Magazine,* Feb., 1991, p. 12; Association of Former Students, "Persian Gulf Resolution," College Station, Mar. 2, 1991; List, "Operation Desert Storm Aggies," Office of the *Texas Aggie,* Feb. 21, 1991; Stewart M. Powell,

"Voices from the War," *Air Force Magazine,* Apr., 1991, pp. 36–42; "Another Vietnam," *Wall Street Journal,* Feb. 6, 1991, p. A12; "Squadron 1 Remembers Cliff Bland '86," *First Call,* fall, 1991, p. 2; Program, Corps Plaza, "Memorial Ceremony," Sept. 28, 1991; Kara Bounds, "A&M Remembers Gulf War Vets with Campus Ceremonies," *Bryan Eagle,* Sept. 29, 1991, p. 1; Maj. Gen. Wallace C. Arnold, letter to Mobley, Mar. 28, 1991, POP. Also see U.S. Senate Committee on Armed Services, "To Recognize and Commend Military Colleges," S. Con. Res. 56, 102nd Cong., 1st sess., July 30, 1991.

24. U.S. General Accounting Office [GAO], "Air Force Academy: Gender and Racial Disparities," NSIAD 93–244, Sept. 24, 1993; GAO, "Military Academy: Gender and Racial Disparities," NSIAD 94–95, Mar. 17, 1994; GAO, "DOD Service Academies: Update on Extent of Sexual Harassment," NSIAD 95–58, Mar. 31, 1995. See also "Many at Academy Believe Women Don't Belong," *Navy Times,* Oct. 22, 1990, p. 4; Sue McMillin, "Sexual Prank at Academy Goes Awry," *Air Force Times,* Oct. 29, 1990; Sally Jacobs, "Dark Tales of Hazing Abound at Alma Mater [Norwich] in Va. Murder," *Washington Post,* May 5, 1991, p. D6; Lisa Leff, "Navy Bias Study Raps Academy for Its Treatment of Women," *Washington Post,* May 5, 1991; Mark Smith, "About-face at West Point," *Houston Chronicle,* circa, 1990; James J. Kilpatrick, "Leave Virginia's [VMI] Male-Only College Alone," *Albuquerque Journal,* June 21, 1991, p. A8.

25. Susan Dodge, "Military Academies Crack Down on Hazing of Freshmen," *Chronicle of Higher Education,* Nov. 13, 1991, pp. A39–40. See also Hank Nuwer, *Broken Pledges: The Deadly Rite of Hazing,* pp. 320–24.

26. Dodge, "Military Academies Crack Down," pp. A39–40; Dennis Steele, "Women at West Point: At First, It Was Just Survival; Now It's Progress," *Army Magazine,* July, 1990, pp. 26–33. See also Wise, *Drawing Out the Man: The VMI Story,* pp. 120–23, 395–401, 422; and Catherine S. Manegold, *In Glory's Shadow: Shannon Faulkner, The Citadel, and a Changing America,* pp. 54–57, 99–102, 229–30, 237–40.

27. Maj. Gen. Thomas Darling, interview, College Station, Apr. 24, 1999; Memorandum, "Meeting, Advisory Panel for a Discrimination-Free Corps of Cadets," Oct. 9, 1990, POP.

28. Jill Butler, "Corps Eliminates Female Units," *Battalion,* Apr. 13, 1990, p. 1.

29. Darling interview, Apr. 4, 1999.

30. Gen. Tom Darling, "Briefing to the Board of Regents," Oct. 4, 1990, POP.

31. Delia M. Rios, "Hard Corps," *Dallas Morning News,* Nov. 24, 1991, p. 1F; Cadet Kurt F. Sauer, "Trip Report for Visit to Virginia Military Institute, 26–31 March 1991," memo to the Commandant, Mar. 31, 1991, POP.

32. "Five Cadets Plead Guilty of Assault at Fall Bonfire," *Bryan Eagle,* Feb. 7, 1987, p. 12D; Incident report, Bonfire Site Discrimination Charge, Oct. 28, 1990. Those involved in the 1990 incident of verbal comments included cadets and civilian students from Walton Hall.

33. "Ordeal in Aggieland," *Texas Aggie,* Dec., 1991, pp. 2–4; *Bryan Eagle* articles: Jade Boyd, "Student Leader Says Media Gave Corps Unfair Treatment," Oct. 25, 1991, p. 1; Kara Bounds, "Corps Panel Reveals Little after Meeting," Oct. 10, 1991, p. 1; and Bounds, "Female Corps Member Reports Second Assault," Oct. 15, 1991, p. 1. See also Martha Villareal, president, and Denise Powers, vice president, TAMU National Organization for Women, letter to Thomas G. Darling, Sept. 23, 1991, POP.

34. "Ordeal in Aggieland," p. 3; Eagle Editorial Board, "Corps Committee Should Conduct Full Investigation," *Bryan Eagle,* Oct. 9, 1991, p. 4; *Battalion* Editorial Board, "Reported Assault Taints Reputation of Corps," *Battalion,* Sept. 23, 1991, p. 11; Jennifer Kerber, "Female Cadet Must Step Forward," *Battalion,* Oct. 18, 1991, p. 11; Robert C. Borden, "Stories of Harassment and Abuse Shine Frightening Light on Corps," *Bryan Eagle,* Sept. 29, 1991, p. 6A; "Female Cadet Quits Corps," *Bryan Eagle,* Oct. 19, 1991, p. 1; Todd Ackerman, "Female Cadet Quits Corps, Sites Harassment," *Houston Chronicle,* Oct. 18, 1991, p. 25A; "Texas A&M Cadets Charge Sex Abuse," *New York Times,* Oct. 7, 1991, p. A10; Kelli Levey, "Texas A&M's Tarnished Corps," *Dallas Morning News,* Oct. 13, 1991, p. 43A.

35. "Ordeal in Aggieland," p. 4; See also letters to Dr. Mobley from Henry C. Wendler '34, Oct. 26, 1991; James E. Richardson '53, Nov. 4, 1991; and Roxie R. Pranglin '78, Oct. 22, 1991, POP.

36. Linda Armstrong and Margaret Hinton, letter to Dr. William Mobley, Oct. 21, 1991, POP. See also letters from A&M Mothers Clubs, A&M Clubs, and former students from across the state in 1991 Corps of Cadets File, POP.

37. Kara Bounds, "Student Lied about Attacks, Officials Report," *Bryan Eagle*, Oct. 23, 1991, p. 1; "Cadet's Stories Fabricated, Leaders Told," *Bryan Eagle*, Oct. 22, 1991, p. 1; Liz Tisch, "Officials Question Cadet," *Battalion*, Oct. 22, 1991, p. 1.

38. Chris Vaughn, "Commandant Expects to Lift Suspension in 2 Weeks," and Karen Praslicka, "Fact-finding Panel Continues to Review Harassment Charges," *Battalion*, Oct. 25, 1991, p. 1; Vaughn, "Committee Reports on Harassment," *Battalion*, Nov. 22, 1991, p. 1; "Ordeal," *Texas Aggie*, Dec., 1991, pp. 3–4; Kara Bounds, "Cavalry Hearings to Include 5 Juniors," *Bryan Eagle*, Oct. 17, 1991, p. 1.

39. Video, Press Conference: Association of Former Students, Oct. 25, 1991, College Station, Texas A&M Archives.

40. E. Dean Gage, interview, Bryan, Tex., Nov. 9, 1999, and Bill Youngkin, interview, Bryan, Tex., Dec. 7, 1999.

41. Manegold, *In Glory's Shadow*, pp. 142–44, 151–52, 200–201, 293.

42. Mark Mattox, "Justice Dept. Counters VMI Appeal," *Lexington News-Gazette*, Nov. 18, 1992, p. 1; "Military College Is Ordered to Admit Women," *New York Times*, Nov. 18, 1993, p. B16; Catherine S. Manegold, "Citadel's Traditions Clash With an Age-Old Issue," *New York Times*, May 29, 1994, p. 4; George Will, "VMI Needs a Victory in This One," *Laredo Morning Times*, Jan. 31, 1993, p. 20; "Women Cadets at Citadel Reportedly Are Set on Fire," *San Antonio Express-News*, Dec. 14, 1996, p. 1A; Linda Kanamine, "Hearings to Begin in Citadel Hazing," *USA Today*, Feb. 21, 1997, p. 2A; Peter Finn, "Women Reach Rat [VMI] Finish Line," *Washington Post*, Mar. 17, 1998, p. 1.

43. John B. Sherman '92, Cadet Colonel of the Corps, "A Rebuttal," Dec., 1991, Corps of Cadets File, Texas A&M Archives. For an opposing view see "Press Conference Shrouded Issues," *Battalion*, Nov. 11, 1991, clipping in Corps of Cadets Files, and Wendy Stock, "TAMU Corps of Cadets Defies Zentgraf Ruling: A Saga of Continued Sexual Harassment and Discrimination," *Touchstone*, Mar., 1992, pp. 3–5.

44. Major Ray, memo, "Female Cadet Interviews," to General Darling, Apr. 2, 1992, POP. See also Mariano Castillo, "Too Pretty for the Corps," *Battalion*, Sept. 20, 1999, p. 15.

45. Texas A&M Office of Public Information press release, Jan. 27, 1992; Carlos Byars, "A&M Corps Reduction Not Linked to Sexual Harassment, Head Says," *Houston Chronicle*, May 3, 1992, p. 8C; Memo, Mobley to Members, Board of Regents, "Corps of Cadets Initiatives on Gender Issues," Jan. 20, 1992, POP; Darling, "Corps Marches on in Proud Tradition," and "Forward, Ho! With the Cavalry," *First Call*, spring, 1992, pp. 4–8: Memo, "Rain Dance Incident of 9 Oct 91," Col. James J. Crumbliss, PAS, to Cadet Bart Lesniewicz, Oct. 28, 1991, POP; Corps of Cadets, Cadet Honor Board, "Summary of Actions of the Cadet Honor Board 1991–92," May 7, 1992.

46. William H. Mobley, letter to Gloria A. Watjus, Mid-Jefferson County A&M Mothers Club, Feb. 3, 1992, POP; Eric W. Trammell, President of the Midland A&M Club, letter to Mobley, Feb. 7, 1992, POP; Mobley, letter to Rod Dockery '66, Jan. 7, 1992, POP; Mimi Swartz, "Love and Hate at Texas A&M," *Texas Monthly*, Feb., 1992, pp. 65–71.

47. Roberto Suro, "Harassment Cited in Cadet Program," *New York Times*, Nov. 25, 1991, p. A17. See also "Texas A&M Cadets Charge Sex Abuse," *New York Times*, Oct. 7, 1991, p. A10.

48. John R. Dunne, USDOJ, letter to Mobley, Dec. 27, 1991, POP; Mobley, letter to Senator Lloyd Bentsen, Aug. 17, 1992, POP; Chris Vaughn, "Committee Reports on Harassment," *Battalion*, Nov. 22, 1991, p. 1; Deval L. Patrick, USDOJ, to Genevieve G. Stubbs, Office of the General Counsel, Texas A&M, Mar. 30, 1994.

49. "President's Cadet Corps Committee Issue Report," *Texas Aggie*, Aug., 1992, pp. 2–3.

50. Ibid.

51. Memo, Darling to Mobley, "Corps Disciplinary Process," Sept. 22, 1992, POP; "Preliminary Report of the President's Ad Hoc Advisory Group on the Corps of Cadets," Dec., 1992, POP. Note: The Ad Hoc Committee members were: Jan Winniford, assistant vice president for Student Services; Kenneth Dirks, director of Student Health Services; Patricia A. Alexander, professor; and Stephen G. Ruth '92, A&M student body president.

52. "Crowd Lauds Dedication of Corps Center," *Texas Aggie,* Nov., 1992, pp. 4–6; Memo, Mobley to Genevieve Stubbs, "Listings of Meetings," Jan. 7, 1993 (see President's Executive Staff Meeting Minutes, Sept. 28, 1992), POP.

53. "Meet the Class of '95," and "'Spend the Night' Visits Increase in Fall '91," *First Call,* spring, 1992, pp. 2, 8; "Contributors Continue Outstanding Support," *First Call,* fall, 1992, p. 11; Don E. Crawford '64, letter to author, Oct. 6, 1998.

54. "The Spirit of Aggieland Abounds on Football Weekend," *First Call,* spring, 1996, pp. 8–10; Joe Fenton, "The Spirit of '02," *First Call,* fall, 1995, p. 10; Marcy Boyce, "Corps'[s] Pumpkin Will Fly Again," *Battalion,* Oct. 29, 1980; "Flight of the Great Pumpkin," *The Saber,* fall, 1987, p. 4; Jeanette Simpson, "Mascot Mourners," *Battalion,* Sept. 13, 1999, p. 1.

55. Gretchen Perrenot, "Corps of Cadets Hosts 14th Military Weekend," *Battalion,* Feb. 24, 1995, p. 1; "Corps of Cadets to March to Brazos for March of Dimes," *Bryan Eagle,* Apr. 16, 1996, n.p.; Gilbert Moreno '96, "Final Review," *Texas Aggie,* June, 1995, p. 25. The above reference to the "14th Military Weekend" is incorrect. Military Weekend was started in 1949 by Col. Haydon Boatner. Due to the annual turnover of writers at the *Battalion,* incorrect data often is passed from year to year.

56. "Corps Academic Support Program Gets an A," *First Call,* spring, 1997, pp. 5–6; "Corps of Cadets Freshman GPR—Fall '87–Spring 97," Corps Academic Center, 1998; "MLSC 489 Senior Course Outline," Office of the Commandant, fall, 1999; Maj. Becky Ray, interview, Nov. 26, 1999, College Station, and Laura Arth '75, interview, May 31, 1999, College Station; Corps of Cadets, *Corps Career Services,* Sept., 1999, pp.

1–2; Corps of Cadets, *Corps Academic Support Programs,* fall, 1999, pp. 1–4; Corps of Cadets, *Corps Recruiting Update,* Apr., 1999, pp. 1–2.

57. E. Dean Gage, provost, letter to Mobley, Feb. 19, 1993, POP.

58. Ibid.

59. Jerry Cooper, "Aiming Toward a Strong Corps," *Texas Aggie,* Nov., 1993, pp. 6–9; Association of Former Students, "Hill Report: Corps of Cadets Recruitment and Retention Study," Mar. 22, 1993; Scott P. Phelan '92, "Marching into the Future: A Proposal for Revitalizing the Corps of Cadets," College Station, n.d. [circa 1992], pp. 1–85; Royce H. Hickman '64, interview, College Station, Oct. 11, 1999.

60. Ad Hoc Advisory Group Memo to Mobley, "Spring Semester Report of the Ad Hoc Advisory Group on the Corps of Cadets," June 11, 1993, POP. The ad hoc committee was dissolved after the June, 1993, report.

61. Ibid.; Rip Woodard, "Commandant Gets Authority to Control Corps Discipline," *Battalion,* Oct. 6, 1988, p. 1; Memo to Mobley from Koldus, "Corps Discipline," June 18, 1993, POP; Gen. Tom Darling, interview, College Station, Oct. 11, 1999.

62. Chip Brown, "Mobley to Become A&M Chancellor in System Shakeup," *Austin American-Statesman,* Aug. 1, 1993, p. B2; Kara Bounds, "Reports: Major Shakeup at A&M," *Eagle,* July 31, 1993, p. 1; "Corps Membership Drops; Ags Worry," *San Angelo Standard-Times,* Dec. 6, 1993, p. 5A; Jerry Cooper, "Aiming Toward a Strong Corps," *Texas Aggie,* Nov., 1993, pp. 6–9; James B. Kelly, "Council Chairman Reflects on Corps's Future," *First Call,* fall, 1993, p. 10.

63. "Joint Statement of Dr. Robert Berdahl, President of UT, and Dr. E. Dean Gage, Interim President of Texas A&M," College Station and Austin, circa Dec. 28, 1993, POP; Memo, General Darling to All Cadet Commanders, Jan. 5, 1994, College Station, POP; Texas A&M, *Texas A&M University Annual Report 1994,* College Station, Oct., 1994, p. 23; Cheryl Heller, "Abduction of Reveille VI Used as Publicity Stunt," *Battalion,* Feb. 23, 1995, p. 1.

64. *Battalion* articles: Craig Lewis, "Corps Focus on Membership Recruitment," July 13, 1994, n.p.;

Katherine Arnold, "Numbers of Freshman, Female Cadets Increase," Aug. 29, 1994; and Amy Collier, "More Recruits, More Dorms for Corps on Quad," Jan. 26, 1995.

65. "Fact and Figures," *The Guidon,* Feb.–Mar., 1995, p. 1; Susan Warren, "Texas A&M, Long an Also-Run, Soars in Popularity and Influence," *Wall Street Journal,* Aug. 13, 1995, p. T1; Gretchen Perrenot, "Tuition in Exchange for Service in National Guard, Bill Proposes," *Battalion,* Feb. 22, 1995, p. 1.

66. "Cadet Leader Returns as University President," *First Call,* fall, 1995, p. 3; "The President's View," *Texas Aggie,* Oct., 1994, pp. 2–5; "Corps Academic Support Program Gets an A," *Texas Aggie,* spring, 1997, pp. 6–7; "Spring 1995 Commandant's Honor Roll," and Laura Arth, "Corps Grades at Record High," *The Guidon,* summer, 1995, pp. 4, 7. See also Paul Burke, "Did You Hear the One about the New Aggies?" *Texas Monthly,* Apr., 1997.

67. Maj. Gen. Ted Hopgood, "Commandant's Report," June 12, 2000, College Station.

68. "Darling Reprises Tough Tenure," *Bryan Eagle,* Apr. 7, 1996, n.p.; Keely Coglan, "A&M Corps Commandant Plans to Step Down June 1," *Dallas Morning News,* Apr. 9, 1996, n.p.; "General Darling Steps Down as Commandant to Lead Endowment Campaign," *First Call,* fall, 1996, pp. 8–9; Gen. Don Johnson, interview, College Station, Oct. 10, 1999.

69. Office of the Commandant, "Memorandum of Understanding Between the Professor of Military Science, the Professor of Aerospace Studies, the Professor of Naval Science, and the President of Texas A&M University." College Station, Apr., pp. 1–5.

70. John Kirsch, "Corps Enrollment on the Rise," *Bryan Eagle,* Dec. 6, 1996, n.p.; "Bill Backs Favoritism for Corps," *Bryan Eagle,* June 6, 1997; "A&M Grads Could Lose Privileged Status in Military," *Houston Chronicle,* Mar. 6, 1997; Richard Stewart, "Gramm Will Battle Army Plan," *Houston Chronicle,* Apr. 4, 1997, p. 27; Office of the Commandant, "Quick Facts: Corps of Cadets—Texas A&M University," College Station, Feb., 1999; Anissa Morton, "Last

Aggies Graduate in G. Rollie White," *Bryan Eagle,* Dec. 21, 1997, p. 1.

71. Condoleezza Rice, "Promoting the National Interest," *Foreign Affairs,* Jan., 2000, p. 51. Rice further stated that the armed forces, operating with fewer resources, were deployed more often between 1992 and 1999 than at any time in the last fifty years—an average of once every nine weeks.

72. Dyan Machan, "We're Not Authoritarian Goons," *Forbes,* Oct., 1994, pp. 246–48; Douglas Donaldson, "ROTC: Cut to the Corps," *American Legion Magazine,* Feb., 1993, pp. 28–29; Editorial, "Ready or Not," *Wall Street Journal,* Oct. 22, 1998, p. 22; Thomas G. Darling, "Forming the Corps of the Future," *Bryan Eagle,* Feb. 25, 1990, p. 13.

73. *Chronicle of Higher Education* articles: Colleen Cordes, "New Law Prohibits Pentagon Grants to Colleges that Bar ROTC," Mar. 8, 1996, p. A28; Cordes, "MIT Tries a New Approach in the Battle over ROTC," May 31, 1996, p. A23; and "Pentagon Acts to Identify 'Anti-military' Colleges," Apr. 25, 1997, p. A33.

74. Darling interview, Apr. 24, 1999; U.S. House of Representatives, *National Defense Authorization Act for Fiscal Year 1998,* H.R. 1119, 105th Cong., Report 105–340, Oct. 23, 1997, n.p.; Public Law 105–85, 10 USC 2111a, Nov. 18, 1997, pp. 342–43. See also Maj. Gen. T. D. Hopgood to Sen. Phil Gramm, May 19, 1998 [re: Naval ROTC scholarship recipients], in the Office of the Commandant; and William C. Moore, "The Military Must Revive Its Warrior Spirit," *Wall Street Journal,* Oct. 27, 1998, p. 22.

75. M. T. "Ted" Hopgood '65, "'Recruit, Retain, and Graduate' Are Key Themes," *First Call,* spring, 1997, p. 2; "Commandant Reviews End of Year, Looks Ahead," *First Call,* fall, 1997, p. 2; Office of the Vice President for Student Affairs, "Status Report of Blue Ribbon Recommendations," College Station, Mar. 15, 1997, pp. 1–12. See also A. Phillips Brooks, "Tradition in the Eye of A&M Storm," *Austin American-Statesman,* June 6, 1996, n.p.; "Commandant Committed to Gender Integration within Corps," *First Call,* fall, 1997, p. 8. Emphasis on

excellence was stressed by the commandant by providing a list of the "Top Ten Books for Cadets to Read"—most of them focusing on leadership.

76. Office of Admissions and Records, "Texas A&M University Profile 1998," 1999, pp. 1–5.

77. Chris Fletcher, "Ruling: Hazing Suspects Can't Be Forced to Report It," *Laredo Morning Times,* Oct. 3, 1999, p. 7; Stuart Hutson, "Corps May Resurrect Drill Team," *Battalion,* Oct. 12, 1999, p. 1.

78. Texas A&M Office of University Relations, press release, "Nine Cadets Suspended from Corps in Alleged Drill Team Hazing," Mar. 31, 1997, and "Corps of Cadets to Have No Drill Team in 97–98," Aug. 11, 1997; "Commandant Issues Statement on Hazing," *First Call,* fall, 1997, p. 8.

79. Commandant's Update, June 12, 2000. The Corps of Cadets' Class of 2003 with 564 cadets had the following class profile: nineteen were student body presidents, twenty-two were National Merit Scholars, 33 percent were in the top 10 percent academically in their graduating class, 55 percent were members of the National Honor Society, 36 percent were Eagle Scouts, 58 percent were high school varsity athletes, and 23 percent were high school varsity captains.

80. Lovell W. Aldrich '65, "Lend a Hand to Help the Corps Grow," *First Call,* spring, 2000, pp. 14–15.

81. Coverage of the 1999 Bonfire tragedy was worldwide news. In addition to Cable News Network, the following papers from November 19–30, 1999, have extensive information: *Battalion, Bryan Eagle, Houston Chronicle, Dallas Morning News,* and *San Antonio Express News* as well as a detailed report in the *Texas Aggie, January, 2000.*

BIBLIOGRAPHY

ARCHIVAL SOURCES

Association of Former Students, Texas A&M University. Clayton Williams Center, College Station, Texas.
    Aggie Ring Collection. Minutes of the Board of Directors of the Association of Former Students, 1880–1999. Muster Files, 1943–2000. Papers of the Association of Former Students. Spanish-American War, WWI, WWII, Korean War, Vietnam War Records, Persian Gulf War, Office of the *Texas Aggie* Magazine. *Texas Aggie* Magazine Collection, 1924–2000.
Biographical Files, Texas A&M University Archives, College Station, Texas.
    Allen, Roderick R.; Anderson, Frank G.; Ashburn, Isaac "Ike"; Avery, Frank R.; Bible, Dana X.; Darling, Thomas G.; Daughtrey, Robert N.; Davis, Eddie Joe; Davis, Joe E.; Dinwiddie, Hardaway; Dooley, Thomas; Downs, Pinckney L. "Pinkie," Jr.; Duke, James H. "Red"; Dunn, J. Harold; Earle, James H.; Edwards, Daniel R.; Eisenhower, Dwight D.; Fenton, Joe; Ferguson, Jim; Fernandez, Raul B.; Foster, Lumpkin L.; Gathright, Thomas S.; Gay, George H.; George, Ray; Giesecke, F. E.; Goehring, Dennis; Guion, John I.; Gutierrez, Hector, Jr.; Halbouty, Michel; Hannigan, James P.; Hardy, G. W.; Harrington, Henry H.; Hohn, Caesar "Dutch"; Hollingsworth, James; Houston, David F.; Houston, Temple Lea; Hutson, Mary and Sophie; James, John G.; Krueger, Carl C. "Polly"; Kruger, Weldon D.; Kyle, Edwin J.; Loupot, J. E.; Luedecke, Alvin R.; McAllister, Martin D.; McCoy, Jim H.; McDavitt, Jerome A.; McInnis, Louis L.; McManus, A. F. W.; McQuillen, Everett E.; Melcher, Robert L.; Meloy, Guy S., Jr.; Miller, Jarvis E.; Miller, John; Milner, Robert T.; Mimms, Marvin; Mitchell, Alva; Mitchell,

Harvey; Mobley, William H.; Moore, George F.; Moore, William T.; Morris, R. P. W.; Moses, Andrew; Myers, Shelley; Mueller, C. H.; Nelson, Charles; Pardee, Jack; Parsons, Thomas R.; Parsons, Walter H.; Proffitt, James A.; Pugh, Marion C.; Rains, Jack M.; Ray, James E.; Rogan, Charles; Rose, Archibald J.; Ross, Lawrence S.; Routt, Joe; Rudder, James Earl; Samson, Charles H.; Sanders, Sam Houston; Sargent, H. H.; Sbisa, Bernard; Schiwetz, E. M. "Buck"; Schriever, Bernard A.; Sharp, John; Sherrill, Jackie; Shuffler, R. Henderson; Simpson, Ormond R.; Sleeper, William M.; Slocum, R. C.; Smart, George; Smith, A. J. "Niley"; Southerland, J. Malon; Stallings, Gene; Stark, J. Wayne; Sterns, Josh B.; Sullivan, James; Tassos '45, Damon; Tate '53, Marvin; Teague '32, Olin E. "Tiger"; Teague, John Olin; Todd '97, Charles C.; Trenckmann '82, William A. Utay '08, Joe.; Vandiver, Frank; Varner '40, Durward B.; Webster 1879, Louis Beauregard; Weirus '42, Richard "Buck"; Welty, Col. Maurice; Weyland '23, Gen. Otto P.; Williams '53, Clayton; Williams, Jack K.; Wilson, James V. "Pinky"; Wipprecht, Walter; Woodall, Col. James; Youngblood, Bonnie; Zentgraf, Melanie; Zinn, Bennie.
Center for American History, University of Texas, Austin, Texas.
    Perry, James Franklin, Papers.
Franklin D. Roosevelt Library, Hyde Park, New York.
    Roosevelt, Franklin D., Papers: 1937–38 Trip files.
Lyndon Baines Johnson Library, Austin, Texas.
    Papers of the Vice President of the United States.
Sam Houston Sanders Corps of Cadets Center, Texas A&M University, College Station, Texas.
    Aggie Ring Collection, 1880–2000. Aggie Uniform and Boot Collection. Allen, Dwight, Papers. Commandant's Office Annual Scrapbook, 1984–98. Cook, Robert, Scrapbook and Papers.

Eikner, Col. James E. Papers (July 12, 1944, After-Action Report—Assault on Pointe du Hoc Battery). Homann '27, Richard E., Papers and 1922–23 Diary. Mahan '20, Jack, Scrapbook and Papers. Parsons '30, Walter H., Papers. Sanders, Sam Houston, Scrapbook and Papers. Thompson, Hazel, Scrapbook. Company W-1 Papers. Wary, William D., Scrapbook.

Texas A&M University Archives, College Station, Texas.
Academic Council Papers. Athletics, Texas A&M, Papers. Bizzell, William, Papers. Bonfire Papers and Bonfire Accident Public Information Log, January 18, 2000. Boldt, Robert B. (Dallas, Tex.), Papers. Bolton, Frank C., Papers. Burges, Austin E., Scrapbook. Cole, James Reid, Papers. Cole, Mary, Diary, September, 1882–September, 1884. Crane, Charles Judson, Papers. Cushing, Edward B., Papers and Scrapbook. Dollar, Fred W., Papers. Futrell, Brent (Dumas, Tex.), Papers. Gathright, Thomas S., Papers. Giesecke, F. E., Papers. Groginski, P. S., Scrapbook. Guyler, Robert W., Papers. Harrell, William G., Papers. Harrington, Henry Hill, Papers. Hartung, L. A., Scrapbook. Heye, William B., Papers. Hill, James E., Papers. Holman, Lucius, Papers. Hughes, Lloyd H., Papers. Hull, Burt, Papers. Korean War Papers. McAdams, Carroll G., Scrapbook. McInnis, Louis L., Papers and Scrapbook. Memorial Student Center Papers. Milner, Robert T., Papers. Mobley, William, Papers. Morgan, David H., Papers. Muster Papers. Nash, W. R., Papers. Pfeuffer, George, Papers. President's Office Papers. Ratchford, Willie R., Papers. Reveille Papers—Mascot Binders 1 and 2. Rogan, Charles, Papers. Roseborough, William D., Papers. Ross, Lawrence S., Papers. Ross Volunteers Scrapbook. Rudder, James Earl, Papers. Russell, Dan, Scrapbook. Stark, J. Wayne, Scrapbook. Teague, Olin E. "Tiger," Papers. Trice, W. P., Scrapbook. Vietnam War Papers. Walton, T. O., Papers. World War II Papers.

Texas A&M University Army ROTC, Military Science Building, College Station, Texas.
Professor of Military Science Papers.

Texas A&M University Corps of Cadets HQ, Dorm 2, College Station, Texas.
Corps of Cadets Collection and Papers. Corps Staff Records and Papers.

MANUSCRIPTS

Adams, John A., Jr. "Governance and Administration in American Land-Grant Universities." College Station, June, 1981.

———. "History of the Corps of Cadets: 100 Years of Heritage." College Station, December 7, 1975.

———. "Out of the Past: Traditions and Past Times of the Texas Aggies." College Station, 1987.

Baker, Dean Paul. "The Partridge Connection: Alden Partridge and Southern Military Education." Ph.D. dissertation, University of North Carolina, 1986.

Barthell, Daniel W. "The Committee on Militarism in Education, 1925–1940." Ph.D. dissertation, University of Illinois, 1972.

Benner, Judith Ann. "Lone Star Soldier: A Study of the Military Career of Lawrence Sullivan Ross." Ph.D. dissertation, Texas Christian University, 1975.

Bradford, Walter H. "Aggie Fashion: A Survey of the Texas A&M Cadet Uniform Through the Years." Texas A&M Archives, 1975.

Brooks, Melvin, and John R. Bertrand. "Student Attitudes toward Aspects of the A. & M. College of Texas: Spring 1954." Texas A&M Archives, June 10, 1954.

Byrns, Robert E. "Lafayette Lumpkin Foster: A Biography." M.A. thesis, Texas A&M University, 1964.

Chelette, Vickie G., ed. "Class of '22: Then and Now." College Station, 1985.

Connally, John B. "Centennial Celebration Keynote Address." College Station, October 4, 1976.

Dillard, Raymond L. "A History of the Ross Family and Its Most Distinguished Member, Lawrence Sullivan Ross." M.A. thesis, Baylor University, 1931.

Dromgoole, Glenn. "Muster Speech: Abilene A&M Club," Abilene, Tex., April 21, 1967. Author's collection.

Garber, L. O. "History and Present Status of Military Training in Land-Grant Colleges." M.A. thesis, University of Illinois, 1926.

Gentry, Mike. "Integrity." Transcript of speech to the Corps of Cadets. College Station, October, 1997.

Gilchrist, Gibb. "Gibb Gilchrist: An Autobiography." Texas A&M Archives, 1991.

Griffith, Marion S. "A Study of ROTC Bands in Colleges and Universities of the United States." M.A. thesis, University of Texas, May, 1954.

Hallmark, James. "Animal A Company." Texas A&M Archives, circa 1985.

———. "The Genealogical History of the Corps Organization at Texas A&M University, 1876–present." College Station, 1992.

Heye, William B. "Remarks: Ross Volunteer Induction Banquet." Mobley Papers, Texas A&M Archives, October 10, 1988.

Kelly, James B. "Muster Speech: Coastal A&M Club." Sinton, Tex., April 21, 1986. Author's collection.

Kerr, Pat W. "As Good As It Gets." College Station, 1998. Author's collection.

Killigrew, John W. "The Impact of the Great Depression on the Army, 1929–1936." Ph.D. dissertation, Indiana University, 1960.

Knippa, Heidi A. "Salvation of a University: The Admission of Women to Texas A&M." M.A. thesis, University of Texas, May, 1995.

Kraus, John D., Jr. "The Civilian Military College." Field study, University of Iowa, June, 1976.

———. "The Civilian Military Colleges in the Twentieth Century: Factors Influencing Their Survival." Ph.D. dissertation, University of Iowa, July, 1978.

Langford, Ernest. "Administrators: Texas A&M University." Texas A&M University Archives, 1971.

———. "Here We'll Build the College." Texas A&M University Archives, 1963.

Love, A. C. "Notes on Railroad Summer Practice, Agricultural & Mechanical College of Texas." M.A. thesis, A&M College of Texas, June, 1917.

Marshall, Elmer Grady. "The History of Brazos County, Texas." M.A. thesis, University of Texas, 1937.

Miller, James E. "A Survey of the Cooperative Housing Plan at the Agricultural and Mechanical College of Texas." Master's thesis, Texas A&M College, 1936.

Morgan, William R. "Rehabilitation." Athens, Tex., June 15, 1975.

Payne, John, Jr. "David Franklin Houston." M.A. thesis, University of Texas, 1953.

Phelan, Scott P. "Marching into the Future: A Proposal for Revitalizing the Corps of Cadets."

College Station, n.d. [circa 1992].

Reed, David. "The State of the Corps—An Analysis." College Station, Tex., January, 1970. Author's collection.

Robin, Doris G. "The Opposition to the Establishment of Military Training in Civil Schools and Colleges in the United States, 1914–1940." M.A. thesis, American University, Washington, D.C., 1959.

Rollins, Charles H. "A Report to the Board of Regents on Hazing." College Station, Tex., December, 1984.

Rush, Lisa. "A. & M. at War: How World War II Affected the Campus of the Agricultural and Mechanical College of Texas." Texas A&M University Archives, circa 1993.

Schneider, Eugene W. "A Survey of the Administration of Air Force Reserve Officers' Training Corps Detachments." M.B.A. thesis, University of Texas, 1954.

Schriever, Bernard A. "ICBM: Step Toward Space Conquest." Speech transcript. San Diego, Calif., February 19, 1957.

Sherman, John B. "A Rebuttal." College Station, December, 1991.

Simpson, Lt. Gen. Ormond R., USMC (Ret.). "Reflections of a Cadet in the Corps—1932–1936." College Station, 1997. Author's collection.

———. "Should I Join the Corps?" College Station, n.d. (circa 1985).

Smith, Robert F. "A Brief Sketch of the Agricultural and Mechanical College of Texas." College Station, 1914 [sic, circa 1904].

Tomlinson, Marie Guy. "The State Agricultural and Mechanical College of Texas, 1871–1879: The Personalities, Politics, and Uncertainties." M.A. thesis, Texas A&M University, May, 1976.

Vandiver, Frank E. "State of the University." Presentation to the Texas A&M Faculty Senate, May 5, 1988.

Ward, Robert D. "The Movement for Universal Military Training in the United States, 1942–1952." Ph.D. dissertation, University of North Carolina, 1957.

Webb, Lester A. "The Origins of Military Schools in the United States Founded in the Nineteenth Century." Ph.D. dissertation, University of North Carolina, 1958.

Whiteside, Myrtle. "The Life of Lawrence Sullivan Ross." M.A. thesis, University of Texas, 1938.

Wiley, James E. "An Aggie's WWII Experiences." SHS Corps Center, 1999.

Ylinen, John A., and August W. Smith. "Preliminary Report: Corps Attitudes/Opportunities Survey." College Station, May, 1978.

MOVIES, FILM FOOTAGE, AND DOCUMENTARIES

Aggie Muster in College Station. April 21, 1986, 1987, 1990, 1993, 1994, 1997, and 1998. Film footage.

Aggie Muster Committee. "Muster Awareness." College Station: April, 1988.

Association of Former Students. Press Conference. College Station: October 25, 1991.

———. "Bonfire 1998." Fox TV. College Station: November 24, 1998.

———. "Proud of the Past . . . Committed to the Future." College Station: 1998.

———. "That Certain Spirit." College Station: April, 1975.

———. "We Are the Aggies." College Station: 1954.

Barker. "Traditions of Aggielend." College Station: 1993.

Burton, Donald. "Corps Briefing." University Center, College Station: January 30, 1986 [audio].

Corps of Cadets. "The Corps of Cadets: The Experience Lasts Forever." College Station: 1999.

Dyal, Donald. "E. B. Cushing: The Untold Story." Houston: 1999 [audio].

Hollingsworth, James. "Dedication of Danger 79er Statue." The Quad, College Station: September 10, 1999 [audio].

Texas A&M University. Commencement. College Station: June, 1987.

———. "Texas A&M University: Inspiration to Greatness." College Station: 1990.

———. Corps of Cadets. "Accept the Challenge." College Station: 1992.

———. "Rough! Tough! Real Stuff!" College Station: circa 1995.

University Relations. Aggie Band Clips, 1938–39. Texas A&M Archives.

———. "Marching to the Beat . . . History of Marching Bands—Featuring the Fightin' Texas Aggie Band." College Station: 1994.

———. 1940 Sugar Bowl Classic, Texas A&M vs. Tulane. New Orleans: 1940. Texas A&M Archives.

———. Visit of President Franklin Roosevelt to Texas A&M, College Station: May 11, 1937. Texas A&M Archives.

Vandiver, Frank. "Address to the Corps of Cadets." University Center: December 5, 1984 [audio].

We've Never Been Licked. Paramount Pictures, 1943.

BOOKS AND ARTICLES

AAF: The Official Guide to the Army Air Force. New York: Army Air Force Aid Society, 1944.

Adams, John A., Jr. "Lawrence Sullivan Ross." Texas Aggie, July, 1979, pp. 8–10.

———. "Softly Call the Muster." Texas Aggie, April, 1992, pp. 2–9.

———. Softly Call the Muster: The Evolution of a Texas Aggie Tradition. College Station: Texas A&M University Press, 1994.

———. "That Undying Aggie Spirit." The Saber, Summer, 1986, p. 7.

———. We Are the Aggies: The Texas A&M University Association of Former Students. College Station: Texas A&M University Press, 1979.

Adams, John C., ed. The Voices of a Proud Tradition: A Collection of Aggie Muster Speeches. Bryan, Tex.: Brazos Valley Printing, 1985.

"Agricultural and Mechanical College." Texas Review 7, March, 1886, pp. 433–37.

"Agricultural and Mechanical College of Texas." ROTC Journal, April, 1954, pp. 3–6.

Allandice, Bruce. "West Points of the Confederacy: Southern Military Schools and the Confederate Army." Civil War History, December, 1997, pp. 310–31.

Allen, Frederick L. The Big Change: America Transforms Itself 1900–1950. New York: Harper and Brothers, 1952.

Ambrose, Stephen E. D-Day—June 6, 1944: The Climactic Battle of World War II. New York: Simon and Schuster, 1994.

———. The Victors. New York: Simon and Schuster, 1998.

Anders, Leslie. The Ledo Road. Norman: University of Oklahoma Press, 1965.

Anderson, Joseph R. *VMI in the World War*. Richmond, Va.: Richmond Press, 1921.

Anderson, Terry N. *A Guide to the Oral History Collection, Texas A&M University*. College Station: Sterling C. Evans Library, 1981.

Andrew, Rod, Jr. *Long Gray Lines: The Southern Military School Tradition, 1839–1915*. Chapel Hill: University of North Carolina Press, 2001.

———. "Soldiers, Christians, and Patriots: The Lost Cause and Southern Military Schools, 1865–1915." *Journal of Southern History* (November, 1998): 677–710.

Anthony, Augusta H. "Lawrence Sullivan Ross: Soldier and Statesman." *The Texas Magazine*, September, 1912, pp. 429–31.

Ashburn, Ike. "D. X. Bible: The Man Who Kept His Feet on the Ground." *Texas Parade*, January, 1955.

———. "Educating for Fullness of Living." *Proceedings of the 1941 Annual Meeting*. Dallas: Assoc. of Texas Colleges, July, 1941.

Association of Former Students. *Alumni Directory*. August, 1917.

———. *Alumni Quarterly*. 1916–21.

———. *A. and M. College Record*. College Station: 1901–1902.

———. *Class of 1913: Historical Directory and Biography 1909–1913*. College Station: [circa 1953].

———. *Directory of Former Students. Bulletin of the Agricultural and Mechanical College of Texas*. 3rd ser., vol. 10, December, 1924.

———. *Directory of Former Students. Bulletin of the Agricultural and Mechanical College of Texas*. 3rd. ser., vol. 15, May, 1929.

———. *Directory of Former Students, 1876–1938*. College Station: 1938.

———. *Directory of Former Students, 1876–1957*. College Station: 1957.

———. *Directory of Former Students, 1876–1962*. College Station: 1962.

———. *Directory of Former Students, 1876–1967*. College Station: 1967.

———. *Directory of Former Students, 1876–1970*. College Station: 1970.

———. *Directory of Former Students, 1876–1973*. College Station: 1973.

———. *1876–1976: Centennial Directory of Former Students*. College Station: 1976.

———. *Directory of Former Students, 1876–1979*. College Station: 1979.

———. *Directory of Former Students, 1876–1982*. College Station: 1982.

———. *Directory of Former Students, 1876–1985*. College Station: 1985.

———. *Directory of Former Students, 1876–1988*. College Station: 1988.

———. *Directory of Former Students, 1876–1997*. College Station: 1997.

———. *Leadership Directory: 1998*. College Station: 1998.

———. *The Reilly Report: Texas A&M Today as seen by High School Seniors, Guidance Counselors, Former Students, and the General Public*. New York: E. L. Reilly Co., November, 1967.

Baritz, Loren. *Backfire: A History of How American Culture Led Us into Vietnam and Made Us Fight the Way We Did*. New York: William Morrow, 1985.

Barnes, Roswell P. *Militarizing Our Youth: Significance of the Reserve Officers' Training Corps*. New York: Committee on Militarism in Education, 1927.

Barr, Alwyn. *Reconstruction to Reform: Texas Politics, 1876–1906*. Austin: University of Texas Press, 1971.

Barron, S. B. (Samuel Benton). *The Lone Star Defenders: A Chronicle of the Third Texas Cavalry, Ross' Brigade*. New York: Neal Publishing, 1908.

Baskir, Lawrence M., and William A. Strauss. *Chance and Circumstance: The Draft, the War, and the Vietnam Generation*. New York: Vintage Books, 1978.

Beman, Lamar T. *Military Training Compulsory in Schools and Colleges*. New York: H. W. Wilson, 1926.

Benedict, Harry Y., ed. *A Source Book on the University of Texas: Legislative, Legal, Bibliographical, and Statistical*. University of Texas Bulletin No. 1757, October 10, 1917.

Benner, Judith Ann. *Sul Ross: Soldier, Statesman, Educator*. College Station: Texas A&M University Press, 1983.

Black, Robert W. *Rangers in World War II*. New York: Ivy Books, 1992.

Bletz, Donald F. *The Role of the Military Professional in U.S. Foreign Policy*. New York: Praeger Publishing, 1972.

Blumenfeld, Charles H. "The Case for ROTC." *Military Review* (n.d.), pp. 1–13.

Bolton, Paul. *Governors of Texas*. Corpus Christi, Tex.: Caller Times Publishing, 1947.

Bond, Charles R., and Terry Anderson. *A Flying Tiger's Diary*. College Station: Texas A&M University Press, 1993.

Bond, O. J. *The Story of the Citadel*. Richmond, Va.: Garrett and Massie, 1936.

Bond, P. S. (Paul Stanley). *Military Science and Tactics*. Washington, D.C.: P. S. Bond Publishing, 1942.

Bond, P. S., et al. *The R.O.T.C. Manual: Freshman Course*. Annapolis, Md.: National Service Publishing, 1921.

Boyne, Walter J. "The Man [Bernard A. Schriever] Who Built the Missiles." *Air Force Magazine*, October, 2000, pp. 80–86.

Bradley, Omar N. *A Soldier's Story*. New York: Henry Holt and Company, 1951.

———. [General of the Army]. "Remarks on the World Military Situation." *Proceedings* 66 (1952): 46.

Brady, Jeff L. "Aggie Spirit in Action: Real Marines Don't Cry Until the Spirit Moves Them." *Zip Tips: Class of '93 Senior Handbook*. July, 1992, pp. 30–31.

Brown, John Henry. *History of Texas, 1685–1892*. 2 vols. 1893. Reprint, Austin and New York: Pemberton Press, 1970.

———. *Indian Wars and Pioneers of Texas*. Austin: L. E. Daniel, 189.

Brown, Preston. "The Genesis of the Military Training Camps." *Infantry Journal*, December, 1930, pp. 609–13.

Bryant, Paul W., and John Underwood. *Bear: The Hard Life and Good Times of Alabama's Coach Bryant*. Boston: Little, Brown, 1974.

Burges, Austin E. *A Local History of A&M College 1876–1915*. Bryan: n.p., 1915.

Burka, Paul. "Did You Hear the One About the New Aggies?" *Texas Monthly*, April, 1997.

Bynum, Mike. *Aggie Pride*. College Station: The We Believe Trust Fund, 1980.

Carlton, John T., and John F. Slinkman. *The ROA Story: A History of the Reserve Officers Association of the United States*. Washington, D.C.: ROA, 1982.

Casey, Paul D. *The History of the A&M College Trouble*. Waco, Tex.: J. S. Hill, 1908.

Cavazos, Amado F. *Zachry: The Man and His Company*. San Antonio: Metro Press, 1994.

Chapman, David L. *Wings over Aggieland*. College Station: Texas A&M Friends of the Library, 1994.

Class of '22. *Memorial Loan Program*. College Station: n.d.

Class of 1942. *Thirtieth Reunion—November 3, 4, 1972*. College Station: 1972.

Cofer, David B. *Early History of Texas A. and M. College through Letters and Papers*. College Station: Association of Former Students, 1952.

———. *First Five Administrators of Texas A. & M. College, 1876–1890*. College Station: Association of Former Students, 1953.

———. *Fragments of Early History of Texas A. and M. College*. College Station: Association of Former Students, 1953.

———. *Second Five Administrations of Texas A&M College*. College Station: Association of Former Students, 1953.

———. *Supplement to First Five Administrators of Texas A. & M. College: James Reid Cole, 1879–1885*. College Station: Association of Former Students, 1955.

Coffman, Edward M. *The Old Army: A Portrait of the American Army in Peacetime, 1784–1898*. New York: Oxford University Press, 1896.

Colby, Eldridge. "Military Training in Land Grant Colleges." *Georgetown Law Journal* 23 (November, 1934): 1–36.

Cole, James R. *Seven Decades of My Life*. Dallas: n.p., 1913.

Colner, Robert C. *James Stephen Hogg: A Biography*. Austin: University of Texas Press, 1959.

Cook, Fred J. *The Warfare State*. New York: Macmillan, 1962.

Cooper, Jerry C., and Henry C. Dethloff. *Footsteps: A Guided Tour of the Texas A&M University Campus*. College Station: Texas A&M University Press, 1991.

Cooper, Lewis P. *The Permanent School Fund of Texas*. Fort Worth: Texas State Teachers Association, 1934.

Cope, Myron. "The Proudest Squares." *Sports Illustrated*, November 18, 1968, pp. 98–110.

Cordes, Colleen. "MIT Tries a New Approach in the Battle over ROTC." *Chronicle of Higher Education*, May 31, 1996, p. 23.

Cosmas, Graham A. "From Order to Chaos: The War Department, the National Guard, and Military Policy, 1898." *Military Affairs* 29 (1965): 105–21.

———. "Military Reform After the Spanish-American War: The Army Reorganization Fight of 1898–1899." *Military Affairs* 35 (February, 1971): 12–17.

Coulter, Ellis M. *College Life in the Old South*. Athens: University of Georgia Press, 1928.

Coumbe, Arthur T., and Lee S. Harford. *US Army Cadet Command: The Ten Year History*. Fort Monroe, Va.: Office of the Command Historian, 1996.

Couper, William. *One Hundred Years at VMI*. 4 vols. Richmond, Va.: Garrett and Massie, 1939.

———. *The V.M.I. New Market Cadets Biographical Sketches*. Charlottesville, Va.: Michie Company, 1933.

Cox, Frank W., III. *I Bleed Maroon: The Life, Traditions, History, and Spirit of Texas A&M University*. Bryan, Tex.: Insite, 1992.

Coyle, Clarence C. "The Story of Texas A. and M." *The Texas Magazine*, September, 1912.

Crane, C. J. *The Experiences of a Colonel of Infantry*. New York: Knickerbocker Press, 1923.

Crawford, Charles W. *One Hundred Years of Engineering at Texas A&M*. College Station: Privately published, 1976.

Cunningham, Bill. "And What a College!" *American Legion Magazine*, March, 1940, pp. 4–5.

Dandorn, David B. "The Agricultural Extension System and the First World War." *The Historian* 41 (February, 1979): 315–31.

Darr, S. C., ed. *The Littlest Aggie: The Story of Texas A&M*. Austin: Littlest Book Co., 1990.

Davis, Bowers. "On the Reserve Officers Training Corps." *Infantry Journal*, September, 1930, pp. 290–94.

Dawes, Charles G. *A Journal of the Great War*. 2 vols. Boston: Houghton Mifflin, 1921.

Dethloff, Henry C. *A Centennial History of Texas A&M University, 1876–1976*. 2 vols. College Station: Texas A&M University Press, 1975.

———. *A Pictorial History of Texas A&M University 1876–1976*. College Station: Texas A&M University Press, 1975.

———. *Suddenly, Tomorrow Came . . . : A History of the Johnson Space Center*. Washington, D.C.: NASA, 1993.

Dinwiddie, H. H. "Industrial Education in Texas." *Texas Review* 7 (March, 1886): 407–16.

Dittmar, Gus C. *They Were First: Recollections of the First Officers Training Camp of Leon Springs, Texas*. Austin: Privately published, 1969.

"The Dixie Classic." *The Collegiate World*. January–February, 1922, pp. 2, 31.

Dodge, Susan. "Military Academies Crack Down on Hazing of Freshmen." *Chronicle of Higher Education*, November 13, 1991, pp. A39–40.

Donaldson, Douglas. "ROTC: Cut to the Corps." *American Legion Magazine*, February, 1993, pp. 28–29.

Donovan, James A. *Militarism, USA*. New York: Scribner's Sons, 1970.

Dugan, Haynes W. *The History of the Great Class of 1934*. College Station: n.p., 1982.

Earle, James H. *We Is the Aggies*. College Station: Privately published, 1954.

———. *Slouch—The Aggies' Aggie*. College Station: Slouch Enterprises, 1961.

Eby, Frederick, ed. *Education in Texas: Source Materials*. New York: Macmillan, 1925.

Eddins, Rufus R., and Hubert R. Voelcker. *Life at A&M As We Knew It: 1905–1909 by the Class of 1909*. College Station: Association of Former Students, 1966.

Eddy, Edward D., Jr. *Colleges for Our Land and Time: The Land-Grant Idea in American Education*. New York: Harper and Brothers, 1957.

———. *The Land-Grant Movement: A Capsule History of the Educational Revolution Which Established Colleges for All the People*. N.p.: Land Grant College Association, 1962.

Eikel, Fred. "An Aggie Vocabulary of Slang." *American Speech*. February, 1946, pp. 29–36.

Ekirch, Arthur A., Jr. *The Civilian and the Military*. New York: Oxford University Press, 1956.

———. "The Idea of a Citizen Army." *Military Affairs* 17 (1953): 30–36.

Elam, Stanley, ed. *The Gallup Polls of Attitudes Toward Education, 1969–1973*. Bloomington, Ind.: Phi Delta Kappa, Inc., 1973.

Ellis, William A. *Norwich University, 1819–1911*. Montpelier, Vt.: Capital City Press, 1911.

Epstein, Cynthia F. "Great Divides: Deceptive Distinctions and Rhetorical Strategies in the VMI and Citadel Cases." *Gender Issues* (Winter, 1998): 34–46.

Eschmann, Karl J. *Linebacker: The Untold Story of the Air Raids over North Vietnam.* New York: Ivy Books, 1989.

Evans, Clement A. *Confederate Military History.* Atlanta: Confederate Publishing Co., 1899.

Evans, Wilbur, and H. B. McElroy. *The Twelfth Man: A Story of Texas A&M Football.* Huntsville, Tex.: Strode Publishing, 1974.

Fallows, James. "Vietnam: Low Class Conclusions." *The Atlantic,* April, 1993, pp. 38–44.

Fehrenbach, T. R. *This Kind of War.* New York: Macmillan, 1963.

Ferguson, Jim. "Texas A&M in the Year 2000." *The Saber,* summer, 1986, pp. 8–12.

Foch, Marshall F. *The Memories of Marshall Foch.* Paris and London: n.p. 1931.

Fowler, Bertram B. "Cooperating Their Way Through College." *Reader's Digest,* June, 1939, pp. 43–46.

Forsyth, John A. *The Aggies and the Horns: 86 Years of Bad Blood and Good Football.* Austin: Texas Monthly Press, 1981.

Frazer, Ronald, ed. *1968: A Student Generation in Revolt.* New York: Pantheon Books. 1988.

Galambos, Louis. *America at Middle Age.* New York: McGraw-Hill, 1986.

Gammel, H. P. N., ed. *The Laws of Texas, 1822–1897.* Austin: Gammel Publishing, 1898.

Gay, George H. *Sole Survivor: A Personal Story about the Battle of Midway.* Midway: Midway Publishers, 1986.

Ginousky, John. "Women's Completion Rate Off Sharply." *Air Force Times,* June 1, 1981.

Ginsburgh, Robert N. *U.S. Military Strategy in the Sixties.* Chicago: Henry Regnery Co., 1970.

Glenn, Nancy. *Texas A&M University—A Legacy of Tradition.* College Station: n.p. 1999.

Glick, Edward Bernard. *Soldiers, Scholars, and Society: The Social Impact of the American Military.* Pacific Palisades, Calif.: Goodyear Publishing, 1971.

Glines, Carroll V. *The Doolittle Raid.* Princeton, N.J.: Van Nostrand Co., 1964.

Goertz, Sharon. *Reveille: The Story of the Texas Aggie Mascot.* Austin: Eakin Press, 1994.

Goldberg, Leonard S. "ROTC and the College: A Complex Relationship." *NASPA Journal* (Fall, 1985): 15–18.

Good, Harry G. *A History of American Education.* New York: Macmillan, 1956.

Gougler, Doyle. "Sul Ross: Indian Fighter, Governor of Texas and Builder of Texas A&M in Its Golden Age." *Cattleman,* August, 1963, pp. 39, 64, 67.

Griffith, Robert K., Jr. "Quality Not Quantity: The Volunteer Army During the Depression." *Military Affairs* (December, 1979): 171–77.

Gutmann, Stephanie. *The Kinder, Gentler Military: Can America's Gender-Neutral Fighting Force Still Win Wars?* New York: Scribner, 2000.

Hagood, Johnson. "R.O.T.C., The Key to National Defense." *Cavalry Journal,* September–October, 1931, pp. 5–9. [reprinted in *Infantry Journal,* September–October, 1931, pp. 403–407.]

Haines, Paul G. *Growing Up in the Hill Country.* Bryan, Tex.: Nortex Press, 1976.

Halberstam, David. *The Best and the Brightest.* New York: Penguin, 1972.

———. *The Fifties.* New York: Villard Books, 1993.

———. *The Reckoning.* New York: William Morrow, 1986.

Halsted, J. Evetts. *Washington, the Ideal of the South: Resurgent in Lee and Ross—Address before the Charles Broadway Rouss Camp of the Sons of Confederate Veterans, February 21, 1899.* Austin: Ben C. Jones and Company, 1899.

Hargrove, Thomas R. *A Dragon Lives Forever: War and Rice in Vietnam's Mekong Delta, 1969–1991, and Beyond.* New York: Ivy Books, 1994.

Harmon, E. N. "Norwich University R.O.T.C. Unit." *Cavalry Journal,* April–October, 1929, pp. 174–77.

Hauser, John N. "The Reserve Officers' Training Corps—Mission and Methods." *Field Artillery Journal,* 1927, pp. 138–45.

Hauser, William L. "Professionalism and the Junior Officer Drain." *Army,* September, 1970.

Hawkes, James R. "Antimilitarism at State Universities: The Campaign Against Compulsory ROTC, 1920–1940." *Wisconsin Magazine of History,* autumn, 1965, pp. 41–54.

Heaton, Dean R. *Four Stars: The Superstars of the United States Military History.* Baltimore: Gateway Press, 1995.

Heinz, W. C. "I Took My Son to Omaha Beach." *Collier's,* June 11, 1954, pp. 21–27.

Herge, Henry C. *Wartime College Training Programs of the Armed Services.* Washington, D.C.: American Council on Education, 1948.

Herring, George C. *LBJ and Vietnam: A Different Kind of War.* Austin: University of Texas Press, 1994.

Hersey, Lewis B. *Selective Service in Peacetime.* Washington, D.C.: GPO, 1941.

Hinshaw, Arned L. *Heartbreak Ridge: Korea, 1951.* New York: Praeger, 1989.

Hobbs, Duke. *I knew I was an Aggie when. . . .* Bryan, Tex.: Insite Group, 1999.

Hodenfield, G. K. "I Climbed the Cliffs with the Rangers." *Saturday Evening Post,* August 19, 1944, pp. 18–19, 98.

Hoffman, Frederick. *The Twenties.* New York: Viking Press, 1955.

Hohn, Caesar. *Dutchman on the Brazos: Reminiscences of Caesar (Dutch) Hohn.* Austin: University of Texas Press, 1963.

Houston, David A. *Eight Years with Wilson's Cabinet, 1913–1920.* New York: Doubleday, 1926.

Hoyle, John. *Good Bull: 30 Years of Aggie Escapades.* Bryan, Tex.: Insite Press, 1990.

———. *Good Bull Number 2: More Aggie Escapades.* Bryan: Insite, 1991.

Huntington, Samuel P. *The Soldier and the State.* Cambridge, Mass.: Harvard University Press, 1957.

———, ed. *Changing Patterns of Military Politics.* New York: Free Press, 1962.

Ilfrey, Jack. *Happy Jack's Go-Buggy: A Fighter Pilot's Story.* Atglen, Pa.: Schiffer Military History, 1998.

Irwin, C. L. "Corregidor in Action." *Coastal Artillery Journal,* January, 1943, pp. 9–12.

Irwin, Michael and Joseph. *Cathedrals of College Football.* Atlanta: Alliance Press, 1999.

Jackson, Grace. *Cynthia Ann Parker.* San Antonio: Naylor, 1959.

James, Edmund J. *The Origin of the Land Grant Act of 1862.* Urbana-Champaign, Ill.: n.p., 1910.

James, Sidney L. "Ensign Gay of Torpedo Squadron 8," *Life Magazine,* August 31, 1942, pp. 70–80.

Janowitz, Morris. *The New Military: Changing Patterns of Organization.* New York: Russell Sage Foundation, 1964.

Jennings, Herbert S. "Stirring Days at A. and M. College." *Southwest Review* 31 (autumn, 1946): 341–44.

Johnson, Edwin C. "Main Issues in the R.O.T.C. Controversy." *Harvard Educational Review,* October, 1939, pp. 450–51.

Johnson, John H. *1918: The Unexpected Victory.* London: Arms and Armour, 1997.

Joulman, George A. "ROTC: An Academic Focus." *Military Review,* January, 1971.

Karnow, Stanley. *Vietnam: A History.* New York: Viking Press, 1983.

Keatley, Robert. "Student Soldiers." *Wall Street Journal,* July 2, 1969.

Kendrick, Alexander. *The Wound Within: America in the Vietnam Years, 1945–1974.* Boston: Little, Brown, 1974.

Kerr, Homer L., ed. *Fighting with Ross's Texas Cavalry Brigade, C.S.A.* Hillsboro, Tex.: Hill Junior College Press, 1974.

Kinnear, Duncan L. *The First 100 Years: A History of Virginia Polytechnic Institute and State University.* Blacksburg, Va.: VPI Education Foundation, 1972.

Kittrell, Norman G. "Address Delivered at the Unveiling of the Monument of Gen. Lawrence Sullivan Ross, Ex-Governor of Texas and Ex-President of the Agricultural and Mechanical College at College Station Friday, May 4, 1919." N.p., n.d.

Kolbe, Parke R. *The Colleges in War Time and After: A Contemporary Account of the Effect of the War upon Higher Education in America.* New York: Appleton, 1919.

Lane, Olivia, ed. *The 1982 Maroon Book: Texas Aggie Football.* Dallas: Taylor Publishing, 1982.

Langford, Ernest. *Getting the College Under Way.* College Station: University Library, 1970.

Langton, Rosalind. "Life of Colonel R. T. Milner." *Southwestern Historical Quarterly,* April, 1941, pp. 407–51.

Lanning, Michael Lee. *The Only War We Had: A Platoon Leader's Journal of Vietnam.* New York: Ivy Books, 1987.

Lawson, Ted W. *Thirty Seconds over Tokyo.* New York: Random House, 1944.

Lederman, Douglas. "Pentagon Acts to Identify 'Anti-military' Colleges." *Chronicle of Higher Education,* April 25, 1997, p. 33.

Leftwich, Bill J. *The Corps at Aggieland.* Lubbock: Smoke Signal Publishing, 1976.

Leuchtenburg, William E. *The Perils of Prosperity, 1914–32.* Chicago: University of Chicago Press, 1958.

Levy, David W. *The Debate over Vietnam.* Baltimore: Johns Hopkins University Press, 1991.

Lindsay, Walter M. "Reserve Officers and Their Institutions." *U.S. Infantry Association Journal,* July 1, 1906, pp. 71–96.

Lippmann, Walter. "Patriotism in the Rough." *New Republic,* October 9, 1915, pp. 277–79.

———. *U.S. War Aims.* Boston: Little, Brown, 1944.

Loupot, J. E. *Aggie Facts and Figures.* College Station: n.p. 1989.

Lyons, Gene M., and John W. Masland. *Education and Military Leadership: A Study of the R.O.T.C.* Princeton, N.J.: Princeton University Press, 1959.

———. "The Origins of ROTC." *Military Affairs,* spring, 1959, pp. 1–12.

MacCloskey, Monro. *Reserve Officers Training Corps: Campus Pathways to Service Commissions.* New York: Richard Rosen Press, 1965.

McCullough, David. *Truman.* New York: Simon and Schuster, 1992.

McGinnis, N. M., ed. *Gold Book: Agricultural and Mechanical College of Texas: A Tribute to Her Loyal Sons Who Paid the Supreme Sacrifice in the World War.* College Station: Alumni Quarterly, August, 1919.

Machan, Dyan. "We're Not Authoritarian Goons." *Forbes,* October 24, 1994, pp. 246–48.

McKeithan, Daniel M., ed. *Selected Letters: John Garland James to Paul H. Hayne and Mary M. M. Hayne.* Austin: University of Texas Press, 1946.

McLean, Malcolm D. "The Significance of San Jacinto Day in Texas." *Texana* 10, no. 2 (1972): 104–15.

Maclear, Michael. *The Ten Thousand Day War: Vietnam, 1945– 1975.* New York: St. Martin's Press, 1981.

McNamara, Robert S. *In Retrospect: The Tragedy and Lessons of Vietnam.* New York: Time Books, 1995.

Malone, Henry O. "Military History, Command Support, and the Mission: The TRADOC Experience." *The Army Historian,* winter, 1986, p. 4.

Manegold, Catherine S. *In Glory's Shadow: Shannon Faulkner, The Citadel, and a Changing America.* New York: Knopf, 2000.

Marston, Anson D. "Wartime Role for Colleges and Universities." *Military Affairs* 18 (1954): 131–44.

Masland, John W., and Laurence I. Radway. *Soldiers and Scholars.* Princeton, N.J.: Princeton University Press, 1957.

Mason, Herbert M. *The Great Pursuit: Pershing's Expedition to Destroy Pancho Villa.* New York: Hawkins and Associates, 1970.

Maysel, Lou. *Here Come the Texas Longhorns.* Fort Worth: Burnt Orange Publishers, 1978.

Meade, Robert D. "The Military Spirit of the South." *Current History,* April, 1929, pp. 55–60.

"Men at War: Two Friends from Texas [re: William G. Harrell]." *Time,* July 30, 1945, p. 30.

Menoher, Pearson. "The Reserve Officers Training Corps." *Cavalry Journal,* April, 1920, pp. 70–80.

Miles, P. L. "Orientation of R.O.T.C. Freshmen." *Infantry Journal,* September, 1931, pp. 439–42.

Miller, Edward A. "The Struggle for an Air Force Academy." *Military Affairs* 27 (1963–64): 163–73.

Miller, Thomas L. *The Public Lands of Texas, 1519–1970.* Norman: University of Oklahoma Press, 1972.

Mitchell, Johnny. *The Secret War of Captain Johnny Mitchell.* Houston: Gulf Publishing Company, 1976.

Moffett, G. C., ed. *YMCA: Students' Hand-Book of A.& M. College for 1914–1915.* College Station: N.p. 1914.

Morrison, Shelly, ed. *Personal Civil War Letters of General Lawrence Sullivan Ross.* Austin: Shelly and Richard Morrison 1994.

Morrow, Lance. "The Long Shadow of Vietnam." *Time,* February 24, 1992, pp. 18–21.

Myers, Fred. "Texas A. & M. Band Puts on Dazzling Show." *Marching Bands,* October, 1955, pp. 14–16.

Newcomb, Richard F. *Iwo Jima.* New York: Holt, Rinehart, and Winston, 1965.

Norman, Geoffrey. "The Boys of New Market." *Quarterly Journal of Military History,* summer, 1997, 19–27.

Nunn, Jack H. "MIT: A University's Contributions to National Defense." *Military Affairs,* October, 1979, pp. 120–25.

Nuwer, Hank. *Broken Pledges: The Deadly Rite of Hazing.* Atlanta: Longstreet Press, 1990.

Nyberg, Kenneth L., and William P. Snyder. "Program Structure and Career Socialization in ROTC: A Bibliographic Note." *Military Affairs,* December, 1976, pp. 179–81.

Odom, William. *The Collapse of the Soviet Military.* New Haven, Conn.: Yale University Press, 1998.

O'Loughlin, W. J. "Anti-Military Training Propaganda." *Infantry Journal,* July, 1926, pp. 44–46.

Ousley, Clarence. *History of the Agricultural and Mechanical College of Texas.* College Station: A&M Press, December, 1935.

Parker, Edwin P. "The Development of the Field Artillery Reserve Officers' Training Corps." *Field Artillery Journal,* July, 1935, pp. 334–42.

Pasco, John. *Fish Sergeant.* College Station: N.p., 1940.

Pate, Steve. *John David Crow: Heart of a Champion.* Indianapolis, Ind.: Masters Press, 1997.

Pattillo, George. "The First Ten Thousand." *Saturday Evening Post,* June 16, 1917, pp. 1–10.

Perry, George Sessions. *The Story of Texas A. and M.* New York: McGraw-Hill, 1951.

Pershing, John J. *My Experiences in the World War.* 2 vols. New York: Frederick A. Stokes Co., 1931.

Peterson, Horace C., and Gilbert C. Fite. *Opponents of War, 1917–18.* Madison: University of Wisconsin Press, 1957.

Philpott, William B., ed. *The Sponsor Souvenir Album and History of the United Confederate Veterans' Reunion, 1895.* Houston: Sponsor Souvenir Co., 1895.

Porter, Jack N. *Student Protest and the Technocratic Society: The Case of ROTC.* Chicago: Adams Press, 1973.

Potter, Waldo C. "Field Artillery Units of the Reserve Officers' Training Corps." *Field Artillery Journal,* January–February, 1919, pp. 17–35.

Quinn, Dan. *The Spirit Within: A True Story of the Spirit and Camaraderie of Life at Texas A&M University.* Bryan, Tex.: Insite, 1992.

Reeves, Ira L. *Military Education in the United States.* Burlington: Free Press, 1914.

Reinert, Al. "Being an Aggie Is No Joke." *Texas Monthly,* January, 1981, pp. 72–80.

Rogan, Octavia F. *Land Commissioner Charles Rogan.* Austin: San Felipe Press, 1968.

Rollins, Joseph G., Jr. *Aggies! Y'all Caught That Dam' Ol' Rat Yet?* San Antonio: Naylor, 1970.

Romanus, Charles F., and Riley Sunderland. *Stillwell's Command Problems: China-Burma-India Theater.* Washington, D.C.: GPO, 1955.

Rose, Victor M. *Ross' Texas Brigade.* 1881. Reprint, Kennesaw, Ga.: Continental Book Co., 1881.

Ross, Earle D. *Democracy's College: The Land-Grant Movement in the Formative Stages.* Ames: Iowa State College Press, 1942.

———. "The Father of the Land Grant College." *Agricultural History,* April, 1938, pp. 151–86.

———. "The Land-Grant College: A Democratic Adaptation." *Agricultural History,* 1941.

"ROTC: The Protester's Next Target." *Time,* March 7, 1969, pp. 54, 59.

R.O.T.C. Manual—Coast Artillery. 7th ed. Harrisburg, Pa.: Military Service Publishing Co., May, 1935.

Rudder, Earl. "What Price Disarmament?" *Ordnance,* November, 1965, pp. 288–90.

Ryan, Cornelius. *The Last Battle.* New York: Simon and Schuster, 1966.

Salmond, John A. *The Civilian Conservation Corps, 1933–1942: A New Deal Case Study.* Durham, N.C.: Duke University Press, 1967.

Sargent, Herbert H. *Napoleon Bonaparte's First Campaign.* 3rd ed. Chicago: A. C. McClurg and Co., 1894.

———. *The Strategy on the Western Front—1914–1918.* Chicago: A. C. McClurg and Co., 1920.

Schaffer, Ronald. "The War Department's Defense of ROTC, 1920–1940." *Wisconsin Magazine of History,* winter, 1969–70, pp. 108–20.

Schmidt, Margaret. *Cynthia Ann Parker: The Life and Legend.* El Paso: Texas Western Press, 1990.

Schoor, Gene. *The Fightin' Texas Aggies: 100 Years of A&M Football.* Dallas: Taylor Publishing, 1994.

Schriever, Bernard A. "The Space Challenge." *Air University Review,* May–June, 1965, pp. 3–5.

Schurman, Jacob G. "Affirmative Discussion: Every College Should Introduce Military Training." *Everybodys,* February, 1915, pp. 59–67.

Schwien, Edwin E. "About the ROTC—Pacifist Agitation in Schools and College." *Infantry Journal,* September, 1927, pp. 277–79.

"Sex and the ROTC." *The Nation,* May 1, 1926, p. 116.

Shannon, Fred A. *The Farmer's Last Frontier.* New York: Holt, 1945.

Sherrill, Robert. "Texas Athletic and Military." *Time,* September 28, 1962, p. 79.

Shirley, Glenn. *Temple Houston: Lawyer with a Gun:* Norman: University of Oklahoma Press, 1980.

Shuffler, Henderson, ed. *Son, Remember . . .* College Station: A&M Press, 1951.

Sinise, Jerry. "Bill Helwig: 'A Blue Chipper.'" *Grain Producers News,* June, 1976, pp. 6–8.

Smith, Edna M., ed. *Aggies, Moms, and Apple Pie.* College Station: Texas A&M University Press, 1987.

Spaw, Patsy McDonald, ed. *The Texas Senate.* Vol. 1. College Station: Texas A&M University Press, 1990.

Spector, Ronald. "'You're Not Going to Send Soldiers Over There Are You!': The American Search for an Alternative to the Western Front 1916–1917." *Military Affairs,* February, 1972, pp. 1–4.

Stancik, William C. "Air Force ROTC: Its Origins and Early Development." *Air University Review,* July–August, 1984, pp. 38–51.

Stoddard, W. L. "For a Citizen Army." *New Republic,* September 4, 1915, pp. 125–27.

Stokes, William N. *Sterling Evans: Texas Aggie, Banker, Cattleman.* College Station: Friends of the Sterling C. Evans Library, 1985.

Stuart, L. L. "The Officers' Reserve Corps." *Infantry Journal,* May, 1931, pp. 281–84.

Summerall, C. P. "The Officers' Reserve Corps." *Infantry Journal,* November, 1930, pp. 461–65.

Tang, Irvin A. *The Texas Aggie Bonfire: Tradition and Tragedy at Texas A&M.* Austin: Morgan Printing, 2000.

Texas A&M University. *Aggieland.* 1941–2000.

———. *The Aggies Are Back.* College Station: Journalism Department, 1968.

———. Biennial Reports. 1880–1980.

———. *The Blue Book.* College Station: 1910–19.

———. *Blueprint for Progress.* College Station: 1962.

———. *Catalogue,* 1876–2000.

———. Century Study. *A Report of the Century Council to the Board of Directors.* College Station: 1962.

———. *College Regulations.* College Station: 1919–63.

———. *Corps Endowment Campaign: Ensuring the Future.* (1996–99 Campaign Summary). College Station: n.d.

———. *Facilities for National Defense.* College Station: 1940.

———. *Fish handbook.* n.p. [College Station], n.d. [fall, 1964].

———. *Fifty Facts about the Agricultural and Mechanical College of Texas.* College Station: Publicity Department, October 11, 1915.

———. *Inauguration of David Hitchens Morgan as Thirteenth President.* College Station: May 20, 1954, pp. 1–44.

———. *Inauguration of Marion Thomas Harrington as Twelfth President.* College Station: November 9, 1950, pp. 1–56.

———. *Inauguration of the State Agricultural and Mechanical College of Texas.* Address of Governor Richard Coke and Thomas S. Gathright. Bryan, Tex.: Appeal and Post, 1876.

———. *InRol at TAMU: Inroad and Outlooks at TAMU.* College Station: 1973.

———. *Longhorn.* 1906–41.

———. *Report of the A. and M. College of Texas Committee on Postwar Planning and Policy.* College Station: A&M College, August, 1944.

———. "Meeting of Corps of Cadets with President of A. & M. College," by David H. Morgan. College Station: March 8, 1956.

———. Meteorological Unit. *32nd Service Co. U.S. Signal Corps.* College Station: 1918, pp. 1–8.

———. *1998–1999 Corps Leadership Outreach: Area Representative's Guide.* College Station: October 9, 1998.

———. *Program of the Semi-Centennial Celebration and the Inauguration of Thomas Otto Walton, LL.D.* College Station: October 15, 1926.

———. *Progress Report for Twelve Years of the Agricultural and Mechanical College of Texas 1925–1937.* College Station: Texas A. and M. Press, 1938, pp. 1–78.

———. *Report of the Committee Appointed to Investigate the Origin of the Fire that Destroyed the Main Building of the Agricultural and Mechanical College of Texas at College Station, Texas. May 27, 1912.* College Station: June 27, 1912, pp. 1–115.

———. *Sixty Facts: About the Agricultural and Mechanical College of Texas.* College Station: circa 1917.

———. "Student Attitudes Toward Aspects of A. & M. College of Texas," by Melvin S. Brooks and John R. Bertrand. June, 1954.

———. *Texas A&M University—Annual Report, 1982.* College Station: November, 1982.

———. *Texas A&M University Student Rules 1999–2000.* College Station: 1999.

———. *Vision 2020: Creating a Culture of Excellence.* College Station: 1999.

———. *Whoop! Start: A Guide to Aggie Life.* College Station: 1999.

Texas A&M Board of Directors. Announcement and Circular of the State Agricultural and Mechanical College, 1876. College Station: 1876.

———. Biennial Report of the Agricultural and Mechanical College of Texas. College Station and Austin: 1885–1919.

———. Minutes of the Board of Directors [Regents]. College Station: May, 1886–2000. Note: Volume I of the board minutes from 1875 to 1886 was destroyed in the fire at Old Main on May 27, 1912.

———. Proceedings of a Hearing by the Board of Directors of the A&M College of Texas, Held at Fort Worth, Texas, February 24 and 25, 1913. College Station: 1913.

———."Proceedings of a Hearing by the Board of Directors of the A.& M. College of Texas: To Inquire into and Receive Complaints Concerning Recent Breaches of Discipline at the A&M College of Texas." Fort Worth: February, 1913.

———. Special Commission on the 1999 Texas A&M Bonfire: Final Report. n.p. [College Station], n.d. [March, 2000].

Texas A&M University Corps of Cadets. Area Representative's Guide: 1999–2000—Corps Leadership Outreach. College Station: August 12, 1999.

———. Articles of the Corps of Cadets. 2nd ed., September 17, 1949. 32 pp.

———. Articles of the Corps of Cadets. September, 1954. 36 pp.

———. Articles of the Corps of Cadets. September 1, 1958. 24 pp.

———. Articles of the Corps of Cadets. September 1, 1961. 79 pp.

———. Articles of the Corps of Cadets. September, 1963. 44 pp.

———. "Award Program Guidelines: Corps Hall of Honor." College Station: 1999.

———. The Cadence—1947–48. College Station: 1948.

———. Cadence. College Station: 1997.

———. Cadre Leadership Training. College Station: 1987.

———. Combat Ball '75: Let's Keep Aggie Traditions. College Station: 1975.

———. "Combined Band Policy—1996–1997." College Station: 1996.

———. Commissioning Exercises. College Station: May 15, 1999.

———. "Corps of Cadets: Texas A&M University" [recruiting brochure]. College Station: 1974.

———. "Corps of Cadets and Vision 2020." College Station: November 16, 1998.

———. Freshman Orientation Week [FOW]: Handbook 1999. College Station: 1999.

———. Office of the Commandant. A Report to the President on the Corps of Cadets. College Station: November 27, 1979.

———. Office of the Commandant. "Memorandum of Understanding Between the Professor of Military Science, the Professor of Aerospace Studies, the Professor of Naval Science and the President of Texas A&M University." College Station, April, 1999, pp. 1–5.

———. Ross Volunteer Company: Policies and Procedures Guide. College Station: April, 1984.

———. The Standard. College Station: 1968; October, 1995; August, 1997; and July, 1999.

———. First Regiment. Standard Operating Procedure. College Station: 1990.

———. Something Extra. College Station: 1975.

———. Texas A&M College: Victory Homecoming—April 19–21, 1946. College Station: 1946.

———. 1998–1999 Corps Leadership Outreach: Area Representative's Guide. College Station: October 9, 1998.

———. The Seventy-fifth Anniversary Committee. Opened October 4, 1876: The Agricultural and Mechanical College of Texas. College Station: 1950.

———. Manual of the Saber. Edited by Jim Ham. College Station: 1972.

———. Yell Book, 1924–25. College Station: n.d. (circa 1924).

"Texas [A&M] Also Has Country's Largest Marching Band." The Baton, fall, 1955, pp. 16–20.

Texas, State of. General and Special Laws of the State of Texas. 48th Leg., 1944.

———. Governor's Messages, Coke to Ross, 1874–1891. Ed. Archives and History Department. Austin: Texas State Library, 1916.

———. Texas Legislative Record. 1881.

———. Journal of the Senate of the Twelfth Legislature of the State of Texas, 1st reg. sess. [n.d.].

Thomas, Lowell. This Side of Hell: Daniel Edwards, Adventurer. New York: Farrer and Rinehart, 1932.

Thompson, W. O. "Military Training at Educational Institutions." Infantry Journal, May, 1929, pp. 496–502.

Thwing, Charles F. *The American Colleges and Universities in the Great War, 1914–1999.* New York: Macmillan, 1920.

Tolson, John J. *Airmobility, 1961–1971.* Washington, D.C.: GPO, 1973.

Tomlinson, Kenneth Y. "ROTC under Attack." *Reader's Digest,* November, 1969, pp. 231–38.

"Top ROTC Grad Honored." *Army ROTC Newsletter,* September, 1976, p. 19.

Tuchman, Barbara. *The Guns of August.* New York: Macmillan, 1962.

Turner, Thomas E., Sr., "A Career That Bred Legends." *Baylor Magazine,* October 30, 1949.

Underwood, Jeffery S. *The Wings of Democracy: The Influence of Air Power on the Roosevelt Administration 1933–1941.* College Station: Texas A&M University Press, 1991.

United States. *Congressional Record,* March 23, 1975.

———. "Detailing Personnel of the Army of the United States to Educational Institutions, Etc." 77th Cong., 2nd sess., House Report No. 1704, January 28, 1942.

———. National Defense Act, 1912. Vol. 35 (1912).

———. National Defense Act, 1916. Vol. 39 (1917).

———. National Defense Act, 1920. Vol. 41 (1920).

———. Selective Training and Service Act. Vol. 44 (1940).

———. *United States Code Annotated.* Title 7, Agriculture.

———. *U.S. Statutes at Large.* Morrill Land Grant Act. Vol. 12, (1862).

United States. Department of the Air Force. *Drill and Ceremonies.* AFM 50–14. Washington, D.C.: GPO, 1971.

———. Department of the Army. *European Theater of Operations: Cross-Channel Attack.* Washington, D.C.: GPO, 1951.

———. Department of the Army. *Truce Tent and Fighting Front.* Washington, D.C.: GPO, 1966.

———. General Accounting Office [GAO]. "Air Force Academy: Gender and Racial Disparities." NSIAD 93–244, GPO, September 24, 1993.

———. GAO. "DOD Service Academies: Problems Limit Feasibility of Graduates Directly Entering the Reserves." NSIAD 96–56, GPO, March 24, 1997.

———. GAO. "DOD Service Academies: Update on Extent of Sexual Harassment." NSIAD 95–58, GPO, March 31, 1995.

———. GAO. "Intercollegiate Athletics: Status of Efforts to Promote Gender Equity." HEHS 97–10, GPO, October 25, 1996.

———. GAO. "Military Academy: Gender and Racial Disparities." NSIAD 94–95, GPO, March 17, 1994.

———. House Committee on Armed Services. "Universal Military Training." H.R. 1752, 82nd Cong., 1st sess., January 23–March 8, 1951.

———. House Committee on Military Affairs. "Abolishment of Compulsory Military Training at Schools and Colleges." Hearings, 69th Cong., 1st sess., 1926.

———. House Committee on Military Affairs. "Abolishment of Compulsory Military Training at Schools and Colleges: Hearing on H.R. 8538." 69th Cong., 1st sess., 1926, pp. 31–32.

———. House Committee on Military Affairs. "Tents for Agricultural and Mechanical College, Texas." 63rd Cong., 2nd sess., April 21, 1914, p. 6559.

———. Office of Education. "Recent Action in the Land-Grant Colleges and State Universities," *Education for Victory.* Washington, D.C.: March 3, 1942.

———. Office of Education and Federal Security Agency. *Education for Victory.* Washington, D.C.: March 3, 1942.

———. Senate. *Medal of Honor Recipients 1863–1973.* 93rd Cong., 1st sess., No. 15, October 22, 1973.

———. Senate Committee on Armed Services. "To recognize and commend military colleges." S. Con. Res. 56, 102nd Cong., 1st sess., July 30, 1991.

———. U.S. Army. Center for Military History. *Correspondence Relating to the War with Spain.* Washington, D.C.: 1993.

———. War Department. *Committee on Education and Special Training—A Review of Its Work During 1918.* Washington, D.C.: circa 1920.

———. War Department. *The ROTC Manual: Basic Course for All Arms.* Harrisburg, Pa.: 1933.

———. War Department. *R.O.T.C. Manual: Coastal Artillery.* 7th ed. Harrisburg, Pa.: 1935.

———. War Department. *War Department Annual Reports, 1919,* Vol. 1, Washington, D.C.: 1920.

———. War Department. Adjutant General's Office. *Annual Report.* Washington, D.C.: 1877–1901.

————. War Department. Office of the Chief of Staff. *Rules of Land Warfare.* Washington, D.C.: GPO, 1914.

United States Military Academy. *The Howitzer.* West Point, N.Y.: 1904, 1909, 1910, 1924.

Utley, Robert M. *Frontier Regulars: The United States Army and the Indian, 1866–1890.* New York: Macmillan, 1973.

Utterback, A. P., Jr. "Cavalry ROTC at Texas A&M." *Cavalry Journal,* May–June, 1942, pp. 82–83.

Vagts, Alfred. *A History of Militarism: Civilian and Military.* New York: Free Press, 1967.

Vandiver, Frank E. "And Now, Mr. President, Bring Back the Draft." *Christian Science Monitor,* March 30, 1982.

————. *Black Jack: The Life and Times of John J. Pershing.* 2 vols. College Station: Texas A&M University Press, 1977.

————. *Blood Brothers.* College Station: Texas A&M University Press, 1992.

Van Overbeek, Will. *Aggies: Life in the Corps of Cadets at Texas A&M.* Austin: Texas Monthly Press, 1982.

Van Schaick, Louis J. "Military Training and Peace." *Infantry Journal,* September, 1926, pp. 252–56.

Virginia Military Institute [VMI]. *Register of Former Cadets.* Lexington: VMI, 1957.

Walker, B. M. "Henry Hill Harrington." *Journal of Mississippi History,* July, 1940, pp. 156–58.

Wall, Ralph, and Ray Herndon, eds. *Southwest Conference Football 1968.* Houston: Football History, Inc., 1968.

Walters, Raymond. "Field Artillery in American Colleges." *Field Artillery Journal,* November–December, 1919, pp. 543–55.

Webb, Lester A. *Captain Alden Partridge and the United States Military Academy 1806–1833.* Northport: American Southern, 1965.

Weeks, John W. "Student Military Training." *Field Artillery Journal,* September–October, 1921, pp. 476–82.

Weems, John E. *To Conquer a Peace.* College Station: Texas A&M University Press, 1984.

Wellman, Paul I. "Cynthia Ann Parker." *Chronicles of Oklahoma,* June, 1934, pp. 163–71.

Wheeler, G. E. "Origins of the Naval Reserve Officers Training Corps." *Military Affairs* 20 (1956): 170–74.

"Why Do We Arm?" *New Republic,* October 30, 1915, pp. 323–24.

Williams, Robert H. "The Case for Peta Nocona." *Texana* 10, no. 1 (1972): 55–72.

Wilson, Charles R. *Baptized in Blood: The Religion of the Lost Cause, 1865–1920.* Athens: University of Georgia Press, 1980.

Wilson, LeRoy. "Rotation of the Cadet Command in the R.O.T.C." *Infantry Journal,* March, 1929, pp. 280–82.

Winstead, G. B. "Texas A. & M. College Blazes the Trail." *The Texas Outlook,* April, 1942, pp. 25–28.

Wise, Henry A. *Drawing Out the Man: The VMI Story.* Charlottesville: University Press of Virginia, 1978.

Wise, Jennings C. *The Military History of the Virginia Military Institute from 1839 to 1865.* Lynchburg: J. P. Bell Co., 1915.

Wood, Leonard. *The Military Obligation of Citizenship.* Princeton, N.J.: Princeton University Press, 1915.

Woolsey, Vernor G. *The Gospel Truth.* San Antonio: Naylor, 1965.

Worthington, Ian. "Socialization, Militarization and Officer Recruiting: The Development of the Officer Training Corps." *Military Affairs,* April, 1979, pp. 90–95.

Wyckoff, Theodore. "Required ROTC: The New Look." *Military Review,* November, 1964, pp. 24–28.

Wythe, George. *History of the 90th Division.* Ninetieth Division Association, 1920.

Young, David M. "ROTC Required or Elective?" *Military Review,* February, 1962, pp. 21–32.

Youngblood, Bonney. *An Economic Study of a Typical Ranching Area on the Edwards Plateau of Texas.* College Station: Texas A&M Experiment Station, 1922.

INDEX